Encyclopedia of Fungicides: Benefits and Drawbacks

Volume II

Encyclopedia of Fungicides: Benefits and Drawbacks Volume II

Edited by **Chris Frost**

New York

Published by Callisto Reference,
106 Park Avenue, Suite 200,
New York, NY 10016, USA
www.callistoreference.com

Encyclopedia of Fungicides: Benefits and Drawbacks
Volume II
Edited by Chris Frost

International Standard Book Number: 978-1-63239-250-3 (Hardback)

Contents

Preface

The book primarily illuminates both the benefits as well as drawbacks of fungicides. Over the past few years, a lot of pesticides have been formulated to kill pests, and fungicides are one of these. They are primarily used to kill fungus, and are mostly utilized in medicine industry, agriculture fields in protection of seeds during storage and prevention of the growth of toxin producing fungi. Due to an increase in the importance of agriculture, the production of fungicides is also growing. Also, because of their harmful effects on humans, birds and fishes; the problems related to safety are also becoming a matter of growing concern. This book is a guide on both the harmful and helpful effects of fungicides. It discusses issues like fungicide resistance, mode of action, management fungal pathogens and defense techniques, side effects of fungicides interfering in the endocrine system, collective use of varied fungicides and the importance of GRAS (generally recognized as safe). This book is useful to almost everyone ranging from post graduate students to environmentalists and even factory owners.

The information shared in this book is based on empirical researches made by veterans in this field of study. The elaborative information provided in this book will help the readers further their scope of knowledge leading to advancements in this field.

Finally, I would like to thank my fellow researchers who gave constructive feedback and my family members who supported me at every step of my research.

Editor

Resistance to Botryticides

Snježana Topolovec-Pintarić

Department for Plant Pathology, Faculty of Agriculture, University of Zagreb
Croatia

1. Introduction

Plant pathogenic fungi and oomycetes as causal of plant diseases are responsible for economical looses in agricultural production worldwide. Therefore, their chemical control by products named fungicides is needed and justified especially when disease tend to became an epidemic. Without fungicides both yield and quality would be severely reduced by the ravages of fungi. The improved performance, specificity and environmental safety of the modern fungicides led to become their ever more widely used. But, as great Renaissance man Leonardo da Vinci said: *"The nature never breaks her own laws"*, the fungi constantly found the new ways to adapt to conditions that human creates and keep existing and living. Fungus develops insensibility to chemical compound aimed to their suppression under constant pressure of often and continuous use of fungicide with specific mechanisms of action. This ability is nothing else than natural phenomenon or evolution. Today this phenomenon is less mysterious than three decades ago when first arise although some new challenges have spring up. Phenomenon of insensibility of fungus to the chemical compound used for controlling it is named resistance. With the increased use and specificity of the product comes a greater risk that resistance will developed because certain members of the target fungal population will not be affected by the product and therefore fungus cannot be controlled adequately any more. That is, they are genetically resistant to it. Although some plant diseases may be managed through the alteration of cultural practices, many diseases are only managed acceptably with the application of fungicides. One of them is grey mould of wine grape caused by ascomycete fungus *Botrytis cinerea* Pers.:Fr. (teleomorph *Botryotinia fuckeliana* (de Bary) Whetzel). Even today the only effective control of *B. cinerea* remains application of fungicides specifically named botryticides. In the past *B. cinerea* has proved to be very prone to resistance development which makes it difficult to control. Those drown attention of scientists and catalyse studies of resistance phenomenon in *B. cinerea*. Furthermore, resistance phenomenon intensified the genetic studies of this fungus because it was assumed that limited understanding of the genetic structure of *B. cinerea* populations is reflecting in difficulties in managing the disease. Despite of gained knowledge about *B. cinerea* resistance and managing solutions the resistance is still an ever present threat with new cases arising and some old problems still continuing. A new segment of the topic becomes issue of multiple drug resistance (MDR). MDR phenomenon is common in human pathogens but it has been rarely described before in field strains of plant pathogenic fungi. Gaining knowledge about MDR revealed existence and involvement of some different mechanisms for resistance development. Fungicide resistance mechanisms can relate to qualitative factors such as absence or presence of a sensitive target site. Beside this, qualitative factors like uptake, transport, storage and metabolism also need to be considered. The MDR phenomenon of *B. cinerea* was firstly recorded in 1998. Since than, more data of MDR monitoring were obtained indicating that *B. cinerea* MDR types in combination with other *B. cinerea* resistant types could represent a significant threat for future chemical control of *B. cinerea*.

2. Bunch rot of grapes: High standard disease of grapevine

Bunch rot of grapes is one of the grapevine diseases of great economical importance because it leads to substantial losses in yield and lowering in quality. Vineyard ecosystem is often difficult to manipulate both the crop and its environment. Also, it is a stage where *B. cinerea* can express its dual nature in causing the destructive bunch rot and, under certain conditions, the non-destructive noble rot, which is not paralleled in plant pathology. Noble rot yields vines of a special quality that are high economical. In the continental climate the bunch rot disease can inflict damages up to 50 or 60 percent and under the Mediterranean climate 3 to 5 percent. The damages are continuing in vine making process. Rotting of grape berries caused by fungus is probably old as winegrowing and some descriptions date from time of Roman Plinius the Older (1. century). Even the genus name *Botrytis* is derived from Latin for "grapes like ashes" by Micheli who erected the genus in 1729. Name of disease, grey mould, actually describes the grey coating spread over the bunch especially beacon before vintage when the most damage is already done. The coating is somatic filamentous body or sporulating mycelia of fungus *B. cinerea*. In grapevine *B. cinerea* causes massive losses of yield and quality of grape berries for vine production especially during cool and wet climatic conditions. This fungus is able to act as saprophyte, necrotroph as well as pathogen. In vineyards *B. cinerea* is present as part of the environmental micro-bionta and predominantly being saprophytic it colonize wounds or senescing tissue. From an economic point of view, only while acting as true pathogen infecting flowers and grape berries are of importance in terms of lowering quantity and quality of yield. Although there are numerous scientific contributions that continue to be published, there are still gaps in our knowledge about the etiology and epidemiology of bunch rot disease in vineyards. Disease starts with infections in flowering and even earlier. Establishment of *B. cinerea* on moribund and injured tissues normally allows pathogen to infect health tissues. Source of inoculum which will initiate further cycle of the disease are sclerotia and mycelium formed in the outer layers of the dead bark of shoots, cane or on plant debris of various origin. The sclerotia may be directly infective as sources of conidia yet some sclerotia are not conidia-bearing but form reproductive body apothecia. The ascospores produced from apothecia can also initiate primary infections although sexual stage is not considered as significant for epidemiology of grey mould yet Anotnin de Bary described easily found apothecia on dead vine leaves in late 1866. Sclerotia are rare in the regions with warm dry summers and therefore it is unlikely apothecia will be found either. Sporulation on sclerotia is repeated and this extend period of conidial production and infection. Rain and splashing water under natural conditions dislodge conidia from germinating sclerotia and conidia are dispersing in air currents, in splashing water droplets and by insects. The "fruit fly" *Drosopilla melanogaster* is considered as plurimodal vector of *B. cinerea*. The concentration of conidia in the air is increasing as the grapevines maturing. The mycelium spread through outer layers of the dead bark of shoots and the bark of invade cane is bleached to almost white colour. *Botrytis* mycelium sometimes invades the nodes and buds on lower parts of the shoots especially if they had bad wood maturation in the autumn. Buds with dormant mycelia will be finally killed and this will reduce bud burst on the basal parts of the fruit canes in the next spring. Sclerotia and mycelium can also exist on various plants surrounding vineyard and from there conidia disperse in air currents are imported to vineyard. Sclerotia and conidia can be developed on pruned cans left *in situ* or on mummified berries. Abundance of described carry over inoculum in the beginning of new vegetation season at pre-flowering stage is quantitatively related and therefore important for flower infections. Infections are favoured by wet period, at least 12 hours duration, and temperature between 15 and 20 °C. Primary infections of grapes occurs at bloom time or at the end of it when *B. cinerea* starts it's life cycle as biotroph infecting flowers through the stigma and style and then into the stylar end to the ovary. Infected flowers are symptomless and only microscopic examination will reveal necrosis of stamens and growth of the pathogen on the style and stigma. These flower infections are invariably followed by a period of latency when fungus remains in a quiescent phase in receptacle area. Flower infection is believed to be an important stage in the epidemiology of *B. cinerea* in grapes. Furthermore, early infections of the generative organs can destroy flower bunches. Infected flowers, also could become potent inocula within developing bunches for berry rot. Because of the abundance of necrotic floral debris in the vineyards, the end of flowering represents an important epidemiological stage for *B. cinerea*. The floral debris provides an excellent nutrient source for the conidia. Floral debris bearing mycelium are dispersed in wind and rain (Jarvis, 1980) onto leaves and berries. After infection at bloom time following symptomless latent phase, generally until berries begin to ripen. Latency could be explained by the ability of the young berries to synthesize stilbenes until veraison (Pezet & Pont, 1992), maintaining the fungus in the receptacle area from where it can spread into the berry during ripening. During the development of berries until veraison, when the berries begin to soften, the berries are resistant to infection. The ripening process corresponds to a senescence process with a degradation of the berry tissues, especially activity allowing disease expression to occur. During this phase, the whole defence mechanisms controlling the pathogen loose their activity, allowing disease expression to occur. Grapevine tissues defend themselves against fungal attack by the accumulation of phytoalexins, like stilbenes, mostly in the green berries but stilbenes appears to be inactive during ripening (Pezet & Pont, 1992; Bais et al., 2000). After veraison the berries become increasingly susceptible to infection. At lower sugar content, less than 13° Brix, the so-called sour rot affects berries and leads very often to a complete loss of attached grapes. Sour rot is favourable with frequent rainfalls. At higher sugar content, attached berries can be processed normally but these forces growers to an earlier harvesting or to picking of moulded grapes. Infections of berries occur at temperature interval between 20 and 25 °C and are accomplished by conidia. Germlings that developed from conidia enter grape berries through different pathways, namely through stigmata, pedicels, natural openings and wounds, or by direct penetration of the cuticle (Coertze *et al.*, 2001; Holz *et al.*, 2003). Conidia are deposited on berry surphace by air, rain or insects. The most prominent symptom of the disease is found on the berries in the ripening period when the disease reaches its highest stage and lasts up to the end of harvest, being marked by softening and decay of grape berries. Infected berries are dark coloured and show the typical greyish, hairy mycelium all over their surface. Especially sporulating mycelium can be seen to grow along cracks or splits on the berries because tufts of condiophores with conida are protrude from stoma and peristomal cracks on the skin of the berry. The *B. cinerea* can also infect young leaves and relatively older leaves. Leaf infections occur occasionally during long rainy periods with continuous leaf wetness over 48 hours and temperature between 15 and 20 °C. Heavy leaves infections are not very common because only long duration of leaf wetness allow mycelium to spread in the mesophyll. Therefore, leaves infections normally take place during rain spring. For the same reason in spring also young shoots can be infected from attached tendrils or small wounds. The quantitative relation between incidence of *B. cinerea* at critical stage in the growth of grapevines; pre-flowering (carry over), flowering and harvest was described by Nair et al. (1995). According to their observation the 50% incidence of *B. cinerea* monitored on grapevine tissues carried over from previous season during pre-flowering can predict 29% primary infections of flowers in the new season.

2.1 Managing grey mould

Although some prognostic models are developed based on etiology and epidemiology of grey mould disease the severity of the grey mould disease in vineyards cannot be easily predicted so therefore control based on prognosis may not be satisfactory. Effective control of grey mould in vineyard is usually based on preventive repeated fungicide applications during the season. Already the Romans used sulphur to control this disease. For the same purpose sulphur and potassium were recommended in 18th century. During the late 1950s fungicides were introduced in viticulture and until 1968 in many countries for *Botrytis* control were used: sulphamides (dichlofluanid), pthalimides (captan, captafol, folpet) and dithiocarbamate (thiram). At this point of time the efficacy of fungicidal treatments for *Botrytis* control ranged between 20 and 50 percent. All this fungicides were multi-site inhibitors, affecting many target sites in fungal cell and therefore acting as general enzyme inhibitors. In 1960s, first fungicides appeared which act primarily at one target site therefore referred to as single-site or site-specific and they more efficiently control pathogen. Today, several families of synthetic site-specific botryticides are available. They can be classified according to their biochemical modes of action into five categories: 1) anti-microtubule toxicants (benzimidazoles); 2) compounds affecting osmoregulation (dicarboximides, fludioxonil); 3) inhibitors of methionine biosynthesis (anilinopyrimidines) and 4) sterol biosynthesis inhibitors (fenhexamid); 5) fungicides affecting fungal respiration (fluazinam, boscalid and multi-site inhibitors). The era of sigle-site or site specific fungicides begun in late 1960s with introduction of benzimidazoles (benomyl, thiophanate-methyl, carbendazim) that improved *Botrytis* control (Dekker, 1977; Georgopoulos, 1979; Beever & O'Flaherty, 1985). Only a few years later the new group of dicarboximides become available and they shadowed all previously used ingredients. Dicarboximides were introduced into the market between 1975 and 1977 primarily for the control of *B. cinerea* in grapes (Beetz & Löcher, 1979). Due to good efficacy they were popularly named botryticides and it seemed that the problem of protection against *Botrytis* had been successfully solved. Dicarboximides or cyclic imides (e.g. chlozolinate, iprodione, procymidone, vinclozolin) are characterized by the presence of a 3,5-dichlorophenyl group. The activity of dicarboximides fungicides was first reported in the early 1970's with the three key commercial products being introduced within three years; iprodione in 1974 (Lacroix et al., 1974), vinclozolin in 1975 (Pommer & Mangold 1975) while procymidone was registered a year later (Hisada et al., 1976). They are typically protectant fungicides and although some claims to systemicity have been made (Hisada et al., 1976) they are best regarded as protectant materials. In the mid-1990s a novel family of botryticides was arose, the anilinopyrimidines, with three representative ingredients: pyrimethanil, cyprodinil and mepanipyrim. Mepanipyrim and pyrimethanil exhibit a high activity against *B. cinerea,* while cyprodinil came in combination with fludioxonil (phenylpyrroles) in protection of grapes. Pyrimethanil and cyprodinil were introduced in French vineyards in 1994 (Leroux & Gredt, 1995) and in Switzerland they were registered since 1995 (Hilber & Hilber-Bodmer, 1998). In Italy cyprodinil was registered in 1997 (Liguori & Bassi, 1998.). Mepanipyrim was in 1995 registered in Switzerland, Japan and Israel (Muramatsu et al., 1996). Mixture of cyprodinil and fludioxonil was firstly introduced in Switzerland in 1995 (Zobrist & Hess, 1996). In Croatia pyrimethanil was acknowledged in 1997 under the commercial name Mythos and cyprodinil came as a mixture with fludioxonil named Switch while mepanipyrim was not registered at all (Topolovec-Pintarić & Cvjetković, 2003). Although anilinopyirimidines showed to be highly effective against *B. cinerea,* a high risk of resistance build up was already evident in the laboratory investigations at preregistration phase (Birchmore & Forster, 1996). In spite of that they have been registered in most European winegrowing countries since 1994 but with recommendations for restricted use: once per season when anilinopyrimidines are applied alone and a maximum of two applications per season is proposed for the mixture cyprodinil + fludioxonil (phenylpyrrol) (Fabreges & Birchmore, 1998; Leroux, 1995). Shortly after introduction of anilinopyrimidines in 1995 fludioxonil (phenylphyroles) start to be used in vineyards against *B. cinerea.* Fludioxonil is synthetic analouge of antibiotic pyrrolnitril (phenylphyrol), an antibiotic compound produced by a number of *Pseudomonas* spp. and is thought to play a role in biocontrol by these bacteria. Fludioxonil belong to class of fungicides affecting osmoregulation and is inhibitor of both spore germination and hyphal growth. In 1999 fluazinam (phenylpyridinamine) was introduced in French vineyards although in Japan has been used since 1990 against grey mould in various crops. Fluazinam belongs to group of fungicides that affecting fungal respiration so, it shows multi-site activity probably related to uncoupling of mitochondrial oxidative phosphorilation. It is highly toxic to spores and mycelia. Any shift of *B. cinerea* toward fluazinam in vineyards has still not revealed. In 1999, firstly in Switzerland, a botryticide with novel botryticidial action was registered, the fenhexamid (Baroffio et al., 2003). Early investigations on the fenhexamid mode of action suggested that it has different mechanism from than of all other botryticides (Rosslenbroich & Stuebler, 2000). Fenhexamid is a 1,4.hydroxyanilide with a high preventive activity against *B. cinerea.* It is easily degraded and therefore presents a favourable toxicological profile and environmental behaviour (Rosslenbroich et al., 1998; Rosslenbroich & Stuebler, 2000). It is characterized by a long duration action. Due to its lipophilic character it shows rapid uptake into the plant cuticle and within the upper tissue layer limited but significant locosystemic redistribution occurs (Haenssler & Pontzen, 1999) and as a result the rain fastness of fenhexamid is very pronounced. Fenhexamid suppresses the germination of spores only at relatively high concentrations but it is highly effective in inhibiting subsequent stages of infections. After the initiation of spore germination the fenhexamid inhibit the germ-tube elongations, germ-tubes collapse and die before they are able to penetrate plant surface. Also, treated hyphae frequently show a characteristic leakage of cytoplasm or cell wall associated material at the hyphal tip area (Haenssler & Pontzen, 1999; Debieu et al., 2001). It is sterol biosynthesis inhibitor and blocks the C4-demethylations (Rosslenbroich & Stuebler, 2000). The lastly released botryticide for use in grapevines, in 2004, is novel ingredient boscalid (syn. nicobifen). Boscalid from carboxamide group is systemic and is the only representative of new generation of fungal respiration inhibitors. It act as inhibitor of fungal respiration morover it is new generation of succinate dehydrogenase inhibitors (SDHIs) which inhibit respirations by blocking the ubiquinone-binding site of mitochondrial complex II. In the future, arrivial of new anilide is expected, still described under code SC-0858.

3. Resistance to botryticides

In *B. cinerea* the resistance phenomenon, as in other plant pathogenic fungi, becomes apparent with the site-specific fungicides. Site-specific or single-site fungicides act primarily at single target under responsibility of single major gene. Thus, just a single gene mutation can cause the target site to alter (monogenic resistance), so as to become much less vulnerable to the fungicide (Brent, 1995). Therefore, within few years of intensive use of such fungicide, in populations of polycyclic pathogen with high propagation rate, can be found a high frequency of resistant mutants. The most common mechanism of fungicide resistance is based on alternations in the fungicide target protein. The resistance to multi-site fungicides, which effect many target sites in fungal cell, has been rarely reported. Multi-site fungicides have been considered as low-risk fungicide from the resistance point of view because they interfere with numerous metabolic steps and cause alternation of cellular structures.

3.1 Retrospective of botryticide resistance

As it was mentioned earlier, the oldest multi-site fungicides used in vineyards against grey mould, were thiram (dithiocarbamate), captan, folpet (chloromethylmercaptan derivates) chlorotalonil (phthalonitrile) and dichlofluanid (phenylsulphamide). This ingredients react with thiol, SH and amino group inducing formation of thiophosgene and hydrogen disulphide. They block several thiol-containing enzymes involved in respiratory processes during spore germination and this multi-site action is believed to prevent the development of resistance (Leroux et al., 2002). Therefore, they have been considered low-risk fungicide from the resistance point of view. But, in the 1980's strains resistant to dichlofluanid and to the chemically related tolylfluanid, chlorthalonil and even to phthalimides, captan and folpet, have occasionally been reported (Malatrakis, 1989; Rewal et al., 1991; Pollastro et al., 1996). Moreover, cross-resistance among captan, thiram, chlorothalonil and related fungicides were identified (Barak & Edington, 1984). Resistance to dichlofluanid is determined by two major genes, named *Dic1* and *Dic2*, probably involved in a detoxifying mechanism and in glutathione regulation (Pollastro et al., 1996; Leroux et al., 2002). The mutation of this genes lead to the formations of two sensitive phenotypes Dic1S and Dic2S, two phenotypes with low level resistance Dic1LR and Dic2LR and one high leveled resistant phenotype Dic1HR. In practice only a few cases of control failure due to dichlofluanid-resistant strains were noted. Although these ingredients are not at risk from resistance development and are still registered their practical use is restricted because they are weak botryticides and their residues can cause problems in vine making process (delay fermentation). First site-specific fungicide used in vineyards since the late 1960's was benzimidazole carbendazim or MBC. But, in the early 1970s, only a few years after commercialization loss of disease control due to resistance was reported in many crops especially in vineyards (Leroux et al., 1998). First report of surprisingly enhanced attacks of *B. cinerea*, rather then suppressed, after benzimidazole treatments was in Germany

(Ehrenhardt et al., 1973; Triphati & Schlosser, 1982; Bolton, 1976) but the outbreak of tolerant strains occurred simultaneously in many winegrowing countries in temperate climate. In Switzerland after only two years of use, in 1973, a complete loss of control by benzimidazole was observed and they were withdrawn (Schuepp & Küng, 1981). In Southern Europe where *B. cinerea* pressure is much lower, resistance appeared more slowly. In Mediterranean climate e.g. Italy satisfactory control was reported until 1977 (Bisiach et al., 1978). In Croatia benzimidazoles was used in protection of vineyards shortly from 1971 to 1974. Primarily they were redrawn from use in vineyards because of toxicological reason (negative residues in must and wine). A decrease in efficacy was in Hungary firstly observed in 1981 and it was confirmed by Kaptas & Dula (1984). In 1987 of special interest become mixture of carbendazime and dietophencarb owing to negatively correlated cross-resistance, allowing destruction of benzimidazole-resistant strains by dietophencarb. Soon, negatively correlated cross-resistance become positive as between 1988 and 1989 an overall increase of resistance from 4 to 22% to both components was detected. An explanation of the quick outcome of benzimidazole-resistance was the local existence of naturally resistant strains in the field population of *B. cinerea* before benzimidazole was introduced and their application acted as selected factor eliminating sensitive strains (Schuepp & Lauber, 1978). Benzimidazole carbendazim (MBC) does not affect spore germination but inhibit germ-tube elongation and mycelial growth at low concentrations. These anti-fungal impacts came from MBC binding to tubulin, which is the main protein in microtubules. Microtubules, one type of cytoskeleton filament, regulate organelle position and movement within the cell. Microtubules consist of long, hollow cylinders of repeating dimers of α- and ß-tubulin. MBC binding to tubulin leads to inhibition of the microtubule assembly (Leroux et al., 2002). Alterations in the binding sites on the ß-tubulin protein are related to benzimidazole-resistance (Leroux & Clearjeau, 1985). Approximately 10 mutations conferring resistance to MBC have been identified in the ß-tubulin gene in laboratory studies with a wide range of different fungi. Benzimidazole-resistance in *B. cinerea* is conferred by polyallelic major gene named *Mbc1* by Faretra & Polastro (1991) with at least four classes of alleles responsible for sensitivity or different levels of resistance variously accompanied by hypersensitivity to *N*-phenylcarbamates (Faretra et al., 1989; Faretra & Pollastro, 1991, 1993a; Pollastro & Faretra, 1992; Yarden & Katan, 1993; Davidse & Ishii, 1995, De Guido et al., 2007). The presumed mutated locus encoded the structural gene for ß-tubulin and single base pair mutations occurred in codons 198 and 200. Two phenotypes exhibiting benzidimadozle-resistance were determined by Leroux et al (2002) in *B. cinerea* populations from French vineyards. Phenotype Ben R1 exhibit high resistance levels (greater then 250) to MBC is simultaneously more sensitive to phenylcarbamate dietophencarb then the wild type strains. The second phenotype Ben R2 was detected after introduction of the mixture carbendazime+ dietophencarb in 1987. Ben R2 is moderately resistant to MBC (levels 100-200) and insensitive to dietophencarb, just like strains sensitive to MBC. In both phenotypes resistance was conferred by alleles of the *Mbc1*: in Ben R1, at position 198 an alanine replaced a glutamate, whereas in Ben R2, at position 200 a tyrosine replaced a phenylalanine (Yarden & Katan, 1993). Resistance to the MBC is a type of 'qualitative' or 'single-step' resistance characterised by a sudden and marked loss of effectiveness, and by the presence of clear-cut sensitive and resistant pathogen populations with widely differing responses (Brent, 1995). Once developed, it tends to be stable, resistant strains have persisted after many years of non-use and sensitivity will usually not be restored by cessation of their use.

Due to stable resistance in vineyards and also for toxicological reason (unwanted toxic residues in vine) MBI were redrawn from use in protection of vineyards.

Benzimidazole carbendazim was followed by dicarboximides which has been available since 1976 (Lorenz & Eichhorn, 1978). Owing to MBI resistance they were welcomed and become recognized as botryticides due to their efficacy superior to formerly used fungicides for that purpose. For almost a decade it seemed that the protection of vineyards against B. cinerea had been successfully solved. The appearance of resistance to dicarboximides did not come as so obvious and sudden loss of efficacy that gave first indication of resistance in the case of MBI. Dicarboximides efficacy was diminishing with time and protection slowly become insufficient. Therefore, resistance to dicarboximides, appears to involve slower shifts toward insensitivity because of multiple-gene involvement. As resistance management strategies were poorly understood at that time this inevitably led to dicarboximides overuse and resistance development. In spite of resistance development no total loss of control occurred so dicarboximides use was continued. Moreover, there were no alternative botryticides at the time and as consequence, the proportion of resistant strains in B. cinerea population increased considerably. Resistance to dicarboximides in vitro was achieved in 1976 (Leroux et al., 1977). Practical dicarboximides-resistance was firstly detected in 1978 in Switzerland (Schüepp & Küng, 1978). The first appearance of resistance in a particular fungicide-pathogen combination in one region has almost always been accompanied, or soon followed, by parallel behaviour in other regions where the fungicide is applied at a similar intensity (Brent, 1995). Thereby, resistance was determined in 1979 in Germany (Holz, 1979) and in Italy (Gulino & Garibaldi, 1979) and in 1982 in France (Leroux & Basselat, 1984; Leroux & Clerjeau, 1985). In Hungary dicarboximides were registered in 1978 and decrease in sensitivity was observed in 1988 and confirmed in 1994 (Dula & Kaptas, 1994). In Slovenian vineyads dicarboximides-resistance was reported (Maček, 1981). In Croatia dicarboximides were introduced in protection of vineyards in 1979. A decrease of efficacy was observed at the end of '80-ties and resistance was proved in 1990 (Cvjetković et al., 1994). Since the beginning of the 1980s, practical resistance to dicarboximides has been related to the selection of moderately resistant strains, named ImiR1 (Leroux & Clerjeau, 1985). Initial studies on dicarboximides-resistance management were started in Germany (Löcher et al., 1985) and France (Leroux & Clerjeau, 1985). To delay the selection of resistant strains during the vegetative period the use of dixarboximides was soon restricted to only two treatments after veraison in Europe (Besselat, 1984; Locher et al., 1987). Unfortunatelly, their efficacy seemed to decrease with infection pressure and goes under 40% and most of the dicarboximides-resistant strains also exhibited high simultaneous resistance to benzimidazoles (Schlamp, 1988). Dicarboximides disturb the synthesis of the cell wall of hyphae by inducing accumulation of glycerol, which burst eventually. A lot of effort was made to investigate primary mode of dicarboximides action. In 1977 was suggested that the primary effect of vinclozolin and iprodion is on DNA production and that lipid metabolism is also affected (Leroux et al., 1977). Following studies showed that dicarboximides have little effect on respiration or the biosynthesis of sterols, nucleic acids, proteins or chitin (Pappas & Fisher, 1979). Edlich & Lyr (1987) described that dicarboximides inactive enzymes are involved in electron transport, causing the production of reactive oxygen products (like O_2^- and H_2O_2) and initiate lipid peroxidation. Moreover, enhanced levels of catalase and superoxide dismutase recorded in some dicarboximides-resistant strains could be responsible for the detoxification of peroxy radicals although a conclusive correlation between amounts of such enzymes and the levels of fungicide resistance has not been found when comparing many field strains and laboratory mutants of *B. cinerea* (Leroux et al., 2002; Edlich & Lyr, 1992). According to Edlich & Lyr (1992) the potential target site of dicarboximides might be a plasma-membrane-bound NADPH-dependent flavin enzyme, inhibition of which would initiate pathological oxidative processes. Therefore, components of glutathione system are targets of dicarboximides. Several findings suggest that they interfere with the osmotic signal transduction pathway consisting of histidine kinase and MAP kinase cascades. Therefore, their primary target sites could be protein kinases involved in the regulation of polyol biosynthesis (Leroux et al., 1999; Schumacher et al., 1997). Set up of target site dicarboximides affecting should enable confirmation of gene responsible for resistance. But, despite of many long-term investigations the mechanism of dicarboximides resistance is not elucidating yet. The most comprehensive data on the genetics of dicarboximides-resistance have been obtained from studies of F. Faretra whose work has clarified the sexual behaviour and matting system of *B. cinerea* and resulted in a reliable technique for obtaining ascospore progeny under laboratory conditions (Faretra & Antonaci, 1987). Resistance to dicarboximides is encoded by a single polyallelic major gene named *Daf1* (Faretra & Antonaci, 1987). Firstly, two alleles of *Daf1* have been recognized (Faretra & Pollastro, 1991): *Daf1* LR and *Daf1* HR responsible for low and high resistance to dicarboximides. Alleles *Daf1* HR also result in hypersensitivity to high osmotic pressure. In further studies conducted with field isolates and laboratory mutants general, was perceived that the resistance mechanism of field isolates differs from that of laboratory isolates. Dicarboximides resistant field isolates were designate as Imi R1 and laboratory mutants as Imi R4 (Leroux et al., 2002). Practical resistance to dicarboximides was only detected with Imi R1 strains (carrying *Daf1* LR alleles) and not with Imi R4 (carrying *Daf1* HR alleles) because of the absence of Imi R4 strains under field conditions. Most dicarboximides-resistant laboratory mutants (Imi R4) acquire high resistance to dicarboximides, but also to aromatic hydrocarbons (AHF) and pheylpyrolles and they are hypersensitive to osmotic stress. High-level dicarboximides-resistant strains of *B. cinerea* have seldom been obtained in the field whereas low- and moderate-level resistant strains (Imi R1) are normally associated with field isolates and are still capable of causing disease control failure. Furthermore, from the field only moderately resistant strains (Imi R1) without osmotic-sensitive phenotypes are recovered (wild type strains are tolerant to osmotic pressure). In addition, dicarboximides-resistant field isolates (Imi R1) show various levels of cross-resistance to aromatic hydrocarbons (AHF) (due to similarity of chemical structure because both have benzene ring in chemical structure) but not to phenylpyrolles (fludioxonil).

Fungicidal toxicity of phenylphyroles is reverted by piperonyl butoxide and α-tocopherol in *B. cinerea*. Different levels of dicarboximides-resistance variously accompanied by resistance to phenylpyrrole fungicides and reduced tolerance to high osmotic pressure point to polymorphism of *Daf1* and with time become evident that there are at least five classes of responsible alleles (Hilber et al., 1995; Faretra & Pollastro, 1991; Faretra & Pollastro, 1993a, 1993b; Vignutelli et al., 2002; Baroffio et al., 2003). Recent studies suggested that an amino acid substitution of serine for isoleucine in the second unit of tandem amino acid repeats on 86 codon of BcOS1p gene is responsible for dicarboximides resistance in the field (Oshima et al., 2002). Preliminary data show that all strains containing a mutation from isoleucine to serine are resistant to dicarboximides without exception. However, some isolates with isoleucine at codon 86 in the second unit are resistant to dicarboximides, suggesting the possibility of other types of resistant strains in the field. Furthermore, Oshima et al. (2002) suggest that most of the mutations within the *BcOS1* gene affect virulence or fitness in *B. cinerea* under field conditions owing to well known fact of dicarboximides-resistant strains rapid decreases after discontinues applications of dicarboximides. According to Leroux (Leroux et al., 2002) dicarboximides-resistant field strains (Imi R1) contained a single base pair mutation at position 365 in a two-component histidine kinase gene, probably involved in the fungal osmoregulation. Dicarboximides-resistant laboratory strains (Imi R4) contained a single base pare mutation on 325 codon in gene also responasble for histidine kinase. In addition, both field strains Imi R1 and laboratory resistant strains Imi R4 showed resistance to the aromatic hydrocarbon fungicides (AHF) and especially to dicloran which is effective against grey mould on lettuces and on fruits during storage. Other *B. cinerea* isolates, Imi R2 and Imi R3, with different patterns of cross-resistance, were also detected in French vineyards (Leroux et al., 1999). Dicarboximides-resistant strains Imi R2 show cross-resistance to both phenylpyrroles and AHFs while Imi R3 are more resistant to dicarboximides then Imi R1 but are weakly resistant to pheylpyrroles. In some *B. cinerea* mutants, fungicide resistance was caused by a mutation in another gene, Daf2, which did not seem to be linked to the Daf1 gene (Faretra & Pollastro, 1993b). Although the primary target site of dicarboximides, phenylpyrroles and AHFs has not been clearly identified, these fungicides are the only commercial ones that seem so far to interfere with plant pathogens through the inhibition of a protein kinase (cit. Leroux et al., 2002). *B. cinerea* practical resistance to phenylpyrroles has not been demonstrated in the vineyards to date.

In the mid-1990s arise a novel family of botryticides, the anilinopyrimidines, with three representative ingredients: pyrimethanil, cyprodinil and mepanipyrim. Although anilinopyirimidines showed to be highly effective against *B. cinerea* a high risk of resistance was already evident in the first laboratory investigations (Birchmore & Forster, 1996) and therefore were put on the market with recommendations for restricted use. In the field pyrimethanil- and cyprodinil-resistant strains of *B. cinerea* were detected during preliminary testing in 1993 and 1994 in French (Leroux & Gredt, 1995) and Swiss vineyards (Forster & Staub, 1996). In Italy resistant strains were detected in 1996 even in vineyards where anilinopyrimidines have never been used before (Gullino & Garibaldi, 1979). Resistance to mepanipyrim was tested only in Japan and was not detected (Muramatsu & Miura, 1996). Organisation FRAC (Fungicide Resistance Action Committee at Global Crop Protection Federation (GCPF)) formed a new working group for anilinopyrimidine-resistance which in 1995 organised "ad hoc EPPO Workshop" in Switzerland and addressed to all winegrowing countries because of: "*... emergent and critical situation of B. cinerea resistance to anilinopyrimidines especially in vineyards...*" Even then was emphasize that efficacy of anilinopyrimidines can be saved and prolonged only with well organized monitoring and antiresistant strategy. Anilinopyrimidines exhibit some systemic translocation in plant tissues, and together with their image of pathogenesis inhibitors they possess protective activity and as it is said also some curative activity. Yet, in order to achieve satisfactory botryticidal effect it is recommended to use them preventively. They do not affect spore germination but germ tube elongation is inhibiting as well as mycelial growth at low concentrations. Under *in vitro* studies toxicity toward mycelial growth depends upon nutrition status of media and is greatly reduced on rich complex media. They posses ability to prevent fungal secretion of hydrolytic enzymes such as protease, cellulase, lipase or cutinase which play an important role in the infection and therefore they are considered as inhibitors of pathogenesis (Miura et al., 1994; Milling & Richardson, 1995). In *B. cinerea* anilinopyrimidines prevent secretion of laccase and in grape treated with pyrimethanil reduce laccase activity can be observed (Dubos et al., 1996). Laccasa is phenol oxidase and causes oxidation of must so reduction of their quantity by pyrimethanil is welcome effect. The exact mechanism of action in the protein secretory pathway is not yet understood; it has been hypothesized that the target of anilinopyrimidines could be a step involving the Golgi complex or a later stage (Milling & Rhichardson, 1995; Miura et al., 1994). Anilinopyrimidines are particularly inhibitors of methionine biosynthesis. Enzymes which are involved in methionine biosynthesis are cystathionine γ-sinthase and cystathionine β-lyase. The late one might be their primary target site (Leroux et al., 2002). Biochemical studies showed that methionine and homocysteine (prior to methionine) were lower in mycelia after treatment by pyrimethanil while slight increase of precursor cystathionine was recored. However, recent enzymatic studies revealed only weak inhibition of cystathionine β-lyase by anilinopyrimidines (Leroux, 1994; Masner et al., 1994) and conclusive results with the isolated enzyme were not given. Moreover, the relevance of the inhibition of methionine biosynthesis in the fungus while it is invading plant tissue (that offers a rich source of methionine) has yet to be elucidated (Rosslenbroich & Stuebler, 2000). According to Rosslenbroich & Stuebler (2000) the inhibition of methionine biosynthesis and the secretion of hydrolytic enzymes may be associated with the mechanism of antifungal action of the anilinopyrimidines but might also be only a secondary effect. The discovery of *B. cinerea* strains that exhibit *in vitro* and *in vivo* resistance to anilinopyrimidines suggests that they do not interfere with pathogenesis alone. Based on long-term monitoring conducted since 1993 in French vineyards Leroux (Leroux et al., 1999) distinguished two groups of strains resistant to anilinopyrimidines: I) highly resistant strains with EC 50 greater than 0.5 mg l-1 and II) less resistant strains with EC 50 lower than 0.4 mg l-1 . All AniR strains were *transposa* types but according to their response to other fungicides they proposed following three anilinopyrimidine-resistant phenotypes: AniR1, AniR2 and AniR3. Practical resistance was observed only with AniR1 strains. Phenotype AniR1 was found in most European countries. It is moderately to highly resistant to anilinopyrimidines and *in vitro* response to new experimental anilide SC-0858 and other fungicidal groups is similar to the wild-type strains. Phenotypes AniR2 and AniR3 are weakly resistant to anilinopyrimidines and resistance was mainly recorded at the germ-tube elongation stage. Most important was founding that AniR2 and AniR3 were cross-resistant to chemically unrelated fungicides: dicarboximides, phenylpyrroles, several inhibitors of sterol biosynthesis, fenhexamide, tolnaftate,14 α-demethylation (DMIs), anilide SC-0858 and cyclohexamide (Leroux et al., 1999; Chapeland et al, 1999). AniR3 isolates were also resistant to azole fungicides. Genetic analysis showed that fungicide resistance in phenotype AniR1 is encoded by one major gene (Chapeland et al., 1999; Hilber & Hilber-Bodmer, 1998) but AniR2 and AniR3 are encoded by two different single major genes. Chapeland et al (1999) hypothesize that AniR2 and AniR3 posse's mechanism of resistance which consist of reduced accumulation of fungicides within mycelium and could be mediated by excretion of toxic molecules. Mechanism is correlated with increased mRNA levels of specific transport genes. Phenomenon of reduced accumulation of fungicides is known as "pleiotropic drug resistance" (PDR) or "multi drug resistance" (MDR) which is discussed later in 3.2. Therefore, as these phenotypes are multi-drug resistant (MDR) Chapeland et al. (1999) renamed AniR2 as MDR1 and determined them as anilniopyrimidine-resistant strains with considerable cross-resistance levels mainly towards fludioxonil, cyprodinil and tolnaftate. Ani R3 become MDR2 which is characterized by increased resistance to fenhexamid, tolnaftate, cycloheximide, and cyprodinil. A third MDR phenotype, MDR3, was first detected in 2001. It is characterized by the highest levels and broadest spectrum of resistance against most fungicides tested (Kretschmer et al., 2009) contraty to MDR1 and MDR2 which have overlapping but distinct profiles. The frequency of MDR strains in the Champagne steadily increased until 2008, when the three MDR phenotypes together represented 55% of the total population. In contrast to the Champagne in German Wine Road region between 2006 and 2009, increasing MDR populations were also observed, but the MDR1 phenotype was clearly dominating (Kretschmer et al., 2009). In Croatia the occurence of resistant phenotype of B. cinerea to pyrimethanil was determined for the first time in 1999 after three years of intensive use of pyrimethanil (Topolovec-Pintarić & Cvjetković, 2002). The cross-resistant strains to cyprodinil were detected also (Topolovec-Pintarić & Cvjetković, 2003). Pyrimethanil- and cyprodinil-resistant strains were also detected in vineyards where anilinopyrimidines had never been used and this strains seemed to be "naturally" resistant and could be of AniR1 type. The growing number of resistant phenotype from the first to the last year of the 3-years trial lead to the conclusion of the appearance of so called "acquired resistance". The testing was conducted *in vitro* by germ tube assay so the resistance was determined in the germ-tube elongation stage. All anilinopyrimidine-resistant strains were *transposa* (Topolovec-Pintarić et al., 2004) so they belong to AniR2 or AniR3. Some of isolates showed cross-resistance to unrelated vinclozolin and fenhexamid (Topolovec-Pintarić, 2009). Described profile imply that this strains could belong to MDR2 type.

In 1999 botryticide with a high preventive activity against B. cinerea, fenhexamide, was introduced into vineyards (Baroffio et al., 2003). Owing to its novel mode of action, differing from all other botryticides it was presumed that resistance to fenhexamid will not occur easily. Analyses of unsaponifiable compounds conducted by Debieu et al. (2001) revealed that fenhexamid inhibited sterol biosynthesis in B. cinerea. The major sterol constitutes in B. cinerea, as well as in most filamentous fungi, is ergosterol. In the presence of fenhexamide ergosterol is reduced while its precursors 4α-methyl and 4-desmethyl 3-keto compounds are accumulated. This indicates that fenhexamid inhibits the 3-keto reductase, one of the four proteins of the enzymatic complex implicated in sterol C-4 demethylation process (Debieu et al, 2001). Thus, inhibition of 3-keto reductase leads to sterol C-4 demethylation inhibition and as a result the 4α-fecosterone and fecosterone are produced. Subsequent isomerization of 4α-fecosterone and fecosterone would give 4α-methylepisterone and episterone. Sterone accumulation is linked to growth inhibition and therefore is responsible for fenhexamide fungi toxicity. High risk of resistance was already evident in the first laboratory investigations. The baseline sensitivity of B. cinerea towards fenhexamide was recorded in 1992 and afterwards resistant strains of B. cinerea were detected in French and Swiss vineyards but so far loss of the fungicide's effectiveness has never been observed (Leroux et al., 1999; Suty et al., 1997; Baroffio et al., 2003). In France were high level fenhexamid-resistant strains collected even before use of fenhexamid. In Switzerland was obtained that fenhexamid-resistance can develop very rapidly, during 4 years from 0% up to 100% (Barofio et al., 2003). Knowledge on the risk of resistance to this fungicide is so far scant, although limiting the number of sprays per season is recommended (de Guido et al., 2007; Fungicide Resistance Action Committee [FRAC], 2006). It seems that fenhexamid resistance is not easily induced in B. cinerea because experiments towards selection of laboratory mutants resistant to fenhexamid (either spontaneous or UV-induced) produced only few mutants often with aberrant morphology and colony growth. Hence, there are an association between fenhexamid-resistance and reduced fitness (de Guido et al., 2007). Genetic analysis indicated that the resistant phenotypes are encoded by single major gene(s) which is/are not linked with *Mbc1* and *Daf1* (de Guido et al., 2007). Fenhexamid resistance is caused by mutations in the erg27 gene encoded 3-keto reductase according to Fillinger et al. (2008). Alberini et al. (2002) described four phenotypes according to their responses towards fenhexamid: HydS, HydR1, HydR2 and HydR3. A HydS type is wild type sensitive to fenhexamid. HydR1 is naturally fenhexamid-resistant strains with negative cross-resistance to other SBIs (sterol biosynthesis inhibitors) such as prochloraz (14α-demethylase inhibitor or DMI) and higher sensitivity to fenpropidin (Δ14-reductase inhibitor). HydR2 and HydR3 are insensitive to fenhexamid and are similar in lower sensitivity to SBIs fungicides and microtubule inhibitors (carbendazim and dietofencarb). They are representatives of acquired resistance differing in response toward fenhexamid in the stage of germ-tube elongation; HydR2 is weakly resistant while HydR3 is highly resistant to fenhexamid. This suggests that distinct mutation in the same locus or in different loci is involved. Because of their high resistance level Hyd R3 strains have to be considered relative to risk of resistance occurrence. But, their poor overwintering capacities suggesting that they probably do not impact fenhexamid field's efficacy as laboratory investigations as well as field trials indicated (Ziogas et al., 2003; Suty et al, 1997; Kretschmer & Hahn, 2008; Topolovec-Pintarić, 2009; Korolev et al., 2011). For example, in long term trial conducted in vineyard by Suty et al. (1999) fenhexamid was used for 3-4 sprays per year and no reduce of effectiveness was observed although *B. cinerea* isolate less sensitive than normal did appear in the field. The Albertini et al. (2002) analyzed gene CYP51 and determined its DNA sequence. The gene CYP51 was highly polymorphic and this allowed distinction among HydR1 and non-HydR1 strains. At HydR1 CYP51 gene show two non-silent mutations: at position 15 expressed is phenylalanin instead of isoleucin and a serine instead of asparagine at position 105. Recently, Billard et al (2011) described that the major mechanism responsible for fenhexamid-resistance at Hyd R2 and Hyd R3 is fenhexamid detoxification by cytochrome P450 named cyp68.4.

Recently released botryticidal ingredient for use in grapevines was boscalid (carboxamide). Preliminary survey of *B. cinerea* populations from Champagne vineyards did not detect any strains moderately or highly resistant to boscalid and showed the absence of cross-resistance with benzimidazoles, phenylcarbamates and anilinopyrimidines (Leroux et al., 2010). The first resistant strains were found in 2006 in French and German vineyards and their number increased till 2008 (Lerox et al., 2010). Boscalid is the succinate dehydrogenase inhibitor and inhibits the fungal respiration by blocking the ubiquinone-binding site at mitochondrial complex II. Therefore, boscalid-resistance is caused by alterations in the respiratory succinate dehydrogenase (Avenot et al., 2008). The six phenotypes were characterized by Laleve et al (2011) according to their resistance pattern: CarR1-CarR4 with low to medium level of resistance and highly resistant CarR5 and CarR6. CarR1 and CarR2 are currently most frequent in France and Germany. For boscalid-resistance seemes to be responsible mutations in SDH proteins. For all except CarR2 phenotype, putative mutatuion occuring in the sdhB gene lead to a specific amino acid change in the *sdhB* gene. Strains of CarR2 phenotype may be distinguished in at least 3 sub-groups: I) point mutation in the *sdhB* gene, II) point mutation in the *sdhD* gene and III) no mutation in any of the four *sdh* genes. To summarize, *B. cinerea* resistant phenotypes in correlation to fungicides *in vitro* effect towards germ-tube elongation and mycelial growth can be separated into three main categories. Phenotypes which exibit resistance at both stages are: I) Ben R1 and Ben R2 resistant to anti-microtubule toxicants; II) Imi R1 resistant to dicarboximides, III) Ani R1 resistant to anilinopyrimidines and IV) Hyd R3 resistant to fenhexamid. Phenotypes which expressed resistance mainly at stage of germ-tube elongation are only anilinopyrimidine-resistant Ani R2 and Ani R3 (also MDR phenotypes). Finaly, only phenotype Hyd R2 expressed fenhexamid-resistance at mycelial growth stage. Leroux et al (2002) stated that only phenotypes whose exibiting in vitro resistance at both development stages of *B. cinerea* will probably lead to practical resistance. This found not to be true for Hyd R3 phenotype as pratical resistance to fenhexamide has never been recoreded naimly to the rarity of Hyd R3 strains in vineyards. Furthermore, three main mechanisms of *B. cinerea* resistance to botryticides are indentified: reduced penetration of toxicants, increased detoxification or decreased conversion to toxic metabolites and reduced sensitivity of the target site.

3.2 Multi drug resistance phenomenon in *Botrytis cinerea*

Multiple drug resistance (MDR) phenomenon imply simultaneous reduced sensitivity to several different unrelated compounds. MDR is known as common in human pathogenic microbes and even cancer cells. In agricultural practice obvious cases of MDR in field strains of plant pathogenic fungi are restricted. The classic example of an MDR phenotype in *B. cinerea* is the cross-resistance to aromatic hydrocarbons, dicarboximides and other fungicides. The actual mechanism for this type of B. cinerea mutants has never been elucidated although many putative mechanisms of resistance were suggested. Recently, de Waard et al. (2006) suggested that drug transporters may have played a role in this case of resistance. But, the first expression of *B. cinerea* MDR phenomenon in the vineyard was detected between 1993 and 1997 from French vineyards located in Alsace, Armagnac, Bordeaux, Champagne and Loire Valley and described by Leroux et al. (1999). In some anilinopyrimidine-resistant phenotypes, named AniR2 and AniR3, they observed resistance extending to dicarboximides, phenylpyrroles, sterol biosynthesis inhibitors (e. g. tolfante, prochloraz, tebuconazole), and finally to hydroxyanilide derivate, fenhexamide. In Italy in 1996 anilinopyrimidine-resistant strains were detected even in vineyards where ingredient from this botryticidal group were never used before. Moreover, one of the isolates showed to be resistant simultaneously to fludioxonil, dixarboximides and benzimidazoles although it's virulence was very low. According to Chapeland et al. (1999) three MDR phenotypes can be distinguished in *B. cinerea*: 1) MDR1 strains show reduced sensitivities against fludioxonil and cyprodinil, 2) MDR2 strains are less sensitive to fenhexamid, cyprodinil and iprodione, 3) MDR3 strains are MDR1xMDR2 recombinants and thus show further reduced fungicide sensitivity. Until today MDR strains with additional boscalid resistance were never observed. Recently description of MDR phenothypes was reported in Germany by Leroch et al. (2011). They found most dominant to be MDR1 phenotype but interestingly, they detected MDR2 in low frequencies (2 %) in 2006 but until 2009 their number escalated (up to 26.7 %). Furthermore, MDR2 strains were carbendasim resistant also. Their hypothesis is that MDR2 strains have migrated eastward from Champagne to Germany based on the investigations conducted by Kretschmeir et al. (2009). MDR monitoring data from various investigations indicate that fungicide resistance patterns in *B. cinerea* are following current fungicide treatments. Development of MDR in human pathogens is explained as consequence of over expression of drug efflux transport proteins. In *B. cinerea*, as in other plant pathogenic fungi this is also one of the various mechanisms that allow them to survive toxic compunds in their environment such as plant defense compounds, antibiotics and fungicides (De Waard et al., 2006). Efflux transporters are plasma membrane factors with low substrate specificity and depended of energy. Mutations leading to over expression of individual transporters can result in increased export of drug molecules back into their outer environment and thereby reduced sensitivity to drug. Thereby, they prevent accumulation of drug up to fungictoxic concentrations at their target sites inside fungal cells, preventing or reducing their toxic action. The major types of drug efflux proteins are ATP binding cassette (ABC) and major facilitator superfamily (MFS) transporters. The role of ABC and MFS transporters in efflux of natural and synthetic toxicants is today well known (De Waard et al., 2006). The genome of *B. cinerea* encodes more than 40 ABC-type, and more than 100 putative MFS-type efflux transporters although most of these transporters are not yet functionally characterized. Several ABC transporter genes have been cloned with variant basal transcript level (Vermeulen et al., 2001): I) undetectable *BcatrC, BcatrJ*, II) low level *BactrA, BcatrB, BcatrE, BcatrG, BcatrK*, and III) high level *BcatrF, BcatrH* and *Bcatrl. BcatrB* and *BctrD* are a true multidrug transporters (De Waard et al., 2006; Hayashi et al., 2002). *BacrB* is a determinant for the anilinopyrimidines, dicarboximides, phenyllpyroles fludioxonil and fenpiclonil and to antibiotic phenazines. *BctrD* affects the sensitivity of *B. cinerea* to azole fungicide oxpoconazol, dicarboximides and benzimidazoles as well as to the antibiotic cyclohexamide. Therefore, it is possible that *BcatrB* and *BctrD* function in the MDR1 and MDR2 isolates. Also, *BcatrK* is a determinant for orgnophosphorus fungicide iprobenfos and antibiotic polyoxin. The ABC transporter AtrB1 has been shown to transport a variety of natural and synthetic drugs (Stefanato et al., 2009). Transcription of *atrB* is induced by various drugs, and requires the zinc cluster transcription factor Mrr1. Factor Mrr is present in nucleus but remain inactive in the absence of inducing drugs. Permanent activation of Mrr1 due to mutations in *mrr1*, resulting in constitutive overexpression of *atrB* and multidrug resistance (MDR) phenotypes. Kretschmer et al. (2009) identified that in MDR1 strains mutations in the transcription factor Mrr1 lead to over expression of the ABC transporters AtrB but in MDR2 strains a promoter rearrangement leads to over expression of the MFS transporters MfsM2.

4. Why is *Botytis cinerea* prone to resistance building?

B. cinerea is pathogen which can inflict an extreme number of plants without apparent specialisation which point to considerable variation. Already in 1922 Brierley nicely expressed that as follows: "*Botrytis cinerea is perhaps the commonest and best known fungus and has been the centre of mycological research since the time of de Bary. The species B. cinerea may be visualized as, at any one moment, a cluster of numerous races or strains morphologically congruent on the host plant but in vitro showing marked and constant cultural differences*". Genetic studies of nature and extent of genetic variability in *B. cinerea* were initiated to provide clues to the mechanisms and adaptability on different hosts. The resistance phenomenon intensified the genetic studies of *B. cinerea* assuming limited understanding of the genetic structure is reflecting in difficulties in managing the disease. Also, behavioral differences towards fungicides have provided additional markers for the wide genetic polymorphism encountered in *B. cinerea* (Giraud et al. 1999; Leroux et al. 1999). Genetic studies showed that this fungus has really great morphological variability and metabolic and genetic diversity but the genetic basis for this variability is not elucidated. As the sexual stages (*Botyotinia fuckeliana*) are rarely observed in vineyards it was generally accepted that sexual recombination played no role in somatic and genetic diversity of *B. cinerea* and that therefore diversity must be due to heterokaryosis and aneuploidy. Hence, the high genetic diversity as well as the equal distribution of the two mating types in field populations indicates that sexual reproduction is a major reason for the heterogeneity of *B. cinerea* population (Choquer et al., 2007). There are also other nonexclusive hypotheses under investigation: occurrence of microviruses among wild populations can be one source of variability and other source can be differences in gene content and gene variability among *B. cinerea* strains. The genetic diversity of *B. cinerea* vineyard populations is so opulent that always seems to exceed the sampling size that can be handled in investigations. Recent genetic analysis showed that one of the causes of *B.cinerea* genotype variability could be transposable elements. Investigation on *B.cinerea* populations in Champagne vineyard showed the existence of two sibling sympatric species which can be distinguished according to the content of transposable elements (Giraud et al., 1999). Transposable elements are parts of DNA molecule, jumping genes which can change their position inside genomes, which means to change their position or locus on the chromosomes causing appearances like: gene inactivation, reactivation of pseudogenes, gene expression disorder, and also mutation type: deletion, insertion, inversion and translocation (Daboussi 1996; Daboussi & Capy, 2003). These sympatric species have been described as *Transposa* and *Vacuma* by Giraud et al. (1999). *Transposa* contains two transposable elements: gypsi-like retrotransposon Boty (Diolez et al. 1995) and transposon Flipper while *Vacuma* strains are devoid of both elements (Levis et al., 1997). *Vacuma* is showing higher diversity than *Transposa*. Newly population studies and phylogenetic analyses in France (Fournier et al., 2003) consider *B. cinerea* as a complex of species containing two genetically distinct populations, namely "Group I" (referred as *Botrytis pseudocinerea*) and "Group II" (*B. cinerea sensu stricto*) that were proposed to be phylogenetic species (Choquer et al., 2007). Groups exhibit difference in ecology and their resistance pattern to fungicides but it should be emphasize that they are unable to cross with each other. Group I is characterized by *vacuma* isolates sensitive to fenhexamide which are genetically isolated from others and therefore form a true phylogenetic species. Group II are all fenhexamid-resistant *vacuma* strains and *transposa*. Therefore, they are three main genetic types: Group I, Group II *vacuma* and Group II *transposa*. In the vineyards, as on various host plant, evident are changes in *B. cinerea* population structure and dynamics between genetic types during season. Group I play only a minor role in the epidemiology of bunch rot and is found only sporadically mostly at flowering on leaves and blossoms. Group II *vacuma* reached maximum on senescing floral caps and then decrease significantly until harvest but increase from autumn to flowering. This type of isolates are mostly isolated from over wintering sclerotia and in order with this it is hypothesized that *vacuma* isolates will express a more ruderal life-strategy with greater saprophytic capability (Furnier et al., 2003). Furthermore, it is regarded as a mix of different migrant populations from other host (Giraud et al., 1999). Group II *transposa* is the most virulent one and seems well-adapted to the grapes cause it is dominant at every phenological stage and isolated from over wintering canes more then other types. As it's showing a peak occurrence in rotted grape berries an epidemic of bunch rot is dominated by Group II *transposa* (Matinez et al., 2005). Group II *transposa* it seams to represent commonly occurring population of *B. cinerea* in European vineyards (Kretschmer & Hahn, 2008; Furnier et al., 2003; Topolovec-Pintarić et al., 2004).

According to Martinez et al. (2005) differences in the saprotrophic and pathogenic ability of the *vacuma* and *transposa* combined with a switch in resource availability from dead to living tissues is the most likely mechanism accounting for the success of *transposa* isolates and the decline of *vacuma* isolates during the course of an epidemic. The isolates that had only Boty and therefore they were named "Boty only" and were detected in France (Giraud et al., 1999) and in Chile (Muñoz et al., 2002). The "Boty only isolates" were frequently isolated on kiwifruit and in lower frequency on grapes and tomato. It is hypothesized that "Boty only isolates" may be result of crosses between *vacuma* and *transposa* and the transposon Boty may even be invading isolates of *vacuma* gropus (Muñoz et al., 2002). The isolates containing only Flipper have been found by Albertini et al. (2002). These isolates were origin from France vineyard and from UK strawberry fields and were resistant to fenhexamide. Considering fact that transposones influence genetic changes, especially expression of genes and gene mutations, it was hypothesized about possible relation to fungicide resistance. Some data about possible influence of transposones on the resistance to botryticides were obtained recently (Giraud et al, 1999). Group I is characterized by natural resistance to fenhexamide (Leroux et al., 2002; Suty et al., 1999; Martinez et al., 2005). Group II *transposa* strains are multiresistant and frequently resistant to vinclozolin and diethofencarb (Giraud et al., 1999; Martinez et al., 2005). Alberini et al. (2002) described four phenotypes according to their responses towards fenhexamid: HydS, HydR1, HydR2 and HydR3. A HydS type is wild type sensitive to fenhexamid and is either *vacuma* or *transposa*. HydR1 is fenhexamid-resistant with negative cross-resistance to other SBIs (sterol biosynthesis inhibitors) such as prochloraz (14α-demethylase inhibitor or DMI) and higher sensitivity to fenpropidin (Δ14-reductase inhibitor). Genetic analyses showed that they are mostly *vacuma* although few was Flipper only type. In comparison to other HydR types conidia of HydR1 are oversized and mycelial growth rate is higher. It should be emphasize that all HydR1 strains are compatible and mating obtains progeny but crosses between HydR1 and HydS failed to obtain progeny. Types HydR2 and HydR3 are of *transposa* type having smaller macroconidia (Giraud et al., 1999) and exhibit slower rates of mycelial extension when grown on highly nutritive agar media at different favorable temperatures (Martinez et al., 2005). They are similar in lower sensitivity to SBIs fungicides and microtubule inhibitors (carbendazim and dietofencarb). They differing in response toward fenhexamid in the stage of germ-tube elongation; HydR2 is weakly resistant while HydR3 is highly resistant to fenhexamid. The Albertini et al. (2002) analyzed gene CYP51 and determined its DNA sequence. The gene CYP51 was highly polymorphic and this allowed distinction among HydR1 and non-HydR1 strains rather then between *vacuma* and *transposa*. At HydR1 CYP51 gene show two non-silent mutations: at position 15 expressed is phenylalanin instead of isoleucin and a serine instead of asparagine at position 105. Absence of genetic exchange between HydR1 and non-HydR1 combined with morphological and somatic incompatibility suggest that these two groups are from distinct genetic entities and might even be non HydR two different species. As types in non HydR group are more variable it is probably composed of different subpopulations whose phylogenetic relationship is still not resolved. The *transposa* isolates resistant to fenhexamid, carbendazim and vinclozolin were detected in other investigations (Martinez et al., 2005; Giraud et al., 1999) and resistance appeared to be associated with an increased virulence. The *transposa* resistant profiles, according to Martinez et al. (2005) are most likely a consequence of population dynamics and are generated by the application scheme of fungicides. Therefore, *transposa* as predominant at veraison is exposed to greater selective pressure of vinclozoline, resulting in a greater frequency of vinclozoline-resistant strains. Also, with speculations about ability of *transposa* isolates to develop fungicide resistance, as well as increased virulence, arises a question if this ability could be based on MDR (multi-drug resistance) systems. The MDR system allows efflux of various cytotoxic drugs such as plant defense compounds (e.g., phytoalexins) and botryticidal coumpound. The role of such mechanisms in the population dynamics of *B. cinerea* genetic types warrants further investigation. At the end, *B. cinerea* can be considered as actually a complex of several different entities (Albertini et al., 2002).

5. Conclusion and future prospect

In order to overcome the problem of *Botrytis* resistance there are continuous world efforts to develop new active ingredients. Monitoring methods were developed and again improving. Antiresistant strategies are applied. However, resistance to botryticides still pursuing to be economically significant problem in *B. cinerea* management as it will be in foreseeable future, mainly due to selection of MDR mutants with high levels of resistance. Ironically, selection of MDR mutant is favoured by some recommendation of antiresistant strategy. In most European vineyards is obey the recommendation to alternate the various groups of botryticides with restriction to one spray per year for each chemical group because they are single-site fungicides and to use fungicide mixtures. However, this antiresistant measure is impeded by the development of MDR phenotypes. This measure delay or prevent the evolution of mutant with target site modifications but presents a multi attack by fungicides which favours stepwise evolution of polygenic resistance related to ABC transporters. On the other hand, gained knowledge of MDR mechanisms could be used as new weapon against not only *B. cinerea* but other phytopathogenic fungi as well. Fungal mutants that lack drug transporters become hypersensitive and can be used as tester strains in the creating new fungicidal compounds. New antifungals perhaps can be compounds that inhibit ABC transporters. Analogues in human medicine exit as blockers or modulators of ABC transporters. Furthermore, these compounds can be synergist of existing fungicides or the ones that annul MDR in phytopathogenic fungus. Another line of new antifungals can be disease control compounds that even do not posses a direct fungitoxic activity and may be considered as inducers of plant defence mechanism. Such compounds can act as inhibitors of ABC transporters involved in fungal virulence which will lead to enhancement of plant defence compounds (phytoalexines, pathogenesis-related proteins) in fungal cells.

Yet after all, the basic for resistance prevention and management remains necessity of updating our knowledge about *B. cinerea*, although it is already abundad and comprehensive. For this goal the data about structure and dynamics of *B. cinerea* populations in commercial vineyards as well as the distribution of different types of fungicide resistance, including MDR types, should be obtained. More attentions should be given to today available alternatives to classical botryticides like biological control, or use of mineral salts and plant activators.

6. Acknowledgements

I am indebted to Edyta Đermić for her helpful contributions and valuable comments during the writing and revision of this test.

7. References

Albertini, C.; Thebaud, G.; Fournier, E.; & Leroux, P. (2002). Eburicol 14α-demethylase gene (*CYP51*) polymorphism and speciation in *Botrytis cinerea*. *Mycological Research*, No. 106, pp. 1171-1178, ISSN 0953-7562

Avenot, H.F.; Sellam, A.; Karaoglanidis, G. & Michailides, T.J. (2008). Characterization of mutations in the iron-sulphur subunit of succinate dehydrogenase correlating with boscalid resistance in *Alternaria alternata* from California Pistachio. *Phytopathology*, No. 98, pp. 736–742, ISSN 0031-949X

Bais, A. J.; Murphy, P. J. & Dry, I. B. (2000). The molecular regulation of stilbene phytoalexin biosynthesis in *Vitis vinifera* during grape berry development. *Austrylian Journal of Physiology*, No. 27, pp. 425-433, ISSN 0310-7841

Barak, E. & Edgington, L.V. (1984). Cross-resistance of *Botrytis cinerea* to captan, thiram, chlorothalonil and related fungicides. *Canadian Journal of Plant Pathology*, No. 6, pp. 318–320, ISSN 0706-0661

Baroffio, C.A.; Siegfried, W. & Hilber, U.W. (2003). Long-term monitoring for resistance of *Botryotinia fuckeliana* to anilinopyrimidine, phenylpyrrole, and hydroxyanilide fungicides in Switzerland. *Plant Disease*, Vol. 87 No., pp. 662-666, ISSN 0191-2917

Beetz, K. J. & Löcher, F. (1979). Botrytisbekämpfung im Weinbau - Versuchserbabrisse aus den Jahren 1973-1978. *Weinberg und Keller*, Vol. 26, No. 25, pp. 238-249, ISSN 0508-2404

Beever, R.H. & O'Flaherty, B.F. (1985). Loe-level benzidmidazole resistance in *B. cinerea* in New Zeland. *New Zeland Journal of Agricultural Research*, No. 28, pp. 289-292, ISSN 0028-823

Besselat, R. (1984). Pourriture grise: Evolution des methods de lutte. *Phytoma*, No. 360, pp. 35-38, ISSN 1164-6993

Billard, A.; Fillinger, S; Lerox, P., Bach, J; Solignac P.; Lanen, C.; Lachaise, H.; Beffa, R. & Debieu, D. (2011). Natural and acquired fenhexamid resistance in Botrytis spp.: What's the difference?. *Fungal Genetics Reports,– Supplement: Proceedings of 26th Fungal genetics conference*, Asilomar, USA, March 15-20,2011. Vol. 58, ISSN 0895-1942

Birchmore, R.J. & Forster, B. (1996). FRAC methods for monitoring sensitivity of *Botrytis cinerea* to anilinopyrimidines. EPPO Bulletin, No. 26, pp. 181-197, ISSN 0250-8052

Bisiach, M.; Minervini, G.; Ferrante, G.& Zerbetto, F. (1978). Ricerche sperimentali sull 'attivita antibotrytica in viticoltora di alcuni derivati della 3,5-dichloranilina, *Vignevini*, No. 5, pp.23-27, ISSN 0390-0479

Bolton, A. T. (1976). Fungicide resistance in *Botrytis cinerea*, the result of selective pressure on resistant strains already present in nature. *Canadian Journal of Plant Science*, No. 56, pp. 861- 864, ISSN 0008-4220

Brent, K. (April, 1995). *Fungicide resistance in crop pathogens: how can it be managed?* (1st edition), Published by GCPF (Brussels), ISBN 90-72398-07-6, Bristol, United Kingdom.

Chapeland, F.; Fritz, R.; Lanen, C.; Gredt, M. & Leroux, P. (1999). Inheritance and mechanisms of resistance to anilinopyrimidine fungicides in *Botrytis cinerea* (*Botryotinia fuckeliana*). *Pesticide Biochemistry and Physiology*, Vol 62., No. 64, pp. 85–100, ISSN 0048-3575

Choquer, M.; Fournier, E.; Kunz, C.; Levis, C.; Pradier, J.M.; Simon, A. & Viaud, M. (2007). *Botrytis cinerea* virulence factors :new insights into a necrotrophic and polyphageous pathogen. *FEMS Microbiologial Letters*, No. 277, pp. 1–10, ISSN 0378-1097

Coertze, S.; Holz, G. & Sadie, A. (2001). Germination and establishment of infection on grape berries by single airborne conidia of *Botrytis cinerea. Plant Disease*, No. 85, pp. 668-677, ISSN 0191-2917

Cvjetković, B.; Topolovec-Pintarić, S. & Jurjević, Ž., (1994). Resistance of *Botrytis cinerea* Pers. ex Fr. to dicarboximides in Croatian vineyards. *Atti Giornate Fitopatologiche*, No. 3, pp. 181-186, ISSN 0567-7572

Daboussi, M. J. (1996). Fungal transposable elements: generators of diversity and genetic tools. *Journal of Genetics*, No. 75, pp. 325-339, ISSN 0022-1333

Daboussi, M. J. & Capy, P. (2003). Transposable elements in filamentous fungi. *Annual Review of Microbiology*, No. 57, pp. 275-299, ISSN 0066-4227

Davidse, L. & Ishii, T. (1995). Biochemical and molecular aspects of benzimidazoles, *N*-phenylcarbamates and *N*-phenilformamidoxines and the mechanisms of resistance to these compounds in fungi. In: *Modern Selective Fungicides*, Lyr H., pp. 305-322, Gustav Fischer, ISBN-10: 3334604551, Jena, Germany.

Debieu, D.; Bach, J.; Hugon, M.; Malosse, C. & Leroux, P. (2001). The hydroxyanilide fenhexamid, a new sterol biosynthesis inhibitor fungicide efficient against the plant pathogenic fungus *Botryotinia fuckeliana* (*Botrytis cinerea*). *Pest Management Science*, No. 57, pp. 1060–1067, ISSN 1526-4998

De Guido, M.A. ; De Miccolis Angelini, R.M.; Pollastro, S.; Santomauro, A. & Faretra, F. (2007). Selection and genetic analysis of laboratory mutants of *Botryotinia fuckeliana* resistant to fenhexamid. *Journal of Plant Pathology*, Vol. 89, No. 2, pp. 203-210, ISSN 0929-1873

Dekker, J. (1977). Resistance, In: *Systemic Fungicides*, Marsh R.W., pp. 176-197. Longman Scientific &Technical, ISBN 0470572507, 9780470572504, London.

De Waard, M.A.; Andrade, A.C.; Hayashi, K.; Schoonbeek, H.J.; Stergiopoulos, I. & Zwiers, L.H. (2006). Impact of fungal drug transporters on fungicide sensitivity, multidrug resistance and virulence. *Pest Management Science*, No. 62, pp. 195–207, ISSN 1526-498X

Diolez, A.; Marches, F.; Fortini, D. & Brygoo,Y. (1995). Boty, a long terminal repeat retroelement in phytopathogenic fungus *Botrytis cinerea. Applied and Environmental Microbiology*, Vol. 61, No. 1, pp. 103-108, ISSN 0099-2240

Dubos, B.; Roudet, J. & Lagouarde, P. (1996). The anti-laccase activity of pyrimethanil. Effect of the anti-*Botrytis* product on harvested grapes. *Phytoma*, No. 483, pp. 47-50, ISSN 0048-4091

Dula, T. & Kaptas, T. (1994). Monitoring study of the resistance of *Botrytis cinerea* to benzimidazole nad dicarboximides fungicidies on grapes in Hungary. *BCPC Monograph, 60: Fungicide resistance*, pp. 239-242. ISSN 0306-3941

Edlich, W. & Lyr H. (1987). Mechanism of action of dicarboximides fungicides. In: *Modern selective fungicides: Properties, Applications, Mechanisms of action*, Lyr H., Fisher G., Verlag J., pp. 107-118., Longman, London.

Edlich, W. & Lyr H. (1992). Target sites of fungicides with primary effects on lipid peroxidation, In: *Target sites of fungicide action*, Koller W., pp. 53–68, CRC Press, Boca Raton, Florida, USA.

Ehrenhardt, H.; Eichron, K. W. & Thate, R. (1973). Zur Frage der Resistenzbildung von *B. cinerea* gegenueber systemichen Fungiziden. *Nachrichtenbllatt des Deutschehen Pfllanzenschutzdienstes (Braunschweig)*, Vol 25, No. 49, ISSN 0027-7479

Fabreges, C. & Birchmore, R. (1998). Pyrimethanil: monitoring the sensitivity of *B. cinerea* in the vineyard. *Phytoma*, No. 505, pp. 38-41, ISSN 0048-4091

Faretra, F. & Antonacci, E. (1987):. Production of apothecia of *Botryotinia fuckeliana* (de Bary) Whetz. under controlled environmental conditions. *Phytopathologia Mediterranea* No. 26, pp. 29-35, ISSN 0031-9465

Faretra, F.; Pollastro, S. & Tonno, A.P. (1989). New natural variants of B.fuckeliana (B.cinerea) coupling benzimidazole-resistance to insensitivity toward the N-phenylcarbamate dietofencarb. *Phytopathologia Mediterranea*, Vol. 28, No. 2, pp. 98-104, ISSN 0031-9465

Faretra, F. & Pollastro, S. (1991). Genetic basis of resistance to benzimidazole and dicarboximides fungicides in *Botroytinia fuckeliana* (*Botrytis cinerea*). *Mycological Research*, No. 95, pp. 943–951, ISSN 0953-7562

Faretra, F. & Pollastro, S. (1993a). Genetics of sexual compatibility and resistance to benzimidazole and dicarboximides fungicides in isolates of *Botryotinia fuckeliana* (*Botrytis cinerea*) from nine countries. *Plant Pathology*, No. 42, pp. 48-57, ISSN 0032-0862

Faretra, F., & Pollastro, S. (1993b). Isolation, characterization and genetic analysis of laboratory mutants of *Botryotinia fuckeliana* resistant to phenylpyrrole fungicide CGA 173506. *Mycological Research*, No. 97, pp. 620-624, ISSN 0953-7562

Fillinger, S.; Leroux, P.; Auclair, C.; Barreau, C.; Al, H.C. & Debieu, D. (2008). Genetic analysis of fenhexamid-resistant field isolates of the phytopathogenic fungus *Botrytis cinerea*. *Antimicrobial Agents and Chemotherapy*, No. 52, pp. 3933–3940, ISSN 0066-4804

FRAC, 2006. Sterol Biosynthesis Inhibitor (SBI) Working Group. In: *Minutes of 2006 annual meeting, Recommendations for 2007*, http://www.frac.info/frac/index.htm.

Forster, B. & Staub, T. (1996). Basis for use strategies of anilinopyrimidine and phenylpyrrole fungicides against *Botrytis cinerea*. *Crop Protection*, Vol. 15, No. 6, pp. 529-537, ISSN 0261-2194

Fournier, E.; Levis, C.; Fortini, D.; Leroux, P.; Giraud, T. & Brygoo, Y. (2003). Characterization of Bc-*hch*, the *Botrytis cinerea* homolog of the *Neurospora crassa* het-c vegetative incompatibility locus, and its use as a population marker. *Mycologia*, No. 95, pp. 251-261, ISSN 0027-5514

Georgopoulos, S. G. (1979). Development of fungal resistance to fungicides. In: *Antifungal Compounds*, Siegel M. R., Sisler H. D, pp. 439-495, Marcel Dekker Inc. New York.

Giraud, T.; Fortini, D.; Levis, C.; Lamarque, C.; Leroux, P.; LoBuglio, K. & Brygoo, Y. (1999). Two sibling species of the *B. cinerea* complex, *transposa* i *vacuma*, are found in sympatry on numerous host plants. *Phytopathology*, Vol. 89, No. 10, pp. 967-973 ISSN 0031-949X

Gullino, M. L. & Garibaldi, A. (1979). Osservazioni sperimentali dalla resistence di isolamenti Italiani di *Botrytis cinerea* a vinclozolin. *La difesa delle piante*, No. 6, pp. 341-350, ISSN 0391-4119

Haenssler, G. & Pontzen, R. (1999). Effect of fenhexamid on the development of *Botrytis cinerea*. *Pflanzenschutz-Nachrichten Bayer (Bayer Crop Science journal)*, No. 52, pp. 158-176, ISSN 0340-1723

Hayashi, K.; Schoonbeek, H. & De Waard, M. A. (2002). Expression of the ABC transporter BcatrD from *Botrytis cinerea* reduces sensitivity to sterol demethylation inhibitor fungicides. *Pesticide Biochemistry and Physiology*, No. 73, pp. 110-121, ISSN 0048-3575

Hilber, U. W.; Schwinn, F.J. & Schüepp, H. (1995). Comparative resistance patterns of fludioxonil and vinclozolin in *Botryotinia fuckeliana*. *Journal of Phytopathology*, No. 143. pp. 423-428 ISSN 0931-1785

Hilber, U. W. & Hilber-Bodmer, M. (1998). Genetic basis and monitoring of resistance of *Botryotinia fuckeliana* to anilinopyrimidines, *Plant Disease*, No. 82, pp. 496-500, ISSN 0191-2917

Hisada, Y.; Maeda, K.; Tottori, N. & Kawase, Y. (1976). Plant disease control by N-(3,5-dichlophenyl)-1,1-dimethyl-cyclopropane-1,2-dicarboxamide. *Journal of Pesticide Science*, No. 1, pp. 145-149, ISSN 1348-589X

Holz, G. (1979). Über eine resistenzerscheinung von *Botrytis cinerea* an Reben gegen die neuen Kontakt-Botrytizide im Gebiet der Mittelmosel. *Weinberg und Keller*, No. 26, pp. 18-25, ISSN 0508-2404

Holz, G.; Gutschow, M.; Coertze, S. & Calitz, F. J. (2003). Occurrence of *Botrytis cinerea* and subsequent disease suppression at different positions on leaves and bunches of grape. *Plant Disease*, No. 87, pp. 351-358, ISSN 0191-2917

Jarvis, W. R. (1980). Epidemiology, In: *The Biology of Botrytis*, Coley-Smith J.R., Verhoeff K. and Jarvis W.R. , pp. 219-250, ISBN 0-12-179850-X, Academic Press, London, UK

Kaptas, T. & Dula, B. (1984). Benzimidazoltipusu fungicideekkel szembeni rezistens *Botrytis cinerea* Pers. törzs kialakulasa szölöben. (Resistance to benzimidazole fungicides built.up in strains of *Botrytis cinerea* Pers. in a vineyard). *Növenyvedelen*, No. 20, pp. 174-182

Korolev, N.; Mamiev, M.; Zahavi, T. & Elad, Y. (2011). Screening of *Botrytis cinerea* isolates from vineyards in Israel for resistance to fungicides. *European Journal of Plat Pathology*, Vol. 129, No. 4, pp. 591-608, ISSN 0929-1873

Kretschmer, M. & Hahn, M. (2008). Fungicide resistance and genetic diversity of Botrytis cinerea isolates from a vineyard in Germany. *Journal of Plant Disease and Protecttion*, No. 115, pp. 214–219, ISSN 1861-3829

Kretschmer, M.; Leroch, M.; Mosbach, A.; Walker A. S.; Fillinger, S.; Mernkel, D.; Schoonbeek H. J.; Pradier, J. M.; Leroux, P.; De Waard, M. A. & Hahn, M. (2009). Fungicide-driven evolution and molecular basis of multidrug resistance in field populations of the grey mould fungus *Botrytis cinerea*. *PLoS Pathogens*, Vol. 5, No. (12), e1000696. doi:10.1371/journal.ppat.1000696

Lacroix, L.; Bic, C.; Burgaud, L.; Guillot, M.; Leblanc, R.; Riottot, R. & Sauli, M. (1974). Etude des properties antifongiques d'une nouvelle famille de derives de l'hydantoine et an particulier du 26 019RP. *Phytiatrie-Phytopharmacie*, No. 23, pp. 165-174, ISSN 00318876

Laleve, A.; Walker, A. S.; Leroux, P.; Toquin, V.; Lachaise, H. & Fillinger, S. (2011). Mutagenesis of *sdhB* and *sdhD* genes in *Botrytis cinerea* for functional analysis of resistance to SDHIs. *Fungal Genetics Reports, Vol. 58 – Supplement 26th Fungal Genetics Conference*, Asilomar, USA, March 15-20, 2011, ISSN 0895-1942

Leroch, M.; Kretschmer, M. & Hahn, M. (2011). Fungicide Resistance Phenotypes of *Botrytis cinerea* Isolates from Commercial Vineyards in South West Germany. *Journal of Phytopatholohy*, No. 159, pp. 63–65, ISSN 0931-1785

Leroux, P.; Fritz, R. & Gredt, M. (1977). Etudes en laboratoire des souches de *Botrytis cinerea* Pers. resistantes a la dichlozoline, au dicloran, au quintozene, a la vinclozoline et au 26019 RP ou glycophene. *Phytopathologische Zeitschrift*, No. 89, pp. 347-, ISSN 0931-1785

Leroux, P. & Basselat, B. (1984). Pourriture grise: La resistance aux fongicides de *Botrytis cinerea*. *Phytoma*, No. 6, pp. 25-31, ISSN 0048-4091

Leroux, P. & Clerjeau, M. (1985). Resistance of *Botrytis cinerea* and *Plasmopara viticola* to fungicides in French vineyards. *Crop proterction*, No. 4, pp. 137-160, ISSN 0261-2194

Leroux, P. (1994). Effect of pH, aminoacids and various organic compounds on the fungitoxicity of pyrimethanil, glufosinate, captafol, cymoxanil and fenpiclonil in *Botrytis cinerea*. *Agronomie*, No. 14, pp. 541-544, ISSN 0249-5627

Leroux, P. & Gredt, M. (1995). Etude *in vitro* de la resistance de *B.cinerea* aux fongicides anilinopyrimidines. *Agronomie*, Vol. 15, No. 6, pp. 367-370, ISSN 0249-5627

Leroux, P.; Chapeland, F.; Arnold, A. & Gredt, M. (1998). Resistance de *Botrytis cinerea* aux fongicides, du laboratoire au vignoble et vice versa. *Phytoma*, No. 504, pp. 62-67, ISSN 0048-4091

Leroux, P.; Chapeland, F.; Desbrosses, D. & Gredt, M. (1999). Patterns of cross-resistance to fungicides in *Botryotinia fuckeliana* (*Botrytis cinerea*) isolates from French vineyards. *Crop Protection*, No. 18, pp. 687–697, ISSN 0261-2194

Leroux, P.; Fournier, E.; Brygoo, Y. & Panon, M.-L., (2002). Biodiversité et variabilité chez *Botrytis cinerea*, l'agent de la Pourriture grise. Nouveaux résultats sur les espèces et les résistances. *Phytoma*, No. 554, pp. 38-42, ISSN 0048-4091

Leroux, P.; Gredt, M.; Leroch, M. & Walker, A.S. (2010). Exploring Mechanisms of Resistance to Respiratory Inhibitors in Field Strains of *Botrytis cinerea*, the Causal Agent of Gray Mold. *Applied and Environmental Microbiology*, Vol 76, No. 19, pp. 6615–6630, ISSN 0099-2240

Levis, C.; Fortini, D. & Brygoo, Y. (1997): Flipper, a mobile Fot 1- like transposable element in *Botrytis cinerea*. *Molecular Genetics and Genomics*, No. 254, pp. 674-680, ISSN 1617-4615

Liguori, R. & Bassi, R. (1998). Cyprodinil (Chorus): nuovo fungicida per la difesa dei frutiferi da tricchiolatura e moniliosi. *Informatore fitopatologico*, No. 4, pp. 43-47, ISSN 0020-0735

Lorenz, D. H. & Eichhorn, K. W. (1978). Untersuchungen zur moeglichen Resistenzbuildung von *B. cinerea* an Reben gegen die Wirkstoffe Vinclozolin und Iprodione, *Die Wein Wissenschaft*, No. 33, pp. 251-255 . ISSN 0375-8818

Löcher, F. J.; Brandes, W.; Lorenz, G.; Huber, W.; Schiller, R.. & Schreber, B. (1985), Entwicklung einer Strategie zur Erhaltung der Wirksamkeit von Dicarboximidesn bei Auftreten von resistenten Botrytis-Stämmen an Reben. *Gesunde Pflanzen*, No. 37, pp. 502-507, ISSN 03674223

Löcher, F. J.; Lorenz, G. & Beetz, K. J. (1987). Resistance management strategies for dicarboximides fungicides in grapes: Results of six years trial work. *Crop protection,* No. 6, pp. 139. ISSN 0261-2194

Maček, J. (1981). O odpornosti sive plesni iz dolenjskih vinogradov proti sistemičnim fungicidom. *Sodijsko kmetijstvo,* Vol. 14, No. 7/8, pp. 293-294,

Malatrakis, N. E. (1989). Resistance of *Botrytis cinerea* to dichlofluanid in greenhouse vegetables. *Plant Disease,* No. 73, pp. 138–141 , ISSN 0191-2917

Martinez, F.; Dubos, B. & Fermaud, M. (2005). The role of saprotrophy and virulence in the population dynamics of *Botrytis cinerea* in vineyards. *Phytopathology,* No. 95, pp. 692-700, ISSN 0031-949X

Masner, P.; Muster, P. & Schmid, J. (1994). Possible methionine biosynthesis inhibition by pyrimidinamine fungicides. *Pesticide Sciience,* No. 42, pp.163-166, ISSN 0031-613X

Milling, R. J. & Richardson, C. J., (1995). Mode of action of the anilinopyrimidine fungicide pyrimethanil. 2. Effects on enzyme excretion in *Botrytis cinerea. Pesticide Science,* No. 45, pp. 43-48, ISSN 0031-613X

Miura, I.; Kamakura, T.; Maeno, S.; Hayashi, S. & Yagamuchi, I., (1994). Inhibition of enzyme secretion in plant pathogens by mepanipyrim, a novel fungicide. *Pesticide Biochemistry and Physiology,* No. 48, pp. 222-228, ISSN 0048-3575

Muñoz, G.; Hinrichsen, P.; Brygoo, Y. & Giraud, T. (2002). Genetic characterisation of *Botrytis cinerea* populations in Chile, *Mycological Research,* Vol. 106, No. 5, pp. 594-601, ISSN 0953-7562

Muramatsu, N. & Miura, I. (1996). Methods for evaluating the sensitivity of *Botrytis cinerea* to mepanipyrim using cucumber cotyledons and paper discs. *Bulletin OEPP,* No. 26, pp. 181-197, ISSN 0250-8052

Nair, N. G.; Guilbaud S.; Barchia I. & Emmett, R. (1995). Significance carryover inoculum, flower infection and latency on the incidence of *B. cinerea* in berries of grapevines at harvest in New South wales, Australia. *Australian Journal of Experimental Agriculture,* No. 35, pp. 1177- 1180, ISSN 0816-1089

Oshima, M.; Fujimura, M.; Banno, S.; Hashimoto, C.; Motoyama, T.; Ichiishi, A. & Yamaguchi, I. (2002). A point mutation in the twocomponent histidine kinase BcOS-1 gene confers dicarboximide resistance in field isolates of *Botrytis cinerea. Phytopathology,* No. 92, pp. 75-80, ISSN 0031-949X

Pappas, A. C. & Fisher D. J. (1979). A comparison of the mechanisms of action of vinclozolin, procimidon, iprodion and prochloraz against *Botrytis cinerea. Pesticide Science,* No. 10, pp. 239-246. ISSN 0031-613X

Pezet, R. & Pont, V. (1992). Differing biochemical and histological studies of two grape cultivars in the view of their respective susceptibility and resistance to Botrytis cinerea. In: Verhoeff, K., Malathrakis, N. E., Wiliamson, B. (Eds.): Recent advances in Botrytis research. Pudoc Scientific Publishers, Wageningen, pp. 93-98.

Pollastro, S. & Faretra, F. (1992). Genetic characterization of *Botryotinia fuckeliana (Botrytis cinerea)* field isolates coupling high resistance to benzimidazoles to insensitivity toward the *N*-phenylcarbamate Diethofencarb. *Phytopathologia Mediterranea,* No. 31, pp. 148-153, ISSN 0031-9465

Pollastro, S.; Faretra, F.; Di Canio, V. & De Guido, A, (1996). Characterization and genetic analysis of field isolates of *Botryotinia fuckeliana (Botrytis cinerea)* resistant to dichlofluanid. *European Journal of Plant Pathology*, No. 102, pp. 607–613, ISSN 0929-1873

Pommer, E. H. & Mangold, D. (1975). Vinclozolin (BAS 352F), ein neuer Wirkstaff zur Bekampfung von *B. cinerea*. *Med. Fak. Landbouwert Rijksuniv. Gent.*, No. 40, pp. 713-722,

Rewal, N.; Coley-Smith, J.R. & Sealy-Lewis, H.M, (1991). Studies on resistance to dichlofluanid and other fungicides in *Botrytis cinerea*. *Plant Pathology*, No. 40, 554–560, ISSN 0032-0862

Rosslenbroich, H. J.; Brandes, W.; Kruger, B. W.; Kuck, K. H.; Pontzen, R.; Stenzel, K. & Suty, A. (1998). Fenhexamid (KBR 2738) – A novel fungicide for control of *Botrytis cinerea* and related pathogens. In: Proceedings of Brighton Crop Protection Conference, Pests and Diseases (pp 327–334) BCPC, Farnham. Surrey, UK

Rosslenbroich, H.J. & Stuebler, D. (2000). *Botrytis cinerea*-history of chemical control and novel fungicides for its management. *Crop Protection*, No.19, pp. 557–561, ISSN 0261-2194

Schlamp, H. A. (1988). Spritzfolgen gegen *Botrytis* im praxis-test. *Der Deutsche Weinbau*, No. 43, pp. 486-489. ISSN: 09443177

Schumacher, M.M.; Enderlin, C.S. & Selitrenniko, C.P. (1997). The osmotic-1 locus of *Neurospora crassa* encodes a putative histidine kinase similar to osmosensors of bacteria and yeast. *Current Microbiology*, No. 34, pp. 340-347, ISSN 0343-651

Schüepp, H. & Lauber, H. P. (1978). Toleranz verhalten der Botrytis-population gegenueber MBC-Fungiziden (Benlate u. Enovit-M mit Wirkstoffen Benomyl u. Methylthiophanate) in den Rebbergen der Nord –u. Ostschweiz. *Schweizerische Zeitschrift für Obst- und Weinbau*, No. 114, pp. 132, ISSN 1023-2958

Schüepp, H. & Küng, M. (1978). Gegenüber Dicarboximid-Fungiziden tolerante Stämme von *B. cinerea* Pers. *Berichte der Schweizerischen Botanischen Gesellschaft*, No.. 88, pp. 63-71. ISSN 0080-7281

Schüepp, H. & Küng, M. (1981). Stability of tolerance to MBC in populations of *B. cinerea* in vineyards of northern and Eastern Switzerland. *Canadian Journal of Plant Pathology*, No. 3, pp. 180-181, ISSN 0706-0661

Stefanato, F.& Abou-Mansour, E.; Buchala, A.; Kretschmer, M. & Mosbach, A. (2009). The ABC-transporter BcatrB from *Botrytis cinerea* exports camalexin and is a virulence factor on Arabidopsis thaliana. *Plant Journal*, No. 58, pp. 499–510, ISSN 09607412

Stellwaag-Kittler, F. (1969). Moeglichkeiten der Botrytisbekaempfung an Trauben unter Bercksichtigung der epidemiologischen Grundlagen. *Weinberg und Keller*, No.16, pp. 109, ISSN 0508-2404

Suty, A.; Pontzen, R. & Stenzel, K (1997). KBR 2738: Mode d'action et sensibilit´e de *Botrytis cinerea*. In: 5th International Conference on Plant Diseases (pp 561–568) ANPP, Paris, France

Suty, A.; Pontzen, R. & Stenzel, K. (1999). Fenhexamid-sensitivity of *Botrytis cinerea*: determination of baseline sensitivity and assessment of the risk of resistance. *Pflanzenschutz-Nachrichten Bayer* No. 52, pp. 145-157, ISSN 0170-0405

Topolovec-Pintarić, S. & Cvjetković, B. (2002). The sensitivity of *Botrytis cinerea* Pers.:Fr. to pyrimethanil in Croatia. *Journal of Plant Diseases and Protection*, 109 (1); 74-79. ISSN 1861-3829

Topolovec-Pintarić, S. & Cvjetković, B. (2003). *In vitro* sensitivity of *Botrytis cinerea* Pers.:Fr. to pyrimethanil and cyprodinil in some Croatian vineyards. *Journal of Plant Diseases and Protection*, Vol. 110, No. 1, pp. 54-58, ISSN 1861-3829

Topolovec-Pintarić, S.; Miličević, T. & Cvjetković, B., (2004). Genetic diversity and dynamic of pyrimethnil resistant phenotype in population of *Botrytis cinerea* Pers.:Fr. in one winegrowing area in Croatia. *Journal of Plant Diseases and Protection*, Vol 111, No. 5, pp. 451-460, ISSN 1861-3829

Topolovec-Pintarić, S. (2009). Resistance risk to new botryticides in *Botrytis cinerea* Pers.:Fr. in vinegrowing areas in Croatia. *Journal of Plant Diseases and Protection*, Vol. 116, No. 2, pp. 73-77, ISSN 1861-3829

Triphati, R. K. & Schlosser, E. (1982). The mechanism of resistance of *B. cinerea* to methylbenzimidazol-2-yl-carbamate (MBC), *Journal of Plant Diseases and Protection*, Vol. 89, No.., pp. 151-156, ISSN 1861-3829

Vermeulen, T.; Schoonbeek, H. & De Waard, M. A. (2001). The ABC transporter BcatrB from *Botrytis cinerea* is a determinant of the phenyllpyrole fungicide fludioxonil. *Pest Management Science*, No. 57, pp. 393-402. ISSN 1526-498X

Vignutelli, A.; Hilber-Bodmer, M. & Hilber, U. W. (2002). Genetic analysis of resistance to the phenylpyrrole fludioxonil and the dicarboximide vinclozolin in *Botryotinia fuckeliana*. *Mycological Research*, No. 106, pp. 329-335, ISSN 0953-7562

Yarden, O. & Katan, T. (1993). Mutation leading to substitutions at aminoacids 198 and 200 of beta-tubulin that correlated with Benomyl-resistance phenotypes of field strains of *Botrytis cinerea*. *Phytopathology*, No. 83, pp. 1478-1483, ISSN 0031-949X

Ziogas, B.N.; Markoglou, A.N. & Malandrakis, A.A. (2003). Studies on the inherent resistance risk to fenhexamid in *Botrytis cinerea*. *European Journal of Plant Pathology*, No. 109, pp. 311-317, ISSN 0929-1873

Zobrist, P. & Hess, E. (1996). Performance of a novel cyprodinil/fludioxonil mixture in an integrated control and resistance management strategy for *Botrytis* in grapes; *Proceedings of XIth International Botrytis Symposium*, 23-27 June, Wageningen, The Netherlands

2

Optimizing Fungicide Applications for Plant Disease Management: Case Studies on Strawberry and Grape

Angel Rebollar-Alviter[1] and Mizuho Nita[2]
[1]Centro Regional Morelia, Universidad Autonoma Chapingo,
[2]Virginia Polytechnic Institute and State University,
Alson H. Smith Jr. Agricutural Research and Extension Center, Winchester, VA,
[1]Mexico
[2]USA

1. Introduction

Fungicides are important tools for management of plant diseases caused by fungal and oomycete pathogens. Without use of fungicides, major crop losses are inevitable, and food supply networks as we know today are most likely not able to sustain itself. As a result, fungicides are applied in regular basis in many parts of the world; however, their applications need to be optimized in order to obtain the best result in disease management due to multiple factors such as fungicide efficacy, the risk of resistance development, environmental concerns, pesticide residue in harvest, impact on beneficial organisms, etc. The ultimate goal is to keep the losses from diseases to a level that does not represent a threat to the crop production and to the economy of the grower while reducing the number of applications as much as possible. In order to achieve this goal, growers commonly employ integrated pest management (IPM) approaches where multiple management options are used together to achieve best efficacy with minimum chemical usage. Especially in environmentally challenging growing areas, use of fungicides is an important component of the IPM approach. Abusive uses of fungicides can cost not only growers' budget, but also cost society and environment. Therefore, fungicide usages need to be carefully planned with a good understanding of plant disease epidemics, their components (host, environment and pathogen), fungicide mode of action (biochemical, biological, physical), risk of resistance development, and host physiology, among other aspects. In this chapter, we will review these components that are involved in decision-making process to optimize fungicide application. The main focus of discussion is on management of diseases of strawberry and grape, because both are high value, intensively managed crops where application of fungicides are conducted on a regular basis.

In both strawberry and grape productions, it is not uncommon to observe an excessive number of fungicide applications, which happens sometimes as a result of the lack of knowledge of the pathogen biology and epidemiology, fungicide mode of action and fungicide residues. Or simply growers do not want take risks because of high costs and values of these crops. Moreover, the availability of several groups and mixtures of fungicides in the market is creating confusion among growers who are constantly in need of learning how to integrate a new chemistry in their plant disease management program. It is further confusing not only to growers but also to educators and researchers as well. Some of new formulations or molecules are simply a mixture of known active ingredients, or a different brand name yet the same active ingredient, or a different chemical name with the same mode of action, or a mixture of known active ingredients with a different percentage, etc.

In some agricultural settings such as the wet areas of the Midwest and Eastern US, tropical and subtropical areas of Central Mexico, the need of fungicide use is continuous during the course of the crop development; therefore it is a challenge for growers to keep their fungicide program season after season. Although it is not always considered, there are many factors that influence the decision making process of a grower to apply fungicides. If you put in a simple sense, what a grower wants is to manage a population of pathogens at the end of the day; however at the same time, he/she needs to be aware of the existence of the right tools that provide her/him an economical, effective, and sustainable (in both economic and environmental sense) solution. In addition, because of social pressure against the use of chemical in agriculture, fungicides applications for plant disease management need to be carefully selected. Since development of any plant disease is a result of a complex interaction among host, pathogen, environment, and sometimes a vector of the pathogen, the optimization of a fungicide application program should be based on the knowledge of disease dynamics, fungicide and mode of action in relation to development of epidemics (Madden 2006; Madden et al. 2007). In order to establish season-long programs to manage key diseases, growers need to learn and understand knowledge of information related to the factors that affect the efficacy of a fungicide, the biology and epidemiology of the disease, and crop physiology and the environment.

In this chapter, we explore the factors that growers, consultants, and researchers need to consider in order to establish optimized season-long programs with ecologically and economically sound approaches. We describe different components that influence the development of epidemics and their impact on crop disease management and the whole season approach to manage diseases, disease epidemics, fungicide resistant and its management, integrated pest management, and uses of disease risk assessment tools. In addition, we present two case studies managing diseases using fungicides based on information considering different tactics and strategies to reduce the number of fungicide application, and risk resistance development on grape and strawberry.

2. Components of epidemics and fungicides

Plant diseases are the result of the interaction among the host, the pathogen and the environment. Plant pathologists often describe this relationship, or model, as a plant disease triangle (Francl 2001; Agrios 2005). Each component of the disease triangle plays an important role on the development of diseases. When there is a compatible interaction between a host and a pathogen (i.e., a pathogen can cause disease on a host), the environment is the element that triggers development of a plant disease. Thus, a basic idea of plant disease management is to break the disease triangle from forming by understanding the role of each element. For instance, planting a disease resistant variety is a way to disturb the disease triangle by eliminating the host so that the triangle cannot be formed.

When we consider the change of plant diseases over time and space, we are dealing with plant disease epidemics (Madden 2006; Madden et al. 2007). Since time is another factor added to the triangle, some use a modified disease triangle, which becomes a tetrahedron (Francl 2001). Sometimes it is a challenging task for agricultural educators (such as crop specialists) to describe the concept because it deals with another dimension (time). However, it is important to inform growers that the disease you see today is a consequence of an infection that happened a certain time ago, or even a consequence of multiple infections that happened over the course of time.

Since we are dealing with the progress of disease(s) over time, we need to understand the life cycle (often referred as a disease cycle) of pathogens, which are divided into two groups, monocyclic (one disease cycle per season) and polycyclic (multiple disease cycles per season). Based on the disease cycle, management strategies can differ. For example, one of strategies of plant disease management can involve elimination or reduction of the amount of primary inoculum, which reduces the rate of infection by reducing the probability of pathogens to find healthy host tissues and/or by limiting the time the pathogen and host populations interact (Nutter 2007). In some monocyclic disease cases, only one application of fungicide might be needed. For example *Fusarium graminearum*, a causal agent of Fusarium head blight of wheat causes infection on kernels during anthesis, therefore, protection of wheat during this stage of development is the key for the management of this disease (Nita et al. 2005).

On the other hand, when a continuous release of pathogen inoculum is occurring and host tissues are susceptible over time, multiple applications might be needed. In order to deal with polycyclic diseases, often several applications are needed to delay the onset of the epidemic. In this case, the impact of fungicide applications will be on the rate of the epidemic by reducing the probability of successful infection and/or successful completion of life cycle (= production of spores) (Fry 1982). Early studies by J. E. Van der Plank (1963) introduced many of these concepts, and it was followed by many plant disease epidemiologists who utilized these concepts to develop plant disease models and management strategies such as a use of disease risk assessment (forecasting or warning) tools (Zadoks and Schein 1979; Zadoks 1984; Madden et al. 2007). Some of disease risk assessment tools aim to determine the critical time when the disease become a threat and/or have an economic impact. Disease risk assessment tools can be very useful to reduce the costs of disease control and increase safety of the produce by helping growers to use fungicides in a timely and more efficient manner (Zadoks 1984; Hardwick 2006; Madden et al. 2007).

3. Fungicide resistance and plant disease management

When we discuss about fungal disease management, discussions on the issue of fungicide resistance cannot be avoided. Development of fungicide resistance fungal isolates was documented as early as 1960's when *Penicillium* spp. on citrus (citrus storage rot) was found to be resistant against Aromatic hydrocarbons (Eckert 1981). The other examples from that decade are resistance to organomercurials by cereal leaf spot and strip caused by

Pyrenophora spp., dodine resistant apple scab (*Venturia inequalis*), and QoI (Quinone-Outside Inhibitor) resistance against grape powdery mildew (caused by *Erysiphe necator*) in the field in Europe, and North America in 1990's to 2000's (Staub 1991; Baudoin et al. 2008).

Fungicide resistance develops when a working mode of action loses its efficacy against target fungal pathogen. When fungicide resistance appears in the field, it is often the case that a particular fungicide (or a mode of action), has been used for a several years or seasons, and growers find that the efficacy of that fungicide has been noticeably reduced or even lost. This type of resistance is often called 'field resistance' or 'practical resistance' in contrast to the cases when the resistance isolate is found only in the laboratory conditions (= laboratory resistance) (Staub 1991). Some of laboratory resistance isolates can only survive under protected conditions because they are not adequately fit to compete and survive in the field thus, the presence of these laboratory resistant isolate may or may not be a threat to the real world. Attempts were made to predict the development of field resistance based on populations of laboratory resistance isolate; however, it has been difficult. For example, although the presence of resistant isolates of *Botrytis* against dicarboximides was found in laboratory, the development of field resistance was slower than expected (Leroux et al. 1988).

The resistant mechanisms, whether a single gene or multi-locus function, maybe present naturally among wild population in a small quantity and a repeated application of a particular mode of action select these rare populations to thrive. In some cases, the development of fungicide resistance appears to be a sudden event. This type of resistance is also called 'qualitative', 'single-step', or 'discrete' resistance (Brent and Hollomon 2007). This qualitative resistance tends to appear relatively soon after the introduction of the compromised mode of action and stay once appeared. One of examples would be benzimidazole resistance of apple scab pathogen (*Venturia inequalis*) where resistant isolates appeared only after two seasons of benzimidazole fungicide application (Shabi et al. 1983; Staub 1991). In some cases, a gradual recovery of sensitivity can happen; however, as soon as an application of the compromised mode of action resumes, the resistance tends to come back quickly as in the example of potato late blight pathogen (*Phytophthora infestans*) to phenylamide fungicides (Gisi and Cohen 1996).

In the other cases, development of fungicide resistance is gradual. This type of resistance is called 'quantitative', 'multi-step' or 'continuous' (Brent and Hollomon 2007). Examples of quantitative resistance are the cases of many fungal pathogens to the DMI (sterol DeMethylation Inhbitors) where the reduction of efficacy can be observed over several years or seasons (Staub 1991). For quantitative resistance, reduced use of fungicides of the same mode of action tends to revert populations back to more sensitive state. This decline of the resistance could be due to incomplete resistance, or lack of fitness, or both (Staub 1991).

Another concern on the resistance is the phenomenon called 'cross-resistance' where resistance to one fungicide translates into resistant to other fungicides, which are affected by the same gene mutation(s). Often time it happens with fungicides that are different in chemical composition, while share the same mode of action. One example is the case of benzimidazole fungicide resistance where pathogen strains that resist benomyl were resistance to carbendazim, thiophanate-methyl, or thiabendazole (Brent and Hollomon 2007). Moreover, in some cases, a fungal strain can be resistant to two or more different mode of action or acquire 'multiple resistance'. For example, *Botrytis cinerea* a causal agent of bunch rot of grape and many other plant diseases, is commonly resistant to both benzimidazole and dicarboximide fungicides (Elad et al. 1992).

As noted previously, repeated application of the same mode of action often increase the risk of development of fungicide resistant population. Because of that, intensively managed agricultural crops, such as wine grapes and strawberries, the risk of fungicide resistance development is higher due to frequent application of fungicides throughout a season. For example, in the eastern US grape growing regions, wine grape growers apply more than 10 applications of fungicide year after year (Wolf 2008), but in other regions such as the Central part of Mexico, more than 20 fungicide applications can be done on strawberries in a growing season. Once fungicide resistance is developed against a certain mode of action, it is not only a loss for growers, but also a huge loss to chemical companies that invested a considerable amount of money and time to develop the product. Currently, there are more than 150 different fungicidal compounds used worldwide (Brent and Hollomon 2007). The total sales of fungicide are estimated to be $7.4 billion in US dollars, and grapes are one of the largest consumers of fungicides.

4. Management of fungicide resistance

There are several tactics to reduce the risk of fungicide resistance development. Common approaches implemented from fungicide manufacturers and regulatory agencies are 1) set a limit on the number of application per year, and 2) production of a pre-mixed material. The aim of setting a limitation or a cap on the number of applications per season is to reduce the rate of shifting from sensitive to resistance population by providing a gap between usages. If fungicide sensitive population is less fit than sensitive population, then the interval will provide a time for sensitive population to take over the resistant population. However, in some cases, the cap on the usage did not help the development of fungicide resistance. For example, Baudoin et al. (2008) found that although growers were following a '3 times per season' cap set by an international organization, Fungicide Resistance Action Committee (FRAC), QoI (e.g. strobirulins) resistant grape powdery mildew appeared in the field after 10-15 applications over several years of use. It seems that the 'cost' of having QoI fungicide resistance (i.e., G143A mitochondorial cytochrome b gene) does not affect the fitness of the fungus. On the other hand, the cap approach could help reducing the risk of DMI resistant isolates since fungal population seems to revert back to be sensitive when DMI is not used in the field (Staub 1991; Brent and Hollomon 2007).

The aim for pre-mixed material is to create a mixture of fungicides with multiple modes of action. There is evidence of reduced rate of fungicide resistance by mixing two (or more) different mode of action. For instance, Stott et al. (1990) compared the population shift of barley powdery mildew (caused by *Erysiphe graminis*) and showed that DMI and ethirimol sensitive populations did not shift to resistant population when both materials were used together. This approach seems to be favored by fungicide manufacturing companies; however, as we noted earlier, these pre-mixed materials can cause confusion among growers especially when seemingly new materials were combinations of previously introduced modes of action in reality.

When a host crop requires intensive disease management, the aforementioned two approaches may not be enough to effectively manage the development of fungicide resistance. For instance, in order to manage grape powdery mildew under eastern US growing conditions, wine grape growers typically use a fungicide (or multiple fungicides) for powdery mildew practically every time they spray (10-15 times per year). If they use a DMI early in the season, they may not have much choice later. Thus, careful planning and execution of plant disease management becomes very important. In order to achieve the goal, many successful growers extensively practice the Integrated Pest Management (IPM) or Integrated Plant Disease Management (IPDM) approach.

5. IPM approaches revisited

A basic concept of IPM is to combine multiple approaches of disease management in order to achieve the best result (Agrios 2005). These approaches are 1) cultural control, 2) use of genetic resistance, 3) biological control, and 4) chemical control. In the case of grape disease management, cultural practice can include (but not limited to) site selection, proper nutrition management, selective pruning of dormant canes, canopy management (shoot training, leaf removal, etc), etc. Genetic resistance can be introduced by selecting disease resistant varieties such as some of French hybrids. Often time the challenge is to select resistant varieties with high market demands. One of success stories of such a case is variety called 'Norton'. This highly disease resistant variety for wine making has gained popularity in the Eastern US grape growing regions since 1990's (Ambers and Ambers 2004). There are several biological agents available for use in grape production; however, none of them seems to produce reproducible results. It is partly due to the fact that growers want to use them as if they were using chemical options. Chemical management approaches should be considered only after these non-chemical approaches are considered. Integration of these approaches not only increases the efficacy of overall management strategies, but also, can reduce the monetary cost associated with chemical management approach (e.g., costs for purchasing chemicals, labor and fuel to apply chemicals, etc).

Even after other IPM approaches are considered, growers often need to resort to chemical management options because of environmental conditions that highly favor disease development. There are a few more items to be considered before application of fungicide in order to increase the efficacy. First of all, growers need to identify target pathogen(s) correctly. Then, growers need to select the best tool for management of the target pathogen(s) based on the situation at hand. In order to guide this complicated decision making process, it is necessary to have a better understanding of pathogen and host biology, as well as awareness on legal requirements.

As with any other pest management, identification of the target organism is a very critical component of plant disease management. For instance, symptoms of downy mildew of grape (caused by *Plasmopara viticola*) and powdery mildew of grape (caused by *Erisyphae necator*) may look similar to untrained eyes; however, materials for downy mildew are most likely not effective against powdery mildew, and *vise versa*. Thus, misidentification of disease symptoms can result in unnecessary application of fungicides.

After correct identification of the target disease(s), growers need to determine the best tool(s) for the situation at hand. Both host crop physiology and pathogen population changes throughout the course of the season, and these changes can influence disease triangle of the target pathogen. As we covered in the previous section, in order for a pathogen to successfully infect a host crop, a susceptible host, a pathogenic pathogen, and a disease-conducive environment have to be present at the same time. However, a pathogen may not produce spores at a right timing or a host may not be susceptible at a certain time of its lifecycle. Even if there are spores and hosts are susceptible, if the environmental conditions are not conducive for infection, disease cannot be developed. Thus, it is very important to understand both pathogen's and host's lifecycles, as well as environmental conditions for infection, so that growers can place fungicide application to efficiently disrupt the formation of the disease triangle without wasting their effort.

Changes in host physiology throughout the season, especially fruit development, can be a key factor to determine when and how fungicide should be applied. For example, it is important to protect flowers of strawberry from Botrytis infection because flower infection result in latent infection on berries later in the season (Mertely et al. 2002). Results from Merteley et al., (2002) indicate that Botrytis fruit rot can be controlled with an application of fenhexamid when applied at anthesis. They were also able to relate a linear regression equation between time of application and Botrytis fruit rot incidence, which can guide growers to adjust their spray timings. Legard et al. (2005) integrated information of the crop physiology, epidemiological information and fungicide efficacy to develop reduced fungicides programs to control Botrytis fruit rot in Florida. Their results indicate that in the early stage of the season low rates of captan were as effective as high rates for disease control, and later in the season the control was significantly improved by applications of fenhexamid at the second bloom peak period. In the case of grape production, ontogenic resistance has been well documented against many of major pathogens such as black rot (Hoffman al. 2004), powdery mildew (Ficke et al. 2002; Gadoury et al. 2003), and downy mildew (Kennelly et al. 2005). Grape berries become resistance against downy mildew, powdery mildew, and black rot approximately 4-5 weeks and 3-4 weeks after bloom for French and for American varieties, respectively. By knowing this information, growers can concentrate their effort to protect berries during this critical period.

In addition to biological factors, legal or legislative factors can influence fungicide application decision-making process. For instance, a product containing mancozeb has a 66-day PHI (pre-harvest interval) set by the EPA (Environmental Protection Agency) for an application on grape in the US. Thus, grape growers need to adjust their spray program against downy mildew or black rot when they are expecting to harvest within 2 months. Also, REI (re-entry interval) can be a limiting factor. A product Topsin-M (thiophanate-methyl) has a REI of 2-days for grapes, and a product Pristine (boscalid + pyraclostrobin) has a warning on the label that growers cannot work on grape canes within 5 days after application. Thus, it is difficult to use either Topsin-M or Pristine when constant canopy management is required. The other factors can influence fungicide application is an incompatibility issue. For example, several fungicides, including chlorothalonil can cause phytotoxicity on 'Concord' and related American grape varieties (Goffinet and Pearson 1991).

6. Physical mode of action of fungicides

There is yet another factor to be considered prior to an application of fungicide, that is, physical mode of action of fungicide. Physical mode of action (PMoA) describes the effect a fungicide with respect to the time of placement of a fungicide in relation to the host-pathogen interaction, that is on pre-infection, post-infection, pre- and post-symptom, and vapor activity (Szkolnik 1981; Wong and Wilcox 2001) and the duration and degree of the fungicides activity (Pfender 2006).

PMoA of protectant fungicides is pre-infection effect. It can reduce the infection efficiency as a result of the placement of a fungicidal material on plant tissues. McKenzie et al. (2009) found that applying captan 2 days before inoculation on strawberry crown rot (caused by *Colletotrichum gloeosporioides*), disease intensity was consistently reduced at the end of the season. Azoxystrobin, pyraclostrobin and thiophanate-methyl performed better if applied 1 day after inoculation, but their effect reducing the disease was variable. Based on these results the recommendation was to spray captan throughout the season in a protectant strategy, and azoxystrobin, pyraclostrobin and thiophanate-methyl if an infection event was present in order to keep the disease at low levels.

On the other hand, systemic fungicides with more curative (eradicating) activity can impact the processes of infection and establishment by pathogen, thus these are post-infection and can be pre- or post-symptom effect. Vapor activity can facilitate pre- and post-symptom effects. A single fungicide can provide both protectant and curative activities. For example, fungicides such as strobilurins (QoI) will mainly impact on spore germination, as they interfere with mitochondrial respiration (Bartlett et al. 2002), giving an excellent protectant activity. At the same time, the QoI can provide good curative activity against rusts such as *Puccinia hemerocallidis* and *Puccinia graminis* subsp *graminicola* (Godwin et al. 1992; Pfender 2006). In some cases such as *Cercospora beticola*, that causes Cercospora diseases on sugarbeet, good post symptom activity (eradicant) an antisporulant activity of this group of fungicides has been reported (Ypema and Gold 1999; Anesiadis et al. 2003). In other cases such as downy mildew of grape, caused by *Plasmopara viticola* (Wong and Wilcox 2001) and *Phytophthora cactorum* on strawberries (Rebollar-Alviter et al 2007) these fungicides provide good protectant activities, but do not perform well in post-infection treatments. Other groups inhibiting the sterol biosynthesis (SBI/DMI) do not have direct effect on spore germination, but impact more directly on mycelial growth. Hoffman et al. (2004) found that a DMI, myclobutanil, provides a better post-infection activity against black rot of grape, compared with azoxystrobin, which provided a slight evidence of a post-infection activity.

7. Fungicide use based on disease risk assessment tools

Now we have covered basics of plant disease development, management approaches, fungicide resistant issues, and physical model of action, the next step is to combine these together. As we briefly touched earlier, one of approaches taken by many researchers and growers are the use of disease risk assessment (or forecasting or warning) tools to minimize the use of fungicides by determining the best timing for application. There are several examples of risk assessment tools used together with the knowledge of the physical mode of action of fungicides. For example, Madden et al. (2000) evaluated an electronic warning system for downy mildew based on infection of leaves of American grapes, *Vitis labrusca*, productions of sporangia and sporangial survival over a period of 7 years. Sprayings were done when the system indicated that environmental conditions were favorable for sporangia production. Their results indicated that during this time the use of the warning system reduced the number of applications of metalaxyl plus mancozeb from one to six applications compared to the standard calendar based program. Wong and Wilcox (2001) evaluated the physical mode of action of azoxystrobin, mancozeb and metalaxyl against *Plasmopara viticola*, the causal agent of grape downy mildew. Azoxystrobin was effective in pre-infection treatments, but was ineffective when applied as a post-infection treatment. However, good effect was observed on reduction of sporulation, and reduction lesion size in post-symptom applications. Mancozeb was also excellent protectant but did not have any effect on post infection applications. Metalaxyl provided good pre-infection, post-infection and eradicant activity. Kennely et al. (2005) indicated that mefenoxam has strong vapor activity against *Plasmopara viticola*, grapevine downy mildew and 48 h of systemic activity in post-infection applications. Caffi et al. (2010) evaluated a warning system to control primary infections of downy mildew on grapevine, and results indicated that the number of applications can be reduced by more than 50% with significant savings in cost per ha without compromising downy mildew control.

Working with anthracnose fruit rot of strawberry, Turecheck et al. (2006) evaluated the pre- and post-infection activity of pyraclostrobin on the incidence of anthracnose fruit rot at different times of wetness periods and temperatures. Results indicated that pyraclostrobin was less effective when applied in post-infection with the longest wetness duration (12 and 24) and high temperature (22 and 30 C). The post-infection application had a significant effect when applications were made within 3 and 8 h after the wetness period. Under field conditions, applications made after 24 h after an infection event were able to successfully control the disease, indicating the possibility to incorporate pyraclostrobin in a disease management program in strawberry in a curative form if infection events occurred in the previous 24 h. In a similar study, Peres et al. (2010) indicated that anthracnose fruit rot was effectively controlled with captan on pre-infection under short wetting period and fludioxonil + ciprodinil was effective when applied in pre-inoculation, but also when applied at 4, 8, and 12 h after inoculation, but the efficacy was higher under short wetting periods (6 o 8 h). These studies indicate that performance of fungicide is strongly influenced by wetness duration regardless of the ability of the fungicide move in plant tissues.

Thus, growers face multiple layers of factors such as host-pathogen dynamics, fungicide resistance, physical and biochemical mode of action, IPM strategies, etc. in order to make decisions on fungicide application. Also, note that we were focus only on biological considerations, but not covering many of social and environmental factors such as society's concerns on fungicide use, issues on waste water management, and so on. In addition, there is a whole art and science of fungicide application technologies that is beyond our scope of this chapter. Instead of widening our topics, we would like to focus on the factors we discussed in this chapter by presenting two case studies that are compilations' of multiple studies to establish an optimal use of fungicide(s).

8. Case study 1: Phomopsis cane blight and leaf spot of grape

A series of studies by Nita et al. (2006a; 2006b; 2007a; 2007b; 2008) showed a multi-prong approach to develop a sound management strategy against Phomopsis cane and leaf spot of grape. Phomopsis cane and leaf spot is a common disease of grape in the U.S. and other grape growing regions around the world (Pearson and Goheen 1988; Pscheidt and Pearson 1989). The fungus, *Phomopsis viticola* (Sacc.), is the causal agent of the disease (Pearson and Goheen 1988). Typical symptoms on leaves are yellow spots, which varies in size (less than 1 mm to a few mm). On canes and rachis, it causes necrotic lesions that can be expanded to cause canker. The infected tissues become weak and prone to be damaged by wind. With heavy infection on rachis, fruit drop can be observed. Infections on fruits cause a fruit rot and thus directly decrease yield and fruit quality. Up to 30% loss of the crop has been reported in the Southern Ohio grape growing regions (Erincik, et al. 2001).

The source of inoculum in a given season consists of canes or trunks that were infected during previous growing seasons (Pscheidt and Pearson 1989). The fungus survives in the infected tissues over the winter, and in the spring, numerous pycnidia arise on infected canes. Conidia from these pycnidia are splashed by rain onto new growth (i.e., canes, leaves, and clusters) to cause infection. According to previous studies, *P. viticola* can be active in relatively cool weather conditions (7-8 C) (Erincik et al. 2003). Since they do not produce new spores during the season, it is considered a monocyclic disease.

In order to evaluate efficacy of currently available fungicides, Nita et al. (2007a) examined several fungicides for their protectant and potential curative activity against Phomopsis cane and leaf spot of grape. Fungicides with variety of mode of action, strobilurin, thiophanate-methyl (benzomidazole), myclobutanil (DMI), EBDC (mancozeb), calcium polysulfide (lime sulfur), were tested as protectant as well as curative application in a controlled environment study. Protectant application was applied a few hours prior to an artificial inoculation of leaves and shoots using spore suspension water that contained 5 x 10^6 spores per ml. Various patterns of post-inoculation (curative) application were tested. The shortest period between inoculation and application of a fungicide was 4 hours and the longest was 48 hours. In addition, a treatment with or without an adjuvant (product name Regulaid or JMS Stylet Oil) was also tested. These adjuvants were added in a hope that it might help facilitate movement of chemical into tissues. In addition, up to 150% of labeled rate of fungicide was examined to see a potential dose effect. Results indicated that all materials tested, regardless of a higher rate and/or a presence of adjuvant, did not show evidence of curative activity. On the other hand, strobilurin, calcium polysulfide, and EBDC showed a good protectant activity, up to >85% disease control [(treatment disease severity-negative (=untreated) control disease intensity)/negative control disease intensity], indicating that the management strategy for Phomopsis cane and leaf spot has to focus on protection of vines.

Then the same group evaluated the effect of dormant season fungicide applications of copper and calcium polysulfide against Phomopsis cane and leaf spot of grape disease intensity and inoculum production (Nita et al. 2006a). These dormant season fungicide applications aimed to reduce the source of inoculum by disturbing fungal colonies surviving on grape trunk tissues. Results indicated that fall and spring and spring applications of calcium polysulfide (10% in volume) provided 12 to 88% reduction in disease intensity and inoculum production. Thus, the reduction of disease intensity was not sufficient. Although inoculum production (the number of pycnidium per square cm) was significantly reduced, none of tested canes had zero pycnidium, indicating that there will be a plenty of inoculum available even with the best treatment. In the same study, the authors examined calendar-based applications of mancozeb or calcium polysulfide (0.5% in volume), which reduced 47 to 100% disease incidence and severity. The result indicated that although sanitation approach against this disease did not provide reasonable reduction in disease development, early season applications of a protectant fungicide (mancozeb or calcium polysulfide) provided a better efficacy. These results confirmed previously discussed management recommendations (Pearson and Goheen 1988; Pscheidt and Pearson 1989).

Nita et al. (2006b) also evaluated a warning system (based on temperature and wetness duration following rain) for Phomopsis cane and leaf spot of grape by applying fungicides based on prediction of infection events considering three criteria for risk: light, moderate and high. The infection condition was determined previously by Erincik et al. 2003. This study was conducted to determine if the warning system would provide a reasonable disease control compared with a calendar-based, 7-day interval protectant fungicide application. The warning-system based approach resulted in two to three times less number of applications while the percentage of control was often not significantly lower than the 7-day protectant schedule based on mancozeb, which constantly provided 70-80% and over 95% disease incidence and severity control, respectively.

The same group expanded this study by examining Phomopsis cane and leaf spot disease survey data using various statistical tools and modeling approach (Nita et al. 2007b; Nita et al. 2008). They found out that the variation of disease incidence observed in 20 different commercial vineyard locations over three consecutive years could be explained by a combination of local weather conditions and fungicide application trends. They further found that growers who had a better early season fungicide program (i.e., a use of dormant application of lime sulfur and/or mancozeb application soon after bud break) tended to have lower disease incidence than others who did not protect their vines during that time.

These series of studies showed that pre-season dormant application does not provide satisfactory reduction of this disease, and there are no potential curative materials; however, a dormant season application can be used in a conjunction with early season protectant fungicide applications, a warning system approach can be a good tool to be used, and more importantly, protection of grape tissues during early part of the season was found to be critical for the management of Phomopsis cane and leaf spot of grape. The Eastern and Midwestern US grape growing regions often receive a series of rains in April to May when new grape shoots are emerging, and pathogen can infect tissues under relatively low temperatures conditions, 7-8 C (Erincik et al. 2003; Nita et al. 2003). Therefore, good protection of newly emerging shoots (when new shoots are about 2.5-7.5 cm in length) using a protectant fungicide is a standard recommendation for this disease (Pscheidt and Pearson 1989; Nita et al. 2007b).

9. Case study 2: Leather rot of strawberry

Crown and root rots, such as those caused by *Colletotrichum* spp, *Phytophthora* spp. and *Verticillium* spp., and fruit rots, such as *Botrytis cinerea*, *Colletotrichum acutatum*, and *Phytophthora cactorum* are among the most important pathogens causing disease on strawberry that cause more losses around the world.

Leather rot caused by *P. cactorum* is one of most common disease on strawberry, especially in systems such as matted row and annual systems. The disease is less severe and not very frequent in high tunnel system, mainly because plastic tunnels prevent rain to reach plants and induce splash dispersal of the pathogen. On strawberry all stages of fruit development may be infected by this pathogen, including flowers. On green fruits dark areas covering the entire fruit may develop which later appear leathery and eventually mummify. Mature fruits do not always show the typical symptoms, except they appear discolored and whitish in some areas. However, diseased fruits are in general easy to distinguish because the bad off-odor and taste, which is caused by phenolic compounds (Jelen et al. 2005). In Ohio, losses over 50% have reported (Ellis and Grove 1983) and in areas with medium to low technology levels in open field strawberry plantings under annual production systems in countries such as Mexico, the disease can be a problem during the rainy season of the year (June to October) where losses can reach up to 30% of production.

Development of leather rot is favored by excessive wet weather, especially on saturated soils with poor drainage. In this pathosystem, oospores represent the primary inoculum, which is a survival structure. With moisture, oospores germinate to produce sporangia. Sporangia can germinate and produce a germ tube for infection, or can give a rise to zoospores that can swim in water. With a rain event, both sporangia and zoospore are splash dispersed to fruits to cause infection. Once established, new sporangia can form on the infected fruit to cause another infection. Thus, it is considered a polycyclic disease. Extensive studies conducted on the epidemiology of the disease in the past decade have shown that wetness duration and temperature (17 to 25 C) are important factors for disease development. Splashing of zoospores and sporangia is caused by rainfall and wetness periods can be as short as 2 h are sufficient for the oomycete to cause infection (Grove et al 1985a; Grove et al. 1985b; Madden et al. 1991). Typically there is a latent period of 5 days for full development of symptoms.

Management of leather rot is based on the use of fungicides and cultural practices such as avoiding saturated soils by proper site selection, improving soil drainage and applying straw mulches between rows. Applying straw mulch between row spaces prevents fruits from touching the soil and standing water, and reduces the splashing of water droplets containing sporangia and zoospores (Madden et al. 1991). Protective fungicide program using captan and thiram are widely used; however, both fungicides are not able to control the disease when weather conditions favor leather rot development. Therefore fungicide with a different biochemical, and physical mode of action with the ability to penetrate plant tissues need to be used.

In order to select the proper fungicide, the efficacy of fungicides was defined in the field (Rebollar-Alviter et al. 2005). During 2003 and 2004, the efficacy of pyraclostrobin, azoxystrobin, potassium phosphite and mefenoxam was evaluated in Wooster Ohio, USA against leather rot of strawberry grown in a matted row system. Treatments were applied as a preventive application at the initiations of bloom. In order to create conditions that favor leather rot development, straw between the rows was removed and then plots were flooded until water puddle between the rows at different times using an overhead irrigation system. Results from these experiments indicated that during the two years of testing, disease incidence on fruits varied from 58 to 67% in the controls. No significant differences were detected among the fungicides treatments. Disease incidences ranged from 0.3 to 0.5% with the QoI fungicides (azoxystrobin and pyraclostrobin), 0.8 to 5.4% with potassium phosphite, and 0.3 to 11% with mefenoxam (Rebollar-Alviter et al. 2005). Interestingly, these experiments showed that both QoI fungicides tested were highly effective for control of leather rot of strawberry. Thus, these QoI fungicides can be used in a disease management program alternating with potassium phosphite and/or mefenoxam, which are known to be efficacious to control the disease (Ellis et al. 1998).

In order to understand some aspects of the physical mode of action of the QoI, potassium phosphite, and mefenoxam fungicides that were tested in the previous work, a greenhouse study was conducted. Fungicides were applied on pre-infection, 2, 4 and 7 days before inoculation with a zoospore suspension (10^5 zoospores/ml) and 13, 24, 36 and 48 h after inoculation. A wetness period of 12 h was applied to plants and fruits either before or after inoculation, and disease incidence was recorded 6 days after inoculation. Results indicated that all fungicides applied in pre-infection provided excellent protection activity against the disease when applied up to 7 days before inoculation. These studies confirmed the protectant activity of all fungicides in previous experiments in strawberry. However, the results when the fungicides were applied in post-inoculation (curative application), both QoI fungicides had some effect 13 h after inoculation reducing disease incidence by 60%. Nevertheless when both fungicides were sprayed 24, 36 and 48 h after inoculation there was no disease control. In contrast, the systemic fungicides potassium phosphite and mefenoxam successfully controlled the disease up to 36 h after inoculation with no significant differences between these two fungicides. At 48 h both fungicides still had some moderate control, but not enough to be considered in a curative strategy for disease management (Rebollar-Alviter et al. 2007a).

These results were then used in conjunction with the previous knowledge on the disease epidemiology in order to evaluate disease management programs and to optimize fungicide application. A 3-year study was conducted in a field to examine efficacy of several modes of action (mefenoxam, phenilamides; azoxystrobin and pyraclostrobin, QoI, and potassium phosphite, phosphonate) against leather rot. In previous studies on a forecasting system for leather rot; occurrence of rain was considered a better indicator of risk of disease development than temperature condition or length of wetness duration (Reynolds et al. 1988; Madden et al. 1991). This is probably because this pathogen requires very short wetness periods (2 h) to infect (Grove et al. 1985a), and it can also infect under a wide range of temperatures. Therefore, specific infection conditions (i.e., temperature or length of wetness duration) would not clearly define the risk conditions. Rather, a detection of individual rainstorm and the amount of rainfall during critical periods is a better indicator for post-infection application of a fungicide. The amount of rainfall is critical because it will be a predictor for the dissemination of the spores to susceptible fruits (Ntahimpera et al. 1998).

Based on previous experiments where post infection activities of mefenoxam and potassium phosphite indicated that this fungicides were able to control the disease up to 36 h after inoculation, and considering that epidemic is basically driven by moderate to heavy rain events (Reynolds et al. 1987; Reynolds et al. 1988), scheduling fungicides after the occurrence of rain events taking in to account fungicide persistence in plant (at least 7 days) and other factors that affect the efficacy of fungicides, as well as weather predictions, it would be possible to reduce the number of applications during the critical time for disease development. These experiments indicated that post infection treatments applied after flooding events were as effective as those applied on a calendar basis, but with 1 to 3 fewer sprayings. One spraying of mefenoxam was sufficient to keep the disease under control when applied within 36 h after a rain event. Similarly, 2 sprayings of potassium phosphite were enough to control the disease when sprayings were done within the same time after the occurrence of a rain event. Whereas in calendar based applications (7 days schedule) four sprayings were necessary to control the disease using programs based on azoxystrobin and potassium phosphite, 1 spraying of mefenoxam and 2 of potassium phosphite were enough to control the disease under high disease pressure (Rebollar-Alviter et al. 2010).

The disease control programs evaluated either as protectant strategy or curative responding to rain events were able to control the disease under weather conditions favoring disease development. Calendar based fungicide applications as well as those responding to rain events take in to consideration the risk of disease development and agree with current recommendation to manage fungicide resistance. Growers have a choice to use the protectant (calendar-based program) or curative strategy under a matted row production system in Ohio and similar strawberry production areas to extend the life of the fungicide by a proper use of fungicide resistance techniques.

As additional factor that contributes to optimize fungicide application for the management of leather rot of strawberry is the distribution of the sensitivity to the tested fungicides. A study was conducted in order to determine the sensitivity of *P. cactorum* to azoxystrobin and pyraclostrobin fungicides among isolates from different parts of the state of Ohio, and other states of the USA, so the risk of resistance development by using these fungicides on strawberry could be determined. The sensitivity of 89 isolates of *P. cactorum* was determined to both fungicides on mycelia and zoospore germination. The results showed that there was a wide distribution of sensitivity to azoxystrobin, indicating a great diversity among the isolates evaluated. Thus, the sensitivity distributions can be used as a baseline sensitivity to monitor shifts in fungicide resistance in *P. cactorum* (Rebollar-Alviter et al. 2007b).

These series of studies showed that both calendar-based and disease risk-based fungicide application can result in a satisfactory disease management. Also a proper combination of protectant and curative approach can extend the life of the fungicide. The results obtained from these experiments are based on growing conditions in the Midwestern US with matted row perennial production; however, it can be also applicable to other type of production systems. For example, in subtropical areas of the central part of Mexico (Michoacan and Guanajuato States), strawberries are grown as an annual crop and season is drastically different from the Midwest; however, rain season coincides with fruit set and first harvest as it is in the Midwestern US. Thus, the same principals for leather rot management can be applied.

10. Concluding remarks

In this chapter, we reviewed major components that are associated with fungicide application decision-making process: basic understanding of disease epidemiology; fungicide resistance and its management; fungicide physical mode of action; and use of plant disease risk assessment tools that can integrate these components. We also discussed two case studies where multiple studies are conducted to develop optimal management recommendations. We believe that this chapter demonstrated the complication involved in an optimization of fungicide uses which growers face every day, and presented some of approaches that can be used to investigate this intriguing study subject.

11. Acknowledgement

Authors are very thankful to our mentors Drs. Laurence V. Madden and Michael A. Ellis of the Department of Plant Pathology at The Ohio State University for their invaluable advice and support during the time experiments from cases 1 and 2 were conducted.

12. References

Agrios, G. N. (2005). Plant Pathology. San Diego, Academic Press.

Ambers, R. K. R. and C. P. Ambers (2004). Daniel Norborne Norton and the origin of the Norton grape. American Wine Society Journal 36(3): 77–87.

Anesiadis, T., G. S. Karaoglanidis and K. Tzavella-Klonari (2003). Protective, Curative and Eradicant Activity of the Strobilurin Fungicide Azoxystrobin against Cercospora beticola and Erysiphe betae. Journal of Phytopathology 151(11-12): 647-651.

Bartlett, D. W., J. M. Clough, J. R. Godwin, A. A. Hall, M. Hamer and B. Parr-Dobrzanski (2002). The strobilurin fungicides. Pest Management Science 58(7): 649-662.

Baudoin, A., G. Olaya, F. Delmotte, C. J. F. and H. Sierotzki (2008). QoI resistance of Plasmopara viticola and Erysiphe necator in the mid-Atlantic United States. Plant Health Progress doi:10.1094: PHP-2008-2211-2002-RS.

Brent, K. J. and D. W. Hollomon (2007). Fungicide resistance in crop pathogens: how can it be managed? Brussels, GCPF.

Caffi, T., V. Rossi and R. Bugiani (2010). Evaluation of a Warning System for Controlling Primary Infections of Grapevine Downy Mildew. Plant Disease 94(6): 709-716.

Elad, Y., H. Yunis and T. Katan (1992). Multiple fungicide resistance to benzimidazoles, dicarboximides and diethofencarb in field isolates of Botrytis cinerea in Israel. Plant Pathology 41(1): 41-46.

Eckert, J. W., Wild, B. L. 1981. Problems of Fungicide Resistance in Penicillium rot of citrus fruits. In : G. P. Georgious and T Saito (Eds.). Pest Resistance to Pesticides. pp. 525-555. Plenum Press, New York.

Ellis, M. A. and G. G. Grove (1983). Leather rot in Ohio strawberries. Plant Disease 67: 549.

Ellis, M. A., W. F. Wilcox and L. V. Madden (1998). Efficacy of Metalaxyl, Fosetyl-Aluminum, and Straw Mulch for Control of Strawberry Leather Rot Caused by Phytophthora cactorum. Plant Disease 82(3): 329-332.

Erincik, O., L. V. Madden, D. C. Ferree and M. A. Ellis (2001). Effect of growth stage on susceptibility of grape berry and rachis tissues to infection by Phomopsis viticola. Plant Disease 85(5): 517-520.

Erincik, O., L. V. Madden, D. C. Ferree and M. A. Ellis (2003). Temperature and wetness-duration requirements for grape leaf and cane infection by Phomopsis viticola. Plant Disease 87(7): 832-840.

Ficke, A., D. M. Gadoury and R. C. Seem (2002). Ontogenic resistance and plant disease management: A case study of grape powdery mildew. Phytopathology 92: 671-675.

Francl, L. (2001). "The disease triangle: a plant pathological paradigm revisited. The Plant Health Instructor DOI:10.1094: PHI-T-2001-0517-2001.

Fry, W. E. (1982). Principals of Plant Disease Management. New York, Academic Press.

Gadoury, D. M., R. C. Seem, A. Ficke and W. F. Wilcox (2003). Ontogenic Resistance to Powdery Mildew in Grape Berries. Phytopathology 93(5): 547-555.

Gisi, U. and Y. Cohen (1996). Resistance to Phenylamide fungicides: A Case Study with Phytophthora infestans Involving Mating Type and Race Structure. Annual Review of Phytopathology 34(1): 549-572.

Godwin, J. R., V. M. Anthony, J. M. Clough and C. R. A. Godfrey (1992). ICIA5504: A novel, broad spectrum, systemic beta -methoxyacrylate fungicide. Brighton Crop Protection Conference, Pests and Diseases -1992. Volume 1. Farnham (United Kingdom), British Crop Protection Council: 435-442.

Goffinet, M. C. and R. C. Pearson (1991). Anatomy of Russeting Induced in Concord Grape Berries by the Fungicide Chlorothalonil. Am. J. Enol. Vitic. 42(4): 281-289.

Grove, G. G., L. V. Madden and M. A. Ellis (1985a). Influence of temperature and wetness duration on sporulation of *Phytophthora cactorum* on infected strawberry fruit. Phytopathology 75: 700-703.

Grove, G. G., L. V. Madden and M. A. Ellis (1985b). Splash dispersal of Phytophthora cactorum from infected strawberry fruit. Phytopathology 75(5): 611-615.

Hardwick, N. V. (2006). Disease Forecasting. The Epidemiology of Plant Diseases. B. M. Cooke, G. J. Jones and B. Kaye. Dordrecht, The Netherlands, Springer.

Hoffman, L. E., W. F. Wilcox, D. M. Gadoury, R. C. Seem and D. G. Riegel (2004). Integrated control of grape black rot: Influence of host phenology, inoculum availability, sanitation, and spray timing. Phytopathology 94: 641-650.

Jeleń, H. H., J. Krawczyk, T. O. Larsen, A. Jarosz and B. Gołębniak (2005). Main compounds responsible for off-odour of strawberries infected by *Phytophthora cactorum*. Letters in Applied Microbiology 40(4): 255-259.

Kennelly, M. M., D. M. Gadoury, W. F. Wilcox, P. A. Magarey and R. C. Seem (2005). Seasonal Development of Ontogenic Resistance to Downy Mildew in Grape Berries and Rachises. Phytopathology 95(12): 1445-1452.

Legard, D. E., S. J. MacKenzie, J. C. Mertely, C. K. Chandler and N. A. Peres (2005). Development of a reduced use fungicide program for control of Botrytis fruit rot on annual winter strawberry. Plant Disease 89: 1353-1358.

Leroux, P., M. Gredt and P. Boeda (1988). Resistance to inhibitors of sterol biosynthesis in field isolates or laboratory strains of the eyespot pathogen Pseudocercosporella herpotrichoides. Pesticide Science 23(2): 119-129.

MacKenzie, S. J., J. C. Mertely and N. A. Peres (2009). Curative and Protectant Activity of Fungicides for Control of Crown Rot of Strawberry Caused by *Colletotrichum gloeosporioides*. Plant Disease 93(8): 815-820.

Madden, L. V. (2006). Botanical epidemiology: some key advances and its continuing role in disease management." European Journal of Plant Pathology 115: 2-23.

Madden, L. V., M. A. Ellis, G. G. Grove, K. M. Reynolds and L. L. Wilson (1991). "Epidemiology and control of leather rot of strawberries. Plant Disease 75(5): 439-445.

Madden, L. V., M. A. Ellis, N. Lalancette, G. Hughes and L. L. Wilson (2000). Evaluation of a Disease Warning System for Downy Mildew of Grapes. Plant Disease 84(5): 549-554.

Madden, L. V., G. Hughes and F. van den Bosch (2007). The study of plant disease epidemics. St. Paul, MN, APS press.

Mertely, J. C., S. J. MacKenzie and D. E. Legard (2002). Timing of Fungicide Applications for Botrytis cinerea Based on Development Stage of Strawberry Flowers and Fruit. Plant Disease 86(9): 1019-1024.

Nita, M., M. A. Ellis and L. V. Madden (2007a). Evaluation of the curative and protectant activity of fungicides and fungicide-adjuvant mixtures on Phomopsis cane and leaf spot of grape: a controlled environment study. Crop Protection 26(9): 1377-1384.

Nita, M., M. A. Ellis and L. V. Madden (2008). Variation in Disease Incidence of Phomopsis Cane and Leaf Spot of Grape in Commercial Vineyards in Ohio. Plant Disease 92(7): 1053-1061.

Nita, M., M. A. Ellis, L. L. Wilson and L. V. Madden (2006a). Effects of Application of Fungicide During the Dormant Period on Phomopsis Cane and Leaf Spot of Grape Disease Intensity and Inoculum Production. Plant Disease 90(9): 1195-1200.

Nita, M., M. A. Ellis, L. L. Wilson and L. V. Madden (2006b). Evaluation of a disease warning system for Phomopsis cane and leaf spot of grape: a field study. Plant Disease 90: 1239-1246.

Nita, M., M. A. Ellis, L. L. Wilson and L. V. Madden (2007b). Evaluations of new and current management strategies to control Phomopsis cane and leaf spot of grape. Online. Plant Health Progress: doi:10.1094/PHP-2007-0726-1006-RS.

Nita, M., L. V. Madden, L. L. Wilson and M. A. Ellis (2003). Evaluation of a disease prediction system for Phomopsis cane and leaf spot of grape. Phytopathology 93(6): S65.

Nita, M., K. Tilley, E. De Wolf and G. A. Kuldau (2005). Effects of moisture during and after anthesis on the development of Fusarium head blight of wheat and mycotoxin production. National Fusarium head blight forum, Wilwaukee, WI.

Ntahimpera, N., M. A. Ellis, L. L. Wilson and L. V. Madden (1998). Effects of a Cover Crop on Splash Dispersal of Colletotrichum acutatum Conidia. Phytopathology 88(6): 536-543.

Nutter, F. F. W. (2007). The Role of Plant Disease Epidemiology in Developing Successful Integrated Disease Management Programs. General Concepts in Integrated Pest and Disease Management. A. Ciancio and K. G. Mukerji, Springer Netherlands: 45-79.

Pearson, R. C. and A. C. Goheen, Eds. (1988). Compendium of Grape Diseases. St. Paul, MN, American Phytopathological Society.

Peres, N. I. A., T. E. Seijo and W. W. Turechek (2010). Pre- and post-inoculation activity of a protectant and a systemic fungicide for control of anthracnose fruit rot of strawberry under different wetness durations. Crop Protection 29(10): 1105-1110.

Pfender, W. F. (2006). Interaction of Fungicide Physical Modes of Action and Plant Phenology in Control of Stem Rust of Perennial Ryegrass Grown for Seed. Plant Disease 90(9): 1225-1232.

Pscheidt, J. W. and R. C. Pearson (1989). Time of infection and control of Phomopsis fruit rot of grape. Plant Disease 73: 893-833.

Rebollar-Alviter, A., L. V. Madden and M. A. Ellis (2005). Efficacy of azoxystrobin, pyraclostrobin, potassium phosphite and mefenoxam for control of strawberry leather rot caused by Phytophthora cactorum. Plant Health Progress doi:10.1094: PHP-2005-0107-2001-RS.

Rebollar-Alviter, A., L. V. Madden and M. A. Ellis (2007a). Pre- and Post-Infection Activity of Azoxystrobin, Pyraclostrobin, Mefenoxam, and Phosphite Against Leather Rot of Strawberry, Caused by Phytophthora cactorum. Plant Disease 91(5): 559-564.

Rebollar-Alviter, A., Madden, L. V., Jeffers, S. N., and Ellis, M. A. (2007b). Baseline and differential sensitivity to two QoI fungicides among isolates of Phytophthora cactorum that cause leather rot and crown rot on strawberry. Plant Dis. 91:1625-1637.

Rebollar-Alviter, A., L. L. Wilson, L. V. Madden and M. A. Ellis (2010). A comparative evaluation of post-infection efficacy of mefenoxam and potassium phosphite with protectant efficacy of azoxystrobin and potassium phosphite for controlling leather rot of strawberry caused by *Phytophthora cactorum*. Crop Protection 29(4): 349-353.

Reynolds, K. M., M. A. Ellis and L. V. Madden (1987). Progress in development of a strawberry leather rot forecasting system. Avd. Strawberry Prod. 6: 18-22.

Reynolds, K. M., M. A. Ellis and L. V. Madden (1988). Effect of weather variables on strawberry leather rot epidemics. Phytopathology 78: 822-827.

Shabi, E., T. Katan and K. Marion (1983). Inheritance of resistance to benomyl in isolates of Venturia inaequalis from Israel. Plant Pathology 32(2): 207-211.

Staub, T. (1991). Fungicide Resistance: Practical Experience with Antiresistance Strategies and the Role of Integrated Use. Annual Review of Phytopathology 29(1): 421-442.

Stott, I. P. H., Noon, R. A., Heaney, S. P. (1990). Flutriafol, ethirimol, and thiabendazole seed treatment - an update on field performance and resistance monitoring. Brighton Crop Protection Conference, Pests and Diseases - 1990. Vol. 3. pp. 1169-1174.

Szkolnik, M. (1981). Physical mode of action of sterol-inhibiting fungicide against apple diseases. Plant Disease 65: 981-985.

Turechek, W. W., N. I. A. Peres and N. A. Werner (2006). Pre- and Post-Infection Activity of Pyraclostrobin for Control of Anthracnose Fruit Rot of Strawberry Caused by *Colletotrichum acutatum*. Plant Disease 90(7): 862-868.

Van der Plank, J. E. (1963). Plant Diseases: Epidemics and Control. New York, Academic Press.

Wolf, T. K., Ed. (2008). Wine grape production guide for eastern North America. Ithaca, N.Y., Natural Resource, Agriculture, and Engineering Service (NRAES) Cooperative Extension.

Wong, F. P. and W. F. Wilcox (2001). Comparative Physical Modes of Action of Azoxystrobin, Mancozeb, and Metalaxyl Against Plasmopara viticola (Grapevine Downy Mildew). Plant Disease 85(6): 649-656.

Ypema, H. L. and R. E. Gold (1999). Kresoxim-methyl: modification of a naturally occurring compound to produce a new fungicide. Plant Disease 83(1): 4-19.

Zadoks, J. C. (1984). A quarter century of disease warning, 1958-1983. Plant Disease 68(4): 352-355.

Zadoks, J. C. and R. D. Schein (1979). Epidemiology and plant disease management. New York, Oxford University Press Inc.

3

Multiple Fungicide Resistance in *Botrytis*: A Growing Problem in German Soft-Fruit Production

Roland W. S. Weber and Alfred-Peter Entrop
Esteburg Fruit Research and Advisory Centre, Jork
Germany

1. Introduction

In the mild and humid climate of northwestern Europe, fungal pre-harvest rots are a major factor limiting the production of soft-fruits such as strawberries and raspberries. By far the most important fungal disease is grey mould caused by *Botrytis cinerea* Pers.:Fr., now known to be an aggregate of at least two distinct species (Fournier *et al.*, 2005; Giraud *et al.*, 1999). Primary infections are initiated at flowering, resulting in a quiescent colonisation of floral organs (Bristow *et al.*, 1986; Puhl & Treutter, 2008). Upon ripening of the infected fruit, a brown rot becomes apparent (Fig. 1A) which rapidly engulfs the entire fruit, culminating in the production of conidiophores (Fig. 1B). If mild and humid conditions prevail at harvest time, conidia released from fruits with primary infections may infect healthy fruits, causing a catastrophic grey mould epidemic (Fig. 1C). Repeated fungicide applications during flowering are therefore essential to control *B. cinerea* in soft-fruit production.

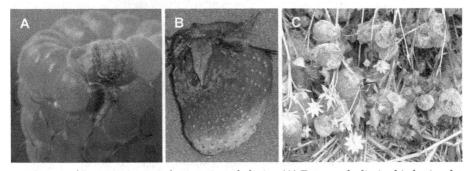

Fig. 1. Stages of *Botrytis cinerea* infections in soft-fruits. (A) Escape of a limited infection from quiescence in a ripening raspberry fruit. (B) Development of a spreading brown rot with the first crop of conidia on an infected strawberry fruit. (C) Occurrence of severe secondary infections during mild and humid weather in a strawberry field in early July 2007.

Northern Germany is a major soft-fruit producing region comprising some 4,000 ha strawberries, 1,500 ha blueberries, 100 ha raspberries and smaller areas dedicated to gooseberry, redcurrant and blackberry production. During the past five years, severe losses to grey mould were recorded by strawberry and raspberry growers. Damage was caused by pre-harvest rot as well as a reduced shelf-life of marketed produce. Although the weather conditions during harvest were often conducive to secondary infections by *B. cinerea*, several fruit farmers also raised concerns about a reduced efficacy of fungicides.

The past four decades have witnessed considerable changes in regional fungicide spray schedules against *Botrytis*. From about 1965 onwards, the sulphamide compounds dichlofluanid and tolylfluanid were widely used as contact (multi-site) fungicides. Although there are literature reports of the development of partial resistance of *B. cinerea* to sulphamides (Malathrakis, 1989; Pollastro *et al.*, 1996; Rewal *et al.*, 1991), no reductions of field efficacy were observed in Northern German soft-fruit production where sulphamides continued to be used as an alternative or complement to a range of specific, single-site fungicides such as benzimidazoles (infrequently used from 1971 to 1976) or dicarboximides (commonly used in 1979-2009). In February 2007, the registration of the last available multi-site fungicide, tolylfluanid, was abruptly withdrawn due to environmental safety concerns. Since then, control of *B. cinerea* has been based solely on compounds with specific (single-site) modes of action, current representatives being fenhexamid (registered since 1999), QoI fungicides (since 2001), boscalid (since 2009), anilinopyrimidines (since 1999) and fludioxonil (since 1999). At present, three widely used botryticides are Switch® (active ingredients cyprodinil + fludioxonil; Syngenta), Signum® (a.i. pyraclostrobin + boscalid; BASF), and Teldor® (a.i. fenhexamid; Bayer CropScience).

Botrytis cinerea is a high-risk pathogen for fungicide resistance development (Brent & Hollomon, 2007; Leroux, 2007), and Northern Germany is a potential 'hot spot' area according to the criteria of Brent & Hollomon (2007). We were concerned about the total reliance on single-site fungicides for grey mould control since 2007, and the lack of any information regarding the current resistance status of regional *Botrytis* populations. In order to address the latter problem, we carried out the surveys and experiments reported here and in previous publications (Weber, 2010b, 2011; Weber & Entrop, 2011; Weber & Hahn, 2011).

2. Fungicide resistance in *B. cinerea*

Botrytis cinerea has earned its reputation as a high-risk pathogen mainly because of its capacity to develop specific resistance to single-site fungicides based on target gene mutations. Specific resistance may emerge within a few years of release of a new fungicide group onto the market, and is usually associated with high resistance factors in laboratory tests. Specific resistance has been described e.g. to benzimidazoles such as benomyl, thiophanate-methyl and carbendazim (Yarden & Katan, 1993; Yourman & Jeffers, 1999), dicarboximides such as iprodione and vinclozolin (Northover & Matteoni, 1986; Yourman & Jeffers, 1999), QoI fungicides (strobilurins; Bardas *et al.*, 2010; Ishii *et al.*, 2009), anilinopyrimidines such as cyprodinil and pyrimethanil (Chapeland *et al.*, 1999; Myresiotis *et al.*, 2007), carboxamides/SDHIs (boscalid; Leroux *et al.*, 2010), and the hydroxyanilide compound fenhexamid (Fillinger *et al.*, 2008; Ziogas *et al.*, 2003). To the best of our knowledge, no reports of specific resistance to the phenylpyrrole compound fludioxonil have been published as yet.

In recent years there have been reports of a non-specific multiple drug resistance (MDR) mechanism based on the activity of ATP binding cassette (ABC) or major facilitator super-family (MFS) transport proteins in *B. cinerea* (Kretschmer *et al.*, 2009; Vermeulen *et al.*, 2001).

MDR phenotypes are usually associated with low or moderate resistance factors, the highest (up to 25-fold relative to baseline sensitivity) being associated with anilinopyrimidines and fludioxonil (Kretschmer et al., 2009).

There is ample evidence that target mutations strongly reduce or even abolish the activity of a fungicide in the field (e.g. Ishii et al., 2009; Myresiotis et al., 2008; Yourman & Jeffers, 1999), and more tentative evidence that this may be the case for MDR phenotypes (Petit et al., 2010; Weber, 2011). Further resistance mechanisms in B. cinerea are the metabolisation of fenhexamid (Billard et al., 2011; Leroux et al., 2002) or the role of alternative oxidases in QoI resistance (Wood & Hollomon, 2003), although the impact of either of these mechanisms on the field efficacy of the respective fungicides is far from clear (Leroux, 2007).

2.1 Development of a resistance assay

An assay of conidial germination and germ-tube growth on the surface of agar media was developed by Weber & Hahn (2011). For thiophanate-methyl, iprodione, fludioxonil and fenhexamid, 1% (w/v) malt extract agar was used. In order to exclude fenhexamid-metabolising isolates from being scored as false positives, germ-tube growth was also examined on tap-water agar containing a critical fenhexamid concentration (Weber, 2010a). For a test of the QoI fungicide trifloxystrobin, 1% malt extract agar was augmented with the alternative oxidase inhibitor salicyl hydroxamic acid (100 µg ml⁻¹). For the boscalid and cyprodinil assays, 0.5% yeast extract agar or 0.5% sucrose agar (respectively) were used instead of malt extract agar.

Pure-culture isolates of B. cinerea were examined for their response to a wide range of fungicide concentrations in order to determine EC_{50} values under the chosen test conditions (Fig. 2; Table 1). For each fungicide Weber & Hahn (2011) also identified discriminatory concentrations at which highly sensitive or baseline (HS) and less sensitive (LS) isolates could be reliably distinguished from those showing moderate resistance (MR) and high resistance (HR). Examples of dose-response patterns to two fungicide groups are given in Fig. 2.

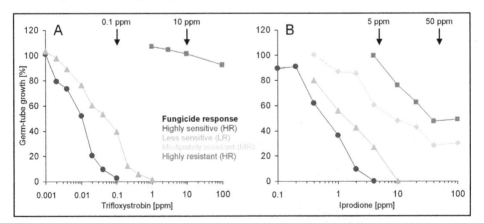

Fig. 2. Responses of selected isolates of Botrytis cinerea to trifloxystrobin (A) and iprodione (B), recorded as germ-tube length after 12 h incubation. Discriminatory concentrations for both assays are indicated by arrows. Adapted from Weber & Hahn (2011).

Based on two discriminatory concentrations for each fungicide (Table 1), a routine test was developed as a decision support tool for fruit farmers and their advisors wishing to optimise their fungicide spray sequence for the control of grey mould. This test had to fulfil certain criteria. Firstly, a relatively high throughput of B. cinerea isolates had to be possible in order to quantify the abundance of resistant strains in a field. Further, the test had to deliver fast results preferably prior to flowering, precluding lengthy isolation and cultivation periods. Thirdly, the test had to be simple so as to be usable with basic equipment and by non-specialised technical staff associated with general microbiology laboratories.

By obtaining conidial suspensions directly from sporulating overwintered plant samples in early spring or from infected fruits at harvest, plating out 10-15 µl drops of suspension onto each of 20 different agar media, and examining germination and/or germ-tube growth with a light microscope after an overnight incubation (Fig. 3), the method was put to practical use during the 2010 and 2011 growing seasons. A high throughput was achieved by accommodating 20-24 drops of different spore suspensions on each agar plate. Samples were collected by the farmers themselves or by their advisors. Results were generated with this method for individual fields, and were communicated to the fruit farmers within 3-7 days of sampling.

Fungicide	Medium	Fungicide concentration [ppm]	EC$_{50}$ [ppm]			
			HS [1]	LS	MR	HR [2]
Thiophanate-methyl	MEA [1]	0, 1, 100	0.047-0.166	-	4.67-16.1	28.0-436
Iprodione	MEA	0, 5, 50	0.399-0.647	0.976-3.51	9.17-23.7	43.9-96.1
Fenhexamid	MEA	0, 1, 50	0.058-0.229	-	-	>100
Trifloxystrobin	MEA +SHAM	0, 0.1, 10	0.007-0.011	0.019-0.056	-	>100
Boscalid	YEA	0, 1, 50	0.041-0.099	0.116-0.346	-	3.43-54.5
Cyprodinil	SA	0, 1, 25	0.028-0.094	0.267-0.482	0.748-1.44	1.89-22.7
Fludioxonil	MEA	0, 0.1, 10	0.013-0.059	0.104-0.178	0.301-0.647	-

[1] Abbreviations: HR (highly resistant), HS (highly sensitive), LS (less sensitive), MEA (1% malt extract agar), MR (moderately resistant), SA (0.5% sucrose agar), SHAM (100 ppm salicyl hydroxamic acid), YEA (0.5% yeast extract agar)
[2] The HR phenotype of the fenhexamid test on MEA was confirmed by full germ-tube growth on tap-water agar containing 10 ppm fenhexamid.

Table 1. Sensitivity categories and their EC$_{50}$ values in the conidial germination assay of 7 fungicides (summarised from Weber & Hahn, 2011).

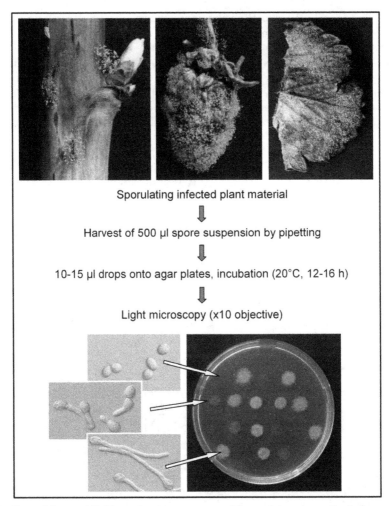

Fig. 3. Outline of the conidial germination assay used for resistance monitoring.

2.2 Reproducibility of the assay method

Overwintered raspberry primocanes infected by *B. cinerea* were collected from selected fields, and incubated in a damp chamber for 2 days. For each cane, conidial suspensions were harvested from three germinated sclerotia separated by a distance of at least 5 cm, and assayed on separate days as described in Fig. 3. The results revealed considerable variations between suspensions obtained from different canes, whereas in most cases the three suspensions collected from any one cane were identical to one another in their fungicide responses (Fig. 4). Further, the observed fungicide resistance phenotypes appeared to be randomly distributed across the field. From these and other results, it was concluded that at least 10 but ideally 15 infected samples should be collected from different parts of the field, and that it was sufficient to analyse one spore suspension from each sample.

Altogether, 93.8% of 1224 conidial suspensions obtained directly from sporulating grey mould lawns on strawberry fruits, raspberry fruits or raspberry canes produced an entirely homogeneous germ-tube growth response to each fungicide, a further 3.4% showing merely minor ambiguities in which less than 10% of conidia differed in their fungicide response from the rest (Weber & Hahn, 2011). The most obvious explanation for such a strong between-sample variation in the near-absence of within-sample variation is that local *B. cinerea* populations may comprise a high genetic diversity with respect to fungicide resistance (Leroch *et al.*, 2011), and that each individual infection is caused by a single germinating conidium.

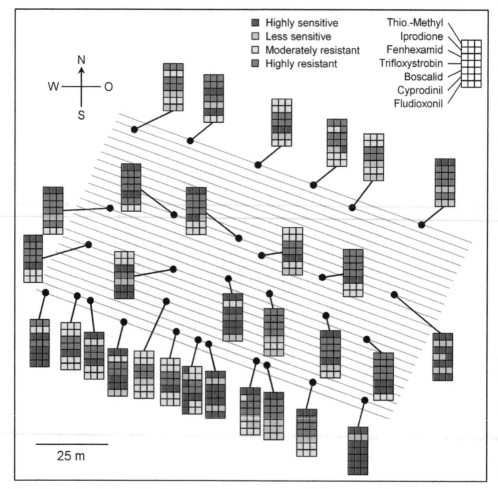

Fig. 4. Reproducibility of the *B. cinerea* resistance assay with 7 fungicides. In April 2010, 30 overwintered infected raspberry canes were collected from a Northern German field at the positions indicated. For each cane, test results of three conidial suspensions obtained from individual germinating sclerotia are shown in columns.

3. Temporal and spatial distribution of fungicide resistance

In a large-scale survey carried out in Northern Germany in 2010, Weber (2011) analysed 353 representative *B. cinerea* conidial suspensions from strawberries, raspberries and other soft-fruits and found that resistance to all 7 fungicides was present at high levels (Table 2). Pure cultures of isolates representing all observed resistance responses to each fungicide were collected. Conidia from MR and/or HR isolates showed a significantly enhanced ability to cause fruit rot in apples pre-treated with the respective fungicides at commercially used concentrations (Weber, 2011), in comparison with sensitive (HS or LS) isolates. In spite of the limitations of the apple inoculation test, these results indicate the possibility that all MR and HR phenotypes identified in the resistance assay might compromise the efficacy of currently registered botryticides in soft-fruit production.

Fungicide	Percent resistant isolates	
	MR	HR
Thiophanate-methyl	18.7%	21.8%
Iprodione	34.8%	29.2%
Fenhexamid	-	45.0%
Trifloxystrobin	-	76.8%
Boscalid	-	21.5%
Cyprodinil	27.2%	14.7%
Fludioxonil	41.1%	-

Table 2. Occurrence of fungicide resistance in Northern German soft-fruit production during 2010 (*n* = 353 isolates tested). From Weber (2011).

3.1 The dimension of time

In Northern Germany, raspberry fields are being cropped for up to 10 years, rendering them useful objects for long-term studies of resistance development. Results of repeated sampling at two commercial sites are presented in Fig. 5, comparing field A under normal management (3-4 annual fungicide applications during flowering) *versus* field B under an intensive crop protection regime with 6-8 applications (Fig. 5A and Fig. 5B, respectively). Both fields had received 1-2 annual applications of fenhexamid during flowering up to and including the year 2008, but none thereafter. Whilst the share of fenhexamid-resistant isolates collapsed in field A, it remained at a high level in field B. On a regional scale, the percentage of fenhexamid resistance in Northern German soft-fruit production has risen from 18.5% in 2009 (Weber, 2010b) to 45.0% in 2010 (Weber, 2011).

Boscalid resistance is another striking case. This compound was not registered for grey mould control in Northern Germany before the 2009 vegetation period, yet by the end of 2010, *i.e.* its second season of use, there was already a 21.5% share of highly resistant isolates across the region (Weber, 2011). Almost simultaneously, boscalid resistant isolates of *B. cinerea* have been described from other regions and other crops (Bardas *et al.*, 2010; Kim & Xiao, 2010; Leroch *et al.*, 2011).

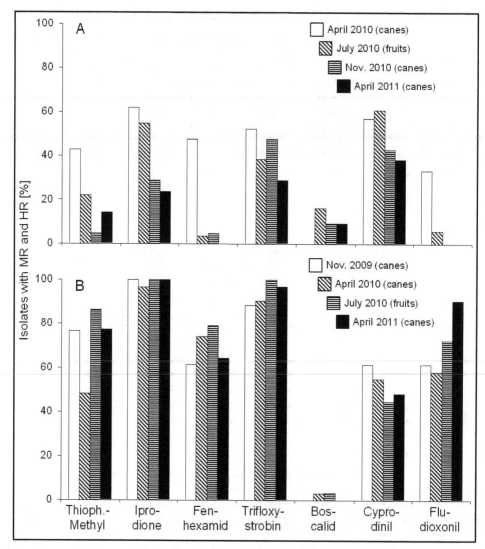

Fig. 5. Percentage of combined moderately and highly fungicide-resistant strains of *B. cinerea* in two Northern German raspberry orchards, one receiving 3-4 annual fungicide sprays at flowering (A), the other 6-8 such treatments (B). In both orchards, fenhexamid was last used in May 2008.

3.2 Selection of resistance by fungicide applications

Tentative correlations between the frequency of fungicide applications and the resulting percentage of resistant strains may be drawn from the results of the 2010 monitoring (Weber, 2011). Unfortunately, only spray schedules for the flowering period immediately preceding sampling were made available by fruit farmers. Nevertheless, for all chemical classes except anilinopyrimidines (cyprodinil, pyrimethanil, mepanipyrim) there was a clear trend towards an elevated share of MR and HR isolates in fields which had received several treatments with the respective fungicides at flowering, as compared to those with the lowest number of applications (Table 3). More detailed analyses incorporating up to 3 years' spray history are being conducted in the course of our 2011 resistance survey.

Fungicide	Percent MR and HR isolates after 0-6 applications					
	0	1	1 or 2	2	3	4-6
Fenhexamid	27.5	46.9	-	59.3	-	-
QoI compounds	-	30.0	-	70.1	91.9	80.8
Boscalid	1.4	20.3	-	25.5	30.4	-
Anilinopyrimidines	-	-	42.7	-	43.1	30.6
Fludioxonil	-	-	14.6	-	55.4	-

Table 3. Percent of fungicide-resistant isolates in comparison with the number of fungicide applications during the flowering period preceding sampling. Evaluation of data collated by Weber (2011).

In order to examine the relationship between fungicide application frequency and resistance development in greater depth, we are performing a series of forced selection experiments (see Brent & Hollomon, 2007). A preliminary trial was carried out in a raspberry orchard during the 2010 season. Variants included (1) a fungicide-free control, (2) four successive sprays of Teldor® as the sole fungicide during flowering, and (3) the farmer's standard spray sequence (Signum® – Switch® – Signum® – Switch®). Resistance tests were conducted with B. cinerea isolates obtained from infected fruits at harvest (Fig. 6).

With only 5.7% grey mould-affected fruits at harvest, the farmer's strategy of alternating treatments with two combination products was successful in terms of crop protection relative to the untreated control (18.8% fruit loss to grey mould). However, as indicated in Fig. 6 there was a significant enrichment of B. cinerea strains with boscalid resistance (16%) as compared to the other two variants (2.3-4.0%). This result substantiates preliminary observations (Table 3) and literature reports of the unusually rapid spread of boscalid resistance following the release of this fungicide in 2009 (see Section 3.1).

Four fenhexamid applications at flowering provided an acceptable control of grey mould (8.2% diseased fruits at harvest) but resulted in a significant enrichment of HR strains (26%) as compared to the other two variants (1.1-3.5%). A sequence of four successive fenhexamid applications may seem excessive but would have been in line with the manufacturer's original recommendation when this fungicide was first registered for the control of grey mould in raspberries (Rosslenbroich, 1999). Clearly, therefore, a strategy based on repeated treatments with the same single-site fungicide(s) may provide good crop protection during the first year(s) of fungicide use but could turn out to be non-sustainable due to the rapid enrichment of resistant B. cinerea strains. In Northern German horticulture there is a veritable history of such boom-and-bust cycles of resistance development within 2-4 years of excessive applications of newly registered fungicides. The best-known examples concern the apple-scab fungus *Venturia inaequalis* which developed resistance to QoI fungicides in 1999 during their third season of intensive use (Palm *et al.*, 2004), and to benzimidazoles in 1974 during their fourth season (Vagt, 1975). Obviously being aware of contemporary reports of benzimidazole resistance in *B. cinerea* from other horticultural situations (Bollen & Scholten, 1971), in a remarkable act of foresight Vagt (1975) cautioned against the continued use of this chemical group for grey mould control in regional soft-fruit production.

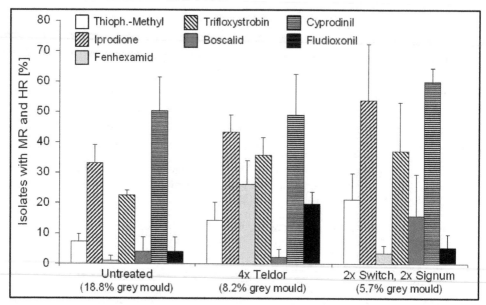

Fig. 6. Forced selection experiment in an 8-year-old raspberry orchard in which plots were left untreated at flowering, or treated with four applications of Teldor® (a.i. fenhexamid) or with an alternating sequence of two Signum® and two Switch® applications. Data are shown as percentage of isolates with MR and HR phenotypes to each of the 7 fungicides tested. Error bars indicate standard deviations of four replicates per treatment. Percentages of fruits with grey mould at harvest are indicated. Altogether 290 isolates were assayed at harvest.

3.3 The issue of multiple fungicide resistance

An important difference between fields A and B shown in Fig. 5 was that strains with multiple resistance were much less common in the former. Thus, only 22.9% of fenhexamid-resistant isolates from field A possessed resistance to more than two of the other four currently used fungicides (QoI, boscalid, cyprodinil and fludioxonil) whereas that share was 49.4% in field B. A situation is therefore conceivable in which strains with resistance to a given fungicide may be enriched or maintained in a field even if that compound is no longer used. Such interactions must be taken into consideration when evaluating the 'fitness' of different fungicide resistance phenotypes.

In general terms, there seems to be a trend in *B. cinerea* towards accumulating resistance, as shown also in the large-scale survey by Weber (2011) who grouped his isolates according to multiples of resistance (0 to 5) to the five currently used fungicide groups, *i.e.* fenhexamid,

QoI compounds, boscalid, anilinopyrimidines, and fludioxonil. There was a significant positive correlation between the number of resistances to current fungicides and the proportion of strains harbouring resistance to iprodione or thiophanate-methyl, two compounds no longer in use. Furthermore, all isolates with quintuple resistance (MR or HR) to current fungicides were obtained from fields with a history of unusually heavy fungicide use.

It is of interest to re-interpret the combinations of resistances in individual isolates (Weber, 2011) in terms of their likely impact on the field efficacy of the commercial products Switch®, Signum® and Teldor®. Thus, even when adding up all fully sensitive isolates as well as those with sensitivity to at least one of the two active ingredients in Signum® or Switch®, merely 38% of all isolates obtained in the 2010 monitoring would have been susceptible to all three products, 40.5% to two, 15% to one, and 6.5% to none of them (Table 4). These data should ring alarm bells with fruit farmers, advisory services and regulatory authorities, especially if a trend towards a further increase in resistance development can be recognised in the results of the ongoing 2011 survey.

Sensitivity[1] to	Percent of isolates[2]
Teldor + Signum + Switch	37.95%
Signum + Switch	24.65%
Teldor + Signum	3.68%
Teldor + Switch	12.18%
Signum only	12.18%
Switch only	2.27%
Teldor only	0.57%
none	6.52%

[1] Meaning sensitivity to at least one of the active ingredients of the commercial product
[2] Total = 353 isolates analysed

Table 4. Proportions of *Botrytis* isolates from Northern German soft-fruit fields with putative field sensitivity to the commercial fungicides Switch® (a.i. fludioxonil and cyprodinil), Signum® (a.i. pyraclostrobin and boscalid) and Teldor® (a.i. fenhexamid), as extrapolated from the resistance tests of Weber (2011).

3.4 Factors favouring the spread of fungicide resistance in *B. cinerea*

From this chapter as well as previous publications (Weber, 2010b, 2011), it is apparent that *B. cinerea* populations associated with Northern German strawberry and raspberry fields harbour an exceptionally high proportion of MR- and/or HR-type resistance against all currently registered fungicide classes, and that resistance development may be on the increase. We are particularly concerned about the possible spread of strains with multiple fungicide resistance because such a development questions the sustainability of the current crop protection practices.

There are several possible reasons to account for the spread of fungicide-resistant strains of *B. cinerea*. Firstly, since February 2007 grey mould control has been based solely on single-site fungicides, all of which except fludioxonil are known to be susceptible to specific resistance development based on target gene mutations. Secondly, the number of fungicide applications made by fruit farmers during flowering has often exceeded regional recommendations (Weber & Entrop, 2011; see Fig. 5). Autumn-cropping strawberries are of particular concern in this respect because their prolonged flowering periods necessitate more numerous fungicide treatments. Thirdly, the short permissible pre-harvest intervals of 3-7 days for the registered fungicides Switch®, Signum® and Teldor® may have encouraged farmers to spray them against ongoing grey mould epidemics between successive fruit pickings at harvest. Fourthly, restrictions by food retail chains concerning the maximum permissible number of detectable pesticide residues in soft-fruits have forced fruit farmers to abandon resistance management strategies in favour of repeated applications of only one or two of the three main botryticides Switch®, Signum® and Teldor®. As we have seen, there is preliminary experimental evidence that such a strategy may lead to a specific build-up of resistance in selected fields within a single vegetation period (Fig. 6). Fifthly, promises of high degrees of efficacy of these botryticides by agrochemical companies may have encouraged fruit farmers to neglect non-chemical crop protection measures related to sanitation, irrigation, aeration and fertilisation.

4. Recommendations

There are several ways to support the activity of fungicides in grey mould control. Most of these complementary approaches are matters of basic horticultural knowledge (see e.g. Sutton, 1998; Weber & Entrop, 2011). All of the suggestions made below may not be applicable to all fruit farmers, but most farmers should be able to implement some of them.

4.1 Fungicide applications

The number of fungicide applications during flowering should be limited to 3-4 sprays at intervals of approx. 7 days. Additional applications do not generally lead to a higher efficacy of grey mould control under Northern German conditions (Faby, 2009) and are counterproductive in terms of resistance management. All three available botryticides should be used, unless site-specific tests have indicated the presence of resistant strains at frequencies at or above 25% (Forster & Staub, 1996; Petit *et al.*, 2010; Weber, 2010a). Switch®, Signum® and Teldor® should not be applied at harvest, as post-harvest treatments e.g. of cane diseases, or indeed at any time of year other than flowering. In view of the widespread resistance to QoI fungicides in *B. cinerea* and their limited efficacy against grey mould, this fungicide class should not be applied singly during flowering.

4.2 Registration of new fungicides

In general terms, broad-spectrum fungicides such as tolylfluanid, chlorothalonil, folpet or thiram possess a lesser efficacy against *B. cinerea* than single-site fungicides (Gullino *et al.*, 1989; Singh & Milne, 1974; Weber & Entrop, 2011), but they are also less susceptible to resistance development (Leroux, 2007). Future experiments should focus on examining their impact on the development of resistance against the current single-site fungicides when applied as tank mixtures. Detailed experiments under regional conditions are required because previous workers have obtained heterogeneous results (Gullino *et al.*, 1989; Hunter *et al.*, 1987; Northover & Matteoni, 1986). If a suitable broad-spectrum fungicide can be found, this should be registered for grey mould control with some urgency.

Similar trials should also be performed with alternative control agents such as inducers of systemic acquired resistance or biological control organisms. As with broad-spectrum fungicides, efficacy should be evaluated as an effect on the spread of fungicide resistance, rather than (or in addition to) grey mould control *per se*.

4.3 Issues of cultivation

Crop cultivation should aim to maximise yields as well as create conditions which optimise fungicide efficacy. Nitrogen fertilisation of strawberries should be reduced to about 60 kg ha^{-1} *per annum* which is sufficient to ensure a good yield whilst avoiding soft plant tissues susceptible to *B. cinerea* infections. Periods of leaf wetness should be reduced to a minimum by using drip irrigation if possible, or alternatively by using rather than extending natural periods of leaf wetness for overhead irrigation. Strawberry or raspberry crops grown in poly-tunnels or under fleece for early cropping should be aerated regularly in order to reduce periods of leaf wetness. Likewise, the planting distance should be sufficient to ensure a good ventilation of leaves and flowers.

Hygiene measures aimed at reducing inoculum should be given a high priority. Thus, rotting or mouldy fruits should be collected separately and removed from the field especially during the first pickings of the season. Likewise, infected raspberry canes should be pruned and removed from the field before bud burst.

4.4 Outlook

In contrast to pome fruit farming, there is no commercially relevant organic soft-fruit production in Germany, the chief reason being dramatic pre- and post-harvest losses to grey mould. Clearly, non-chemical crop protection measures alone are insufficient to provide acceptable control of *B. cinerea* in Northern Germany. However, as we have discussed in the present chapter, a similar situation may soon hold for the current crop protection strategy based on specific fungicides. Fungicide resistance development in *B. cinerea* should catalyse a change of paradigm towards a truly integrated crop protection concept embracing both chemical and supplementary non-chemical measures. There should be ample rewards for fruit farmers willing to embark on such a concept because of a stable demand for fresh and regionally produced soft-fruits on the market.

5. References

Bardas, G.A., Veloukas, T., Koutita, O. & Karaoglanidis, G.S. (2010). Multiple resistance of *Botrytis cinerea* from kiwifruit to SDHIs, QoIs and fungicides of other chemical groups. *Phytopathology* Vol.98, No.4, pp.443-450, ISSN 0031-949X.

Billard, A., Fillinger, S., Leroux, P., Bach, J., Solignac, P., Lachaise, H., Beffa, R. & Debieu, D. (2011). Fenhexamid resistance in the complex of *Botrytis* species, responsible for grey mould disease. In: *Fungicides/Book 2*, N. Thajuddin (Ed.), pp.XX-YY, InTech, ISBN 979-953-307-554-8, Vienna, Austria.

Bollen, G.J. & Scholten, G. (1971). Acquired resistance to benomyl and some other systemic fungicides in a strain of *Botrytis cinerea* in cyclamen. *Netherlands Journal of Plant Pathology* Vol.77, No.3, pp.83-90, ISSN 0028-2944.

Brent, K.J. & Hollomon, D.W. (2007). *Fungicide Resistance: The Assessment of Risk* (second edition). Croplife International, Brussels, Belgium. Avaliable from http://www.frac.info/frac/publication/anhang/FRAC_Mono2_2007.pdf

Bristow, P.R., McNicol, R.J. & Williamson, B. (1986). Infection of strawberry flowers by *Botrytis cinerea* and its relevance to grey mould development. *Annals of Applied Biology* Vol.109, No.3, pp.545-554, ISSN 1744-7348.

Chapeland, F., Fritz, R., Lanen, C., Gredt, M. & Leroux, P. (1999). Inheritance and mechanisms of resistance to anilinopyrimidine fungicides in *Botrytis cinerea* (*Botryotinia fuckeliana*). *Pesticide Biochemistry and Physiology* Vol.64, No.2, pp.85-100, ISSN 0048-3575.

Faby, R. (2009). Termine und Häufigkeit der Botrytis-Anwendungen in Erdbeeren. *Mitteilungen des Obstbauversuchsringes des Alten Landes* Vol.64, No.4, pp.148-153, ISSN 0178-2916.

Fillinger, S., Leroux, P., Auclair, C., Barreau, C., al Hajj, C. & Debieu, D. (2008). Genetic analysis of fenhexamid-resistant field isolates of the phytopathogenic fungus *Botrytis cinerea*. *Antimicrobial Agents and Chemotherapy* Vol.52, No.11, pp.3933-3940, ISSN 0066-4804.

Forster, B. & Staub, T. (1996). Basis for use strategies of anilinopyrimidine and phenylpyrrole fungicides against *Botrytis cinerea*. *Crop Protection* Vol.15, No.6, pp.529-537, ISSN 0261-2194.

Fournier, E., Giraud, T, Albertini, C. & Brygoo, Y. (2005). Partition of the *Botrytis cinerea* complex in France using multiple gene genealogies. *Mycologia* Vol.97, No.6, pp.1251-1267, ISSN 0027-5514.

Giraud, T., Fortini, D., Levis, C., Lamarque, C., Leroux, P., LoBuglio, K. & Brygoo, Y. (1999). Two sibling species of the *Botrytis cinerea* complex, *transposa* and *vacuma*, are found in sympatry on numerous host plants. *Phytopathology* Vol.89, No.10, pp.967-973, ISSN 0031-949X.

Gullino, M.L., Aloi, C. & Garibaldi, A. (1989). Influence of spray schedules on fungicide resistant populations of *Botrytis cinerea* Pers. on grapevine. *Netherlands Journal of Plant Pathology* Vol.95, Suppl.1, pp.87-94, ISSN 0028-2944.

Hunter, T., Brent, K.J., Carter, G.A. & Hutcheon, J.A. (1987). Effects of fungicide spray regimes on incidence of dicarboximide resistance in grey mould (*Botrytis cinerea*) on strawberry plants. *Annals of Applied Biology* Vol.110, No.3, pp.515-525, ISSN 0003-4746.

Ishii, H., Fountaine, J., Chung, W.-H., Kansako, M., Nishimura, K., Takahashi, K. & Oshima, M. (2009). Characterisation of QoI-resistant field isolates of *Botrytis cinerea* from citrus and strawberry. *Pest Management Science* Vol.65, No.8, pp.916-922, ISSN 1526-4998.

Kim, Y.K. & Xiao, C.L. (2010). Resistance to pyraclostrobin and boscalid in populations of *Botrytis cinerea* from stored apples in Washington State. *Plant Disease* Vol.94, No.5, pp.604-612, ISSN 0191-2917.

Kretschmer, M., Leroch, M., Mosbach, A., Walker, A.-S., Fillinger, S., Mernke, D., Schoonbeek, H.-J., Pradier, J.-M., Leroux, P., De Waard, M.A. & Hahn, M. (2009).

Fungicide-driven evolution and molecular basis of multidrug resistance in field populations of the grey mould fungus *Botrytis cinerea*. *PLoS Pathogens* Vol.5, No.12, e1000696.

Leroch, M., Kretschmer, M. & Hahn, M. (2011). Fungicide resistance phenotypes of *Botrytis cinerea* isolates from commercial vineyards in South West Germany. *Journal of Phytopathology* Vol.159, No.1, pp.63-65, ISSN 1439-0434.

Leroux, P. (2007). Chemical control of *Botrytis* and its resistance to chemical fungicides. In: *Botrytis: Biology, Pathology and Control*, Y. Elad, B. Williamson, P. Tudzynski & N. Delen (Eds.), pp.195-222, Springer Verlag, ISBN 978-1-4020-6586-6, Dordrecht, Netherlands.

Leroux, P., Fritz, R., Debieu, D., Albertini, C., Lanen, C., Bach, J., Gredt, M. & Chapeland, F. (2002). Mechanism of resistance to fungicides in field strains of *Botrytis cinerea*. *Pest Management Science* Vol.58, No.9, pp.876-888, ISSN 1526-4998.

Leroux, P., Gredt, M., Leroch, M. & Walker, A.-S. (2010). Exploring mechanisms of resistance to respiratory inhibitors in field strains of *Botrytis cinerea*, the causal agent of gray mold. *Applied and Environmental Microbiology* Vol.76, No.19, pp.6615-6630, ISSN 0099-2240.

Malathrakis, N.E. (1989). Resistance of *Botrytis cinerea* to dichlofluanid in greenhouse vegetables. *Plant Disease* Vol.73, No.2, pp.138-141, ISSN 0191-2917.

Myresiotis, C.K., Karaoglanidis, G.S. & Tzavella-Klonari, K. (2007). Resistance of *Botrytis cinerea* isolates from vegetable crops to anilinopyrimidine, phenylpyrrole, hydroxyanilide, benzimidazole, and dicarboximide fungicides. *Plant Disease* Vol.91, No.4, pp.407-413, ISSN 0191-2917.

Myresiotis, C.K., Bardas, G.A. & Karaoglanidis, G.S. (2008). Baseline sensitivity of *Botrytis cinerea* to pyraclostrobin and boscalid and control of anilinopyrimidine- and benzimidazole-resistant strains by these fungicides. *Plant Disease* Vol.92, No.10, pp.1427-1431, ISSN 0191-2917.

Northover, J. & Matteoni, J.A. (1986). Resistance of *Botrytis cinerea* to benomyl and iprodione in vineyards and greenhouses after exposure to the fungicides alone or mixed with captan. *Plant Disease* Vol.70, No.5, pp.398-402, ISSN 0191-2917.

Palm, G., Kuck, K.-H., Mehl, A. & Marr, J. (2004). Aktueller Stand der Strobilurin-Apfelschorf-Resistenz an der Niederelbe. *Mitteilungen des Obstbauversuchsringes des Alten Landes* Vol.59, No.8, pp.291-295, ISSN 0178-2916.

Petit, A.-N., Vaillant-Gaveau, N., Walker, A.-S., Leroux, P., Baillieul, F., Panon, M.-L., Clément, C. & Fontaine, F. (2010). Determinants of fenhexamid effectiveness against grey mould on grapevine: Respective role of spray timing, fungicide resistance and plant defences. *Crop Protection* Vol.29, No.10, pp.1162-1167, ISSN 0261-2194.

Pollastro, S., Faretra, F., Di Canio, V. & De Guido, A. (1996). Characterization and genetic analysis of field isolates of *Botryotinia fuckeliana* (*Botrytis cinerea*) resistant to dichlofluanid. *European Journal of Plant Pathology* Vol.102, No.7, pp.607-613, ISSN 0929-1873.

Puhl, I. & Treutter, D. (2008). Ontogenetic variation of catechin biosynthesis as basis for infection and quiescence of *Botrytis cinerea* in developing strawberry fruits. *Journal of Plant Diseases and Protection* Vol.115, No.6, pp.247-251, ISSN 1861-3829.

Rewal, N., Coley-Smith, J.R. & Sealy-Lewis, H.M. (1991). Studies on resistance to dichlofluanid and other fungicides in *Botrytis cinerea*. *Plant Pathology* Vol.40, No.4, pp.554-560, ISSN 1365-3059.

Rosslenbroich, H.-J. (1999). Efficacy of fenhexamid (KBR2738) against *Botrytis cinerea* and related fungal pathogens. *Pflanzenschutz-Nachrichten Bayer* Vol.52, No.2, pp.127-144, ISSN 0340-1723.

Singh, G. & Milne, K.S. (1974). Field evaluation of fungicides for the control of chrysamthemum flower blight. *New Zealand Journal of Experimental Agriculture* Vol.2, No.2, pp.185-188, ISSN 0301-5521.

Sutton, J.C. (1998). Botrytis fruit rot (gray mold) and blossom blight. In: *Compendium of Strawberry Diseases* (second edition), J.L. Maas (Ed.), pp.28-31, APS Press, ISBN 0-89053-194-9, St. Paul, USA.

Vagt, W. (1975). Die Schorfsituation 1974 und unsere Spritzempfehlungen für 1975. *Mitteilungen des Obstbauversuchsringes des Alten Landes* Vol.30, No.3, pp.76-80, ISSN 0178-2916.

Vermeulen, T., Schoonbeek, H. & de Waard, M. (2001). The ABC transporter BcatrB from *Botrytis cinerea* is a determinant of the activity of the phenylpyrrole fungicide fludioxonil. *Pest Management Science* Vol.57, No.5, pp.393-402, ISSN 1526-4998.

Weber, R.W.S. (2010a). Schnelle und einfache Methode zum Nachweis der Fenhexamid-Resistenz bei *Botrytis*. *Erwerbs-Obstbau* Vol.52, No.1, pp.27-32, ISSN 0014-0309.

Weber, R.W.S. (2010b). Occurrence of Hyd R3 fenhexamid resistance among *Botrytis* isolates in Northern Germany. *Journal of Plant Diseases and Protection* Vol.117, No.4, pp.177-179, ISSN 1861-3829.

Weber, R.W.S. (2011). Resistance of *Botrytis cinerea* to multiple fungicides in Northern German small fruit production. *Plant Disease* Vol.95, No.10, pp.1263-1269, ISSN 0191-2917.

Weber, R.W.S. & Entrop, A.-P. (2011). Auftreten, Bedeutung und Vermeidung von Fungizid-Resistenzen bei *Botrytis* an Erdbeeren und Himbeeren. *Mitteilungen des Obstbauversuchsringes des Alten Landes* Vol.66, No.5, pp.136-144, ISSN 0178-2916.

Weber, R.W.S. & Hahn, M. (2011). A rapid and simple method for determining fungicide resistance in *Botrytis*. *Journal of Plant Diseases and Protection* Vol.118, No.1, pp.17-25, ISSN 1861-3829.

Wood, P.M. & Hollomon, D.W. (2003). A critical evaluation of the role of alternative oxidase in the performance of strobilurin and related fungicides acting at the Qo site of complex III. *Pest Management Science* Vol.59, No.5, pp.499-511, ISSN 1526-4998.

Yarden, O. & Katan, T. (1993). Mutations leading to substitutions at amino acids 198 and 200 of beta-tubulin that correlate with benomyl-resistance phenotypes of field strains of *Botrytis cinerea*. *Phytopathology* Vol.83, No.12, pp.1478-1483, ISSN 0031-949X.

Yourman, L.F. & Jeffers, S.N. (1999). Resistance to benzimidazole and dicarboximide fungicides in greenhouse isolates of *Botrytis cinerea*. *Plant Disease* Vol.83, No.6, pp.569-575, ISSN 0191-2917.

Ziogas, B.N., Markoglou, A.N. & Malandrakis, A.A. (2003). Studies on the inherent resistance risk to fenhexamid in *Botrytis cinerea*. *European Journal of Plant Pathology* Vol.109, No.4, pp.311-317, ISSN 0929-1873.

4

Fenhexamid Resistance in the *Botrytis* Species Complex, Responsible for Grey Mould Disease

A. Billard[1,2] et al[*]

[1]*INRA UR 1290 BIOGER-CPP Thiverval-Grignon,*
[2]*Bayer SAS, Bayer CropScience, Research Center La Dargoire, Lyon,*
France

1. Introduction

1.1 Chemical control of grey mould in French vineyards

The three major fungal pests of grapevine, powdery and downy mildew and grey mould are mostly controlled through the application of fungicides. Some of those are particularly active against the grey mould agent *Botrytis cinerea*. The panel of fungicides authorized in France comprise since many years anilinopyrimidines, benzimidazoles, dithiocarbamates, dicarboximides, phenylpyrroles and pyridinamines. Lately, the panel has been completed by fenhexamid, a sterol biosynthesis inhibitor (SBI) and the pyridine boscalid, a succinate dehydrogenase inhibitor (SDHI). In addition, a biofungicide based on the *Bacillus subtilis* strain QST 713 (Serenade) has been authorized in 2010 against grey mould in vineyards. The compounds authorized against grey mould in French vineyards are listed in Table 1.

Since grey mould may infect grapevine from flowering until harvest, optimal protection needs to be obtained during this period. Nowadays up to three treatments per season are recommended in vineyards, corresponding to the stages A-C (A: flower cap falling – B: bunch closure – C: veraison). To reduce pesticide applications along with the general trend of reduction of chemical inputs in agriculture chemical treatments against grey mould are positioned according to epidemiological and meteorological parameters. Grey mould not only affects quantity of harvest but also the wine quality. Therefore, treatments also depend on the economic value of the wine. The number of treatments is variable between regions and years according to the factors cited above.

In those regions with regular applications of anti-*Botrytis* fungicides, especially in the Northern regions, resistant strains have been selected which can ultimately lead to treatment failure. In order to reduce the risk of specific resistance development each anti-*Botrytis* mode-of-action is limited to one application/season in France since the 90's, involving alternations of different chemical families to combat grey mould in the vineyards.

[*] S. Fillinger[1], P. Leroux[1], J. Bach[1], C. Lanen[1], H. Lachaise[2], R. Beffa[3] and D. Debieu[1]

1 *INRA UR 1290 BIOGER-CPP Thiverval-Grignon, France,*
2 *Bayer SAS, Bayer CropScience, Research Center La Dargoire, Lyon, France,*
3 *Bayer CropScience AG, Frankfurt, Germany.*

Chemical families	Fungicides	Timing of application
Anilinopyrimidines	Pyrimethanil, Mepanipyrim, Cyprodinil	A, B or C
Benzimidazoles	Thiophanate-methyl	A or B
Carboxamides	Boscalid	A, B or C
Dithiocarbamates	Thiram	A
Hydroxyanilides	Fenhexamid	A, B or C
Dicarboximides	Iprodione	A, B or C
Phenylpyrroles	Fludioxonil	A or B
Pyridinamines	Fluazinam	A, B or C
Bacillus subtilis (biofungicide)	Serenade	up to 8 applications/year

Table 1. Active substances approved against grey mould of grapevine in France and recommended stages of applications (French National *Botrytis* Note - Vigne 2010). For further explanations, see text.

1.2 Fenhexamid: A single mode of action

Fenhexamid is a narrow spectrum fungicide (Rossenbloich and Stuebler, 2000) belonging to the chemical family of hydroxyanilides inhibiting sterol biosynthesis (SBI)(Debieu *et al.*, 2001). Bayer CropScience has introduced it to the French fungicide market more than 10 years ago as a specific anti-*Botrytis*. It is principally used in vineyards, but also on tomato and strawberry cultures up to two treatments per season (Couteux & Lejeune, 2003). Fenhexamid differs from other known SBIs (allylamines, azoles, imidazoles and morpholines/amines) by the enzymatic step that it inhibits. Sterols are essential lipid compounds found in all eukaryotes. They are principally localized to cytoplasmic or endo-membranes, although the latter are composed predominantly of phospholipids. Sterols participate in membrane permeability and rigidity, in the formation of lipid rafts and, eventually, in post-translational control of membrane proteins (Hac-Wydro *et al.*, 2007; Yeagle, 1990; Epand, 2008; Gilbert, 2010; Paila & Chattopadhyay, 2010). The major fungal sterol is ergosterol and specific to this kingdom. Therefore, many attempts to develop targets for antifungal treatments have been orientated towards the ergosterol biosynthesis pathway. Noticeably, all SBIs used against fungal diseases target enzymatic steps that are conserved among all kingdoms and not fungal specific.

The major class of SBIs used in agriculture represented by azoles, imidazoles and triazolinethions inhibits the 14α-demethylase enzyme therefore called DMIs (demethylation inhibitors), amines inhibit the Δ14-reductase and/or the Δ8→Δ7-isomerase. The youngest SBI used against grey mould, fenhexamid, blocks the demethylation at C4 of sterols. This important process involves three different enzymes: Erg25 (C4-methyle oxidase), Erg26 (C3-dehydrogenase, C4-decarboxylase) and Erg27 (3-ketoreductase). The first two enzymes eliminate methyl groups at C4 leading to a keto group at C3. The Erg27 enzyme then reduces this keto group to a hydroxyl (Gachotte *et al.*, 1999) leading to a functional sterol.

This last step is the target of fenhexamid. Its mode-of-action has been uncovered in 2001, by the characterization of sterol composition of fenhexamid treated *B. cinerea* strains. Debieu and co-workers, observed an accumulation of compounds with a keto group at C3, suggesting the inhibition of the 3-ketoreductase step (Debieu *et al.*, 2001). Since this enzymatic step is conserved in all euascomycetes, one may expect fenhexamid to be a large spectrum fungicide as are the DMIs instead of its narrow spectrum (see paragraph 2.3).

1.3 Resistance to fenhexamid in the *Botrytis* species complex

The activity spectrum of a new fungicide is generally established by biological tests on diverse fungal species allowing distinguishing sensitive and tolerant species. While naturally resistant - or tolerant species - are insensitive to a given fungicide prior to its introduction and defining the fungicide's spectrum of activity, sensitive fungal species can evolve and select less susceptible - or resistant - isolates. Natural polymorphism or randomly occurring mutations may confer a certain level of resistance to some strains. Through the application of the corresponding fungicide, these isolates are preferentially selected and their proportion within the population will increase (selection pressure of the fungicide). The differentiation between natural and acquired resistance can sometimes lead to the identification of new fungal species; e.g. *Oculimacula yallundae* and *O. acuformis*: the second species is naturally resistant to triazole SBIs. Another example is *Rhizoctonia solani* which is subdivided into different anastomosis groups some of which are tolerant to pencycuron, an anti-microtubular agent (Campion *et al.*, 2005).

	Botrytis cinerea					*B. pseudocinerea* HydR1
	EC50[a] HydS	Resistance Factor				
		HydR2	HydR3⁻	HydR3⁺	MDR2	
Germ tube	≈ 0.05	≈ 2	≈ 75	> 350	≈ 10	≈ 5 – 10
Mycelial growth	≈ 0.015	≈ 50	≈ 20 - 30	> 350	≈ 10	> 350

[a] concentration of fenhexamid (mg l⁻¹) inhibiting growth at 50%.

Table 2. Sensitivities of resistant phenotypes towards fenhexamid on germ tube elongation and mycelium growth stages. HydS= sensitive to fenhexamid; HydR= resistant to fenhexamid; MDR= multidrug resistant.

Also in the case of grey mould, it became clear that the species later on named *Botrytis pseudocinerea* a close relative of *B. cinerea*, was naturally resistant to fenhexamid at the stage of mycelial growth (Fournier *et al.*, 2005, Walker *et al.*, 2011). Later, resistant isolates were selected from the *B. cinerea sensu stricto* populations, initially sensitive to fenhexamid. Until now, four different fenhexamid resistance categories could be identified among grey mould populations (Table 2): HydR1, corresponding to the naturally resistant species *B. pseudocinerea* and the *B. cinerea* phenotypes MDR, HydR2 and HydR3 (Leroux *et al*, 2002).

For the phenotypes HydR1 (*B. pseudocinerea*) and HydR2 (*B. cinerea*), high to moderate resistance levels respectively, are observed nearly exclusively at the level of mycelial growth, whereas the other phenotypes display similar resistance levels at both developmental stages, germ tube elongation and mycelial growth. In addition, *B. pseudocinerea* (HydR1) presents increased sensitivities to other fungicides, such as DMIs, fenproprimorph, fenpropidine as well as to some SDHIs (Leroux *et al.*, 2002). Only few HydR2 strains have been isolated and to our knowledge, they have not yet been found in French vineyard populations. On the opposite, HydR3 phenotypes are regularly found since 2003. This category has been subdivided into two sub-classes, HydR3- or HydR3+ with moderate or high resistance levels respectively.

MDR2 (and MDR3) phenotypes display weak to moderate resistance levels (resistance factor <15) to various unlinked mode-of-actions including fenhexamid, SDHIs, dicroboximides, anilinopyrimidines etc. This multidrug resistance (MDR) is linked to the over expression of a membrane transporter leading to increased drug-efflux outside the fungal cell (Kretschmer *et al.*, 2009). Anti-resistance strategies involving systematic alterations of different mode-of-actions seem to favor the selection of these MDR phenotypes instead of specific resistances, in particular in regions with more than one treatment per season such as the Champagne wine region.

2. Acquired resistance to fenhexamid in *Botrytis cinerea*

2.1 Target alterations cause reduced affinity of fenhexamid and are responsible for resistance

In agriculture, acquired resistance towards agrochemical products is frequently conferred by target site modifications leading to decreased inhibitor affinities. Target overproduction is rarely found. Pesticide detoxification is a mechanism particularly important for insecticide resistance involving diverse enzymes such as hydrolases, cytochrome P450 monooxygenases, glutathione-S-transferases or glycosyl transferases. In rare cases, resistance results from non-activation of pro-pesticides. Concerning antifungal compounds, the principal resistance mechanisms are target site modifications and increased efflux, in the agronomical field as well as in the medical field (Ma & Michaelides, 2005; Sanglard *et al.*, 2009).

With respect to the *B. cinerea* HydR phenotypes, target site modifications were found in the HydR3 (+ and -) isolates, but not in HydR2 phenotypes (Albertini & Leroux, 2004; Fillinger *et al.*, 2008). In *B. pseudocinerea* the fenhexamid target encoding gene, *erg27* shows a high degree of polymorphism compared to the sensitive *B. cinerea* allele. Its implication in fenhexamid resistance is described in paragraph 3.1.

Eight different amino acid replacements were found in *erg27* alleles from HydR3- field strains either as single or as double mutations (L195F, N196Y, V309M, A314V, S336C, N369D, L400F/S, et L501W). HydR3+ field isolates display a single mutation of the phenylalanine at positions 412 (F412S, F412I or F412V) (Fig. 1; Fillinger *et al.*, 2008).

Other *erg27* mutations have been found in fenhexamid resistant strains selected after chemical mutagenesis (G23S, C53R, T63I, K73E, H105Y, K159N, L195S, T273A, S310P, I397V, I411V, H423R, A452P, Q495R and C516R) (Saito *et al.*, 2010). However, their involvement in resistance to fenhexamid remains to be shown.

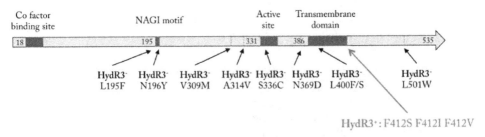

Fig. 1. Protein structure and functional domains of the 3-ketoreductase Erg27 of *B. cinerea*. The active site catalyzing the formation of a hydroxyl group on position C3 of sterols is located near the transmembrane domain allowing the anchorage of the enzyme to the endoplasmic reticulum. The position of the mutations detected in HydR3- resistant isolates are indicated by black arrows and concerning HydR3+ isolates by a red one.

We analyzed the involvement of the natural HydR3 mutations in fenhexamid resistance by site directed mutagenesis. The *erg27* allele - encoding 3-ketoreductase – was replaced in a sensitive strain by the *erg27* allele of a HydR3 strain by homologous recombination. Two mutations identified from HydR3- strains (L195F et V309M) in addition to the three HydR3+ alleles were studied. The fenhexamid sensitivity profiles of the generated mutants were similar to those from the field strains (Table 3). Indeed the transgenic *erg27L195F* or *erg27V309M* mutants display low to moderate resistance levels according to the developmental stage, whereas the transgenic HydR3+ mutants (F412S, F412I, F412V) present high resistance levels at both growth stages, as do the highly resistant HydR3+ field strains.

Fenhexamid susceptibilities	EC50 Sensitive strain (mg l^{-1})	Resistance level (RL)				
		HydR3$^-$ phenotype		HydR3$^+$ phenotype		
		Erg27::L195F	*Erg27::V309M*	*Erg27::F412S*	*Erg27::F412I*	*Erg27::F412V*
Germ tube	0.05	**MR**	**MR**	**HR**	**HR**	**HR**
Mycelial growth	0.015	**LR/MR**	**LR/MR**	**HR**	**HR**	**HR**

Table 3. **Sensitivity to fenhexamid of transgenic mutants.** LR=low resistance; MR=moderate resistance; HR=high resistance

In addition, the inhibition of 3-ketoreductase activity by fenhexamid is reduced in the artificial mutants (Billard *et al.*, unpublished), strongly suggesting a decrease in fenhexamid's affinity to its target 3-ketoreductase in HydR3 mutants. These results prove the direct correlation between the amino acid substitution and the specific resistance to fenhexamid in the *B. cinerea* HydR3+ and HydR3- phenotypes. In addition they show target site modification as sole resistance mechanism in these isolates.

In contrast to HydR3 phenotypes, neither target site modification nor over expression seem to be responsible for fenhexamid resistance in HydR2 phenotypes. The only hint towards the potential resistance mechanism is an observed synergy between fenhexamid and DMIs suggesting fenhexamid metabolization involving a cytochrome P450 monooxygenase enzyme. Which mechanism precisely is at work in HydR2 strains needs yet to be identified.

We have modeled the *B. cinerea* Erg27 protein after alignment with the peptide sequences of various dehydrogenases/reductases whose crystal structures have been established. The homology-based model presented in Fig. 2 monitors the position of the HydR3 mutations relative to the substrate- and cofactor binding sites. None of the mutations are in the vicinity of these docking sites, although it cannot be excluded that the L195F and N369D replacements modify the Erg27 protein structure leading to different substrate- and cofactor affinities. Mutations of F412 however, potentially interfere with the helical structure of the transmembrane helix.

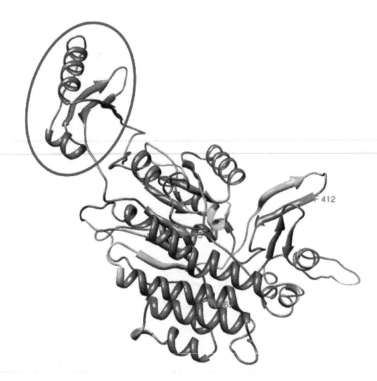

Fig. 2. Predicted 3D structure of the 3-ketoreductase (Erg27) of *Botrytis cinerea* based on the alignement with Salutaridine Reductase From *Papaver Somniferum* (Higashi *et al.*, 2011). The cofactor (NADHP) binding site and the active sites are highlighted in cyan and purple respectively. The green ribbon represents the transmembrane domain. The mutated residues responsible for specific fenhexamid resistance (HydR3) are highlighted in red. The blue circle designates the extension of *Botrytis* and *Sclerotinia* Erg27 proteins compared to other fungal orthologs.

2.2 Why fenhexamid has a narrow spectrum?

Fenhexamid is the sole SBI with a narrow antifungal spectrum. Its activity is restricted to close relatives of *B. cinerea*, e.g. *Botrytis* and *Monilinia spp.*, *Sclerotinia sclerotiorum* (Rosslenbroich & Stuebler, 2000). We therefore raised the question if 3-ketoreductase of ascomycetes naturally resistant to fenhexamid is insentitive to inhibition exerted by fenhexamid. The alignment of Erg27 to orthologous peptide sequences show that both 3-ketoreductase sequences, those from *B. cinerea* and *S. sclerotiorum*, harbor three additional fragments compared to the other fungal proteins, including one extension of 34 amino acids (Albertini & Leroux, 2004; Fillinger, unpublished; Fig. 2). In addition, we have shown by biochemical assays that 3-ketoredcutase activity of insensitive fungal species, is barely inhibited by fenhexamid, suggesting that fenhexamid has a strong affinity towards 3-ketoreductase of *Botrytis* related species only (Debieu *et al.*, submitted). In how far the above mentioned protein extensions and/or specific residues of the *B. cinerea* Erg27 protein sequence are involved in this affinity remains to be established.

2.3 The HydR3$^+$ resistance acquisition entails a cost in controlled conditions

The fitness corresponds to the capacity of one individual (strain) to survive among others under the same conditions. In the absence of selective pressure exerted by the application of fungicides, the relative fitness determines the persistence of a resistant isolate among a natural fungal population. Acquiring a fungicide resistance linked to a high fitness cost is a clear disadvantage for the fungal strain, as it is counter selected in absence of the corresponding fungicide from a mixed population with sensitive and resistant strains. In natural grey mould populations, this seems to be the case for the phenylpyrrol fungicides for which no (or only few) specifically resistant strains were detected up to now (Leroux *et al*, 2002; Moyano *et al.*, 2004). The efficacy of such fungicides can be maintained over long periods. On the opposite, resistant mutations that are not associated to a fitness cost, may threaten the "life-time" of the corresponding fungicide. Indeed, such kind of resistances without associated fitness costs were obtained after few years only with the strobilurins used against cereal fungal pathogens or - on different crops and different pathogens - with the benzimidazoles. Resistance against either or both fungicides is generalized in many pests.

Concerning fenhexamid, the frequency of the HydR3$^+$ strains of the greatest concern in practice, is in progression since 2003 and has reached now non-negligible levels, e.g. 27% in Champagne. This increase remains weak in regions despite annual applications and not comparable to the rapid generalization of strobilurin resistance. This evolution indicates that probably an effective anti-resistance strategy combined to a potential fitness cost linked to high levels of fenhexamid resistance limits the progression of HydR3$^+$ frequencies among grey mould populations.

In order to evaluate the fitness cost linked to the HydR3$^+$ phenotypes, we measured divers physiological parameters under controlled conditions. The general problem for fitness measurements in natural *B. cinerea* strains is the phenotypic variability between strains. Different growth parameters cannot be easily correlated to resistance due to different genetic backgrounds. We circumvented this problem by the construction of HydR3$^+$ (F412S, F412I and F412V) mutants using site directed mutagenesis. The strains were isogenic except for the *erg27* allele providing the ideal material for fitness analyses. Particular attention was paid to temperature and nutritional factors in the measurement of the fitness parameters as well as the quantification of developmental stages important for survival and dispersion (sclerotia and conidia respectively). Statistically significant decreases were observed in the mutants for mycelial growth – especially in extreme conditions (restricted nutrients at low temperatures), for conidia- and sclerotia-production – the last parameter especially at low temperatures, and for mycelial outgrowth from frozen sclerotia. Theses results obtained with isogenic strains, like those reported by Ziogas *et al* (2003) and Saito *et al.* (2010) for chemically resistant mutants of *B. cinerea*, indicate that the acquisition of high-level specific resistance to fenhexamid is associated with a decrease in fitness (Billard *et al.*, submitted). Our data show that *in vitro* pathogenicity is not affected (measured on detached bean leaves), whereas stages of the fungal life cycle during which there is no selection pressure seem to be the most affected.

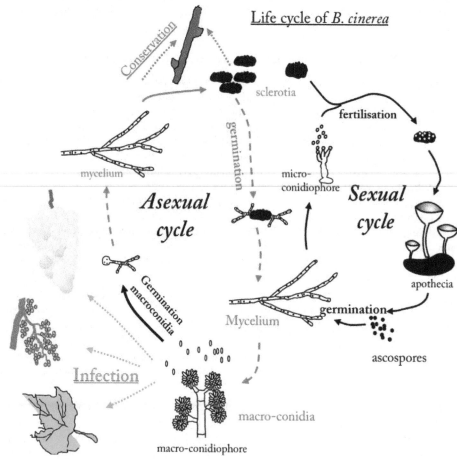

Fig. 3. **Life cycle of *Botrytis cinerea*.** In laboratory conditions, a fitness cost of the HydR3[+] resistance was observed in isogenic mutants at different phases of the asexual cycle (in red) of *B. cinerea*. However, the pathogenicity of these mutants, measured on beans, remained unchanged from the parental strain sensitive to fenhexamid (green).

These findings for isogenic *erg27*[F412] mutants suggest that the HydR3+ phenotype does not persist in the absence of the fungicide and that this phenotype has a moderate impact on the efficacy of fenhexamid treatments for controlling grey mould disease in vineyards. However long-term studies in vineyards are required to confirm these laboratory observations.

2.4 Fenhexamid monitoring in fields: A molecular approach

Monitoring of this resistance is a crucial area of research where our knowledge of the field distribution, evolution and impact of fungicide resistance depends on. In most cases, the degree of sensitivity of fungal populations to one or more fungicides is assessed by biological methods. These bioassays conducted, *in vitro* or *in vivo*, have been miniaturized (i.e. microtiter plate methods), but nonetheless consume considerable resources and remain time-consuming. When the molecular mechanisms of resistance are known (e.g. target mutation, target over expression, increased drug efflux), and particularly when the underlying DNA polymorphisms (single-nucleotide polymorphism (SNPs), deletions or insertions) have been defined, various molecular methods can be used to monitor antimicrobial resistance. The principal methods for quantifying resistance are based on real-time PCR (polymerase chain reaction) technology. Generally, studies have been made on isolated and purified strains using qualitative methods like RFLP, AS PCR or Real time PCR (Benzimidazole resistance in *Botrytis cinerea* Banno *et al.*, 2008, *Venturia inequaelis* Koenraadt *et al.*, 1992, *Rhyncosporium secalis* Wheller *et al.*, 1995, *Helminthosporium solani* McKay *et al.*, 1998; Strobilurines resistance on *Erisiphe graminis* Baümler *et al.*, 2003, *Plasmopara viticola* Furuya *et al.*, 2009, *Alternaria spp.* Ma & Michailides, 2003, *Mycosphaerella graminicola* Fraaije *et al.*, 2007, *Mycosphaerella fijiensis* Sierotzki *et al.*, 2000) and on the opposite some of them on populations, where resistance allele is quantify in gDNA pools by using Real time PCR technologies (Benzimidazole resistance in *Sclerotinia sclerotiorum* Chen *et al.*, 2009 ; Strobilurins resistance on *Blumeria graminis* Fraaije *et al.*, 2002 and *Pyrenophora teres* Kianianmomeni *et al.*, 2007). However, one limitation of this method concerns the nonspecific amplification of alleles, which may affect precision. This limitation does not generally hinder detection of the polymorphism, but it may affect quantification capacity, particularly for mutated alleles with a low abundance.

Taking these results into account, we investigated the development of a new technique for quantifying, with a high precision, the three different *erg27* alleles from the HydR3+ phenotype. The underlying DNA polymorphism is the modification of the TTC codon encoding the F412 residue, which is converted into a TCC (serine), GTC (isoleucine) or ATC (valine) (Fillinger *et al.*, 2008) codon. The best result was obtained with a non multiplexed method combining four allele-specific MGB Taqman® probes and four mismatched specific primers. This technique was named the *Allele Specific Probe and Primer Amplification Assay* (ASPPAA PCR), (Fig. 4; Billard *et al.*, submitted).

The sensitivity of ASPPAA PCR is sufficiently good to quantify a SNP at a rate of 1% in a DNA pool. In the future, the multiplexing, in the same run, of the analysis of several polymorphisms at different genomic loci with probes picked up in different fluorophore channels is conceivable and would be expected to decrease the time required for monitoring, and its cost, significantly.

The principle disadvantage of these molecular quantification methods is that they do not allow to detect emerging resistances in contrast to biological methods. One alternative molecular method to be considered is the HRM (high resolution melt curve) analysis (Pasay *et al.*, 2008). HRM offers a fast and convenient method of assessing the presence of mutations without sequencing in a short (< 400 bp) defined genomic region. Using this tool, the identification of new mutations in the *erg27* gene is possible.

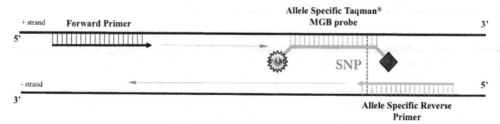

Fig. 4. Principle of ASPPAA PCR for quantitative SNP detection. The high robustness of this new allele quantification method using real-time PCR is caused by the specificity of amplification of the mutation by allele specific probes and primers present on both DNA strands.

3. Natural resistance to fenhexamid in *Botrytis pseudocinerea*

The genus *Botrytis* (Ascomycota) contains 22 highly specialised species and one hybrid. A multiple-gene gene genealogy study recently showed that this genus could be subdivided into two categories, one consisting of *Botrytis* species acting as pest on monocots, and the other containing *Botrytis* species acting as parasites on eudicots (Staats *et al.*, 2005). Within this second category, *Botrytis cinerea* has the widest host range, being able to infect more than 220 types of eudicot, including grapevine and many fruit and vegetable crops. However, *B. cinerea* has recently been shown to be a complex of two sibling species living in sympatry. *B. cinerea sensu stricto* is the predominant species (Fournier *et al.*, 2005). The other species, called *Botrytis pseudocinerea*, has been found at low frequency in French populations. This species is morphologically indistinguishable from *B. cinerea sensu* stricto, but Walker *et al* (2011) established several molecular markers to distinguish both species. In addition *B. pseudocinerea* has a different pattern of fungicide susceptibility (phenotype HydR1), displaying natural resistance to fenhexamid and hypersensitivity to fenpropidin and edifenphos (Leroux *et al*, 2002).

In contrast to the *B. cinerea* fenhexamid resistant phenotypes (HydR2 and HydR3), *B. pseudocinerea* was resistant to fenhexamid prior to its introduction, therefore displaying a natural resistance to fenhexamid.

3.1 Role of the natural target polymorphism in fenhexamid resistance

The genetic polymorphism of *B. pseudocinerea* compared to *B. cinerea* is also visible on the 3-ketoredcutase encoding gene *erg27*. Twelve codon modifications lead to amino acid replacements in the Erg27 protein (N93V, D146L, I211V, I215V, M218T, V234A, I235V, D261G, S264T, P269L, A285T and Q354K) (Albertini & Leroux, 2004). None of theses substitutions corresponds to any of the HydR3 mutations. In order to evaluate the impact of the *B; pseudocinerea erg27* allele in fenhexamid resistance we replaced the *erg27wt* allele in a sensitive *B. cinerea* strain by that of a HydR1 "resistant" *B. pseudocinerea* strain. The gene replacement strategy is explained in Fig. 5.

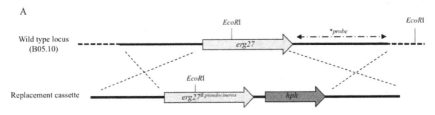

Fig. 5. Gene replacement strategy of the *erg27* allele by that of *B. pseudocinerea* in a sensitive strain of *B. cinerea*. The *hph* gene conferring resistance to hygromycine was used as transformation selection marker.

The transformants harboring the *B. pseudocinerea* allele instead of the *B. cinerea* allele showed a slightly increased resistance to fenhexamid (RF<10) compared to the parental *B. cinerea* strain (Billard *et al.*, unpublished). Those results indicate that the major part of fenhexamid resistance in *B. pseudocinerea* is conferred by a mechanism independent of the target.

3.2 Characterization of an unusual resistance mechanism towards fungicides: Detoxification

Indications about the second mechanism were obtained more than ten years ago by two studies. Suty and co-workers (1999) performed fenhexamid metabolization assays on *B. cinerea* and *B. pseudocinerea* and observed that naturally resistant species displayed a metabolization pattern different from *B. cinerea* suggesting fenhexamid degradation as possible resistance mechanism. These results were supported by the synergy between DMIs and fenhexamid observed only for *B. pseudocinerea* (Leroux *et al.*, 2002). Together these results suggested for *B. pseudocinera* an enzyme(s) similar to Cyp51 (the target of DMIs) to degrade fenhexamid and therefore conferring resistance to fenhexamid.

We have searched the *B. cinerea* genome database (Amselem *et al.*, 2011) (http://urgi.versailles.inra.fr/Species/Botrytis) for cytochrome P450 proteins similar to Cyp51 and compared their expression profiles between *B. cinerea* and *B. pseudocinerea*. The candidate gene showing the highest similarity to *cyp51* (19 % protein identity, 37 % protein similarity) showed increased expression in *B. pseudocinerea* (Billard *et al.*, unpublished). According to the nomenclature of P450s, after phylogenic analysis against all P450s of the ascomycete *Aspergillus nidulans* (Kelly *et al.*, 2009) the corresponding gene was named *cyp684*. We created *cyp684* deletion mutants in a *B. pseudocinerea* strain through a gene replacement strategy. The Δ*cyp684* mutants showed a 200 fold reduction in fenhexamid resistance and a simultaneous reduction in DMI-fenhexamid synergy (Table 4, Fig. 6). Although the *B. pseudocinerea* Δ*cyp684* mutants do not reach *B. cinerea* sensitivity levels, theses results show that the cytochrome P450 encoding gene *cyp684* is responsible for the biggest part of *B. pseudocinerea's* natural resistance to fenhexamid. Natural polymorphism between the *B. cinerea/B. pseudocinerea cyp684* orthologues may account for the different phenotypes. It remains to be shown if *cyp684* overexpression and/or differential enzyme parameters are involved.

In *B. pseudocinerea* however the *cyp684* deletion abolishes the hypersensitivity to the phosphothiolate edifenphos in addition to the loss of fenhexamid resistance (Table 6) suggesting that Cyp684 could have a similar enzyme activity on fenhexamid as on edifenphos. Phosphorothiolates (e.g. edifenphos) are profungicides, nearly exclusively used against the rice-blast disease caused by *Magnaporthe oryzae*.

	B. cinerea	*B. pseudocinerea* (RL)	
		wt	*Δcyp684*
Fenhexamid (Hydroxyanilides)	S	HR	**MR**
Fenhexamid+DMI	Ind.	Syn.	Ind.
edifenphos (Phosphorothiolates)	S	HS	S

RL= resistance levels, S= sensitive, HS=hypersensitive, HR=highly resistant, MR=moderately resistant, Ind.= independent, Syn.= synergistic

Table 4. Sensitivity profiles to fenhexamid and edifenphos (phosphorothiolate) of *B. cinerea*, *B. pseudocinerea* and *Δcyp684* mutants of *B. pseudocinerea*.

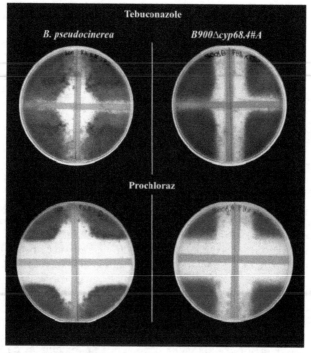

Fig. 6. Joint action between fenhexamid (vertical) and prochloraz or tebuconazole (horizontal strip) on the *B. pseudocinerea* B900 strain and a B900*Δcyp684* deletion mutant. The curves around paper crosses indicate strong synergism between chemicals (in the left panel). In the *B. pseudocinerea Δcyp684* mutant growth inhibition is typical of independent actions (right panel).

These compounds need to be modified in order to become active. In the case of edifenphos, the active metabolite is produced by the cleavage of a phosphor-sulfur bridge (Uesugi, 2001). As for *B. pseudocinerea*, a strong synergy was observed between phosphorothiolates and DMIs. Moreover negative cross-resistance between phosphorothiolates and phosphoroamidates in *M. oryzae* allowed the identification of the biochemical reactions involved (Uesugi & Takenake, 1992 ; Fig. 7). The enzymes involved are yet unknown, but negative cross resistance between BPA and edifenphos on one hand, and the synergy between edifenphos and DMIs on the other, suggest the involvement of a cytochrome P450 accepting as a substrate edifenphos as well as BPA. The comparable situation observed in *B. pseudocinerea* is in favor of a comparable enzymatic reaction involved in the detoxification of fenhexamid. We are currently analyzing the metabolites produced by *B. pseudocinerea* from fenhexamid in order to unravel the reaction mechanism.

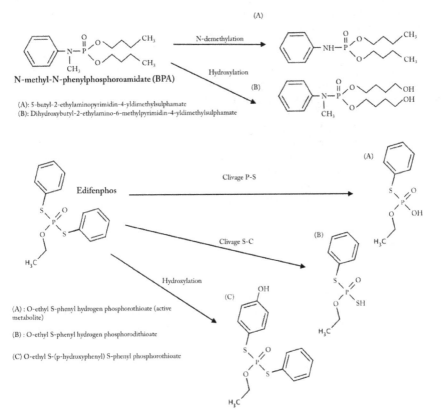

Fig. 7. Metabolic pathways of phosphorothiolates (edifenphos) and phosphoroamidates (BPA) in susceptible and resistant strains of *Magnaporthe oryzae*. (from Uesugi & Takenaka, 1992, modified).

4. Conclusions

We and others have identified and characterized various phenotypes displaying reduced sensitivity to fenhexamid among natural grey mould populations. In the case of the phenotypes with low resistance at the spore germination stage (HydR1, HydR2, MDR2 et MDR3 phenotypes), Petit *et al.*, 2010 have shown these have only a limited impact on fenhexamid's efficacy in artificial inoculation studies and in vineyards studies. In the same line, monitoring started before the introduction of fenhexamid, show that the variations in frequencies of *B. pseudocinerea* – naturally resistant to fenhexamid – in grey mould populations, seem to be independent of selection by fenhexamid. *B. cinerea* and *B. pseudocinerea* are found in sympatry on the same hosts, but clearly differ in their phenology, demonstrating differences in ecological niche (Walker *et al.*, 2011): although present at low frequencies (0-15 %), *B. pseudocinerea* is predominantly found in spring populations and has only a reduced impact on grey mould epidemics.

Finally, the HydR3 phenotype comprises strains with moderate (HydR3-) to high (HydR3+) resistance levels at all growth stages. This resistance is conferred by target site modifications. HydR3+ strains are the predominant strains resistant to fenhexamid in French vineyards and those of greatest concern in practice, because they threaten the sustainability of fenhexamid in combating grey mould. Frequencies of HydR3+ strains are slowly but steadily increasing in French vineyard populations. Even if their mean frequencies stayed moderately low (e.g. 27 % in Champagne in 2010, Fig. 8), their proportion within same plots has strongly increased in the last years. Although one can observe over 50 % highly resistant strains at the end of the season after fenhexamid treatment, our multi-annual monitoring survey did not reveal sites with such important frequencies over several years. In addition, we are not aware of any loss-of-efficacy observed for fenhexamid due to the selection of resistant populations. Several hypotheses can be put forward to explain the situation: (1) the use of fenhexamid limited to one application per year (as for other botryticides), (2) the fitness cost of the HydR3+ mutations and (3) the predominance of MDR strains at least in Champagne (55 % mean frequencies). Indeed, strains combining MDR (multiple drug resistance) and specific resistance to fenhexamid are rarely found.

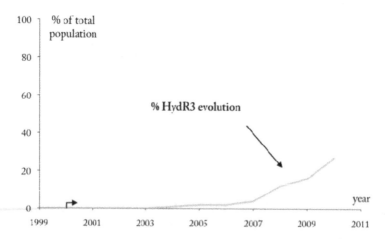

Fig. 8. Evolution of the HydR3 phenotype in Champagne since the introduction of fenhexamid (source : CIVC, AS Walker INRA).

We have developed a molecular tool allowing the rapid detection and quantification of the three major HydR3 alleles from a mixture of *B. cinerea* spores. (ASPPAA PCR). This tool will make large scale field assays realistic to evaluate the percentage of HydR3+ strains among populations subjected to different fungicide pressures in order to precise the best strategy for fenhexamid treatment and the best state for its application. Moreover, this method may monitor frequencies of HydR3+ strains after stopping fenhexamid treatments and therefore allows to analyze the persistence of such resistance phenotype under real conditions.

5. References

Albertini, C. & Leroux, P. (2004). A *Botrytis cinerea* putative 3-keto reductase gene (ERG27) that is homologous to the mammalian 17β-Hydroxysteroid Dehydrogenase type 7 gene (17β-HSD7). *European Journal of Plant Pathology* 110(7): 723-733.

Amselem, J. ; Cuomo, C.; van Kan, J.A.L.; Viaud, M., *et al.,* (2011). Genomic analysis of the necrotrophic fungal pathogens *Sclerotinia sclerotiorum* and *Botrytis cinerea*. *PLoS Genetics* in press.

Banno, S.; Fukumori, F.; Ichiishi, A.; Okada, K.; Uekusa, H.; Kimura, M. & Fujimura, M. (2008). Genotyping of benzimidazole-resistant and dicarboximide-resistant mutations in *Botrytis cinerea* using real-time polymerase chain reaction assays. *Phytopathology*. 98(4): 397-404.

Baümler, S.; Sierotzki, H.; Gisi, U.; Mohler, V.; Felsenstein, F.G. & Schwarz, G. (2003). Evaluation of *Erysiphe graminis f sp tritici* field isolates for resistance to strobilurin fungicides with different SNP detection systems. *Pest Manag. Sci.* 59:310–314

Billard, A.; Fillinger, S.; Leroux, P.; Lachaise, H.; Beffa, R. & Debieu, D. (2011). Strong resistance to the fungicide fenhexamid entails a fitness cost in *Botrytis cinerea*, as shown by comparisons of isogenic strains. *Pest Manag. Sci.*, in press.

Billard, A.; Laval, V.; Fillinger, S.; Leroux, P.; Lachaise, H.; Beffa, R. & Debieu, D. (2011). Allele Specific Probe & Primer Amplification Assays (ASPPAA): a new real time PCR method for fine SNP quantification in DNA pool. Submitted to *Antimic. Agents Chem.*

Breitling, R.; Krazeisen, A.; Möller, G. & Adamski, J. (2001). 17 beta-hydroxysteroid dehydrogenase type 7 an ancient 3-ketosteroid reductase of cholesterogenesis. *Mol. Cell. Endocrinol.* 171(1-2): 199-204.

Campion, C.; Chatot, C.; Perraton, B. & Andrivon, D. (2003). Anastomosis groups, pathogenicity and sensitivity to fungicides of *Rhizoctonia solani* isolates collected on potato crops in France. *European Journal of Plant Pathology*. 109(9): 983–992.

Chen, Y. & Ming-Guo, Z. (2009). Characterization of *Fusarium graminearum* isolates resistant to both carbendazim and a new fungicide JS399-19. *Phytopathology* 99: 441.

Couteux, A & Lejeune,V. (2003). Index phytosanitaire Acta 2003 – 38e edition. ACTA, Paris, France

Debieu, D.; Bach, J.; Hugon, M.; Malosse, C. & Leroux, P. (2001). The hydroxyanilide fenhexamid, a new sterol biosynthesis inhibitor fungicide efficient against the plant pathogenic fungus *Botryotinia fuckeliana* (*Botrytis cinerea*). *Pest Manag. Sci.* 57(11): 1060-1067.

Epand, R.M. (2008). Proteins and cholesterol-rich domains. *Biochim. Biophys. Acta.* 1778(7-8): 1576-1582.

Fillinger, S.; Leroux, P.; Auclair, C.; Barreau, C.; Al Hajj, C. & Debieu, D. (2008). Genetic analysis of fenhexamid-resistant field isolates of the phytopathogenic fungus *Botrytis cinerea*. *Antimicrob. Agents Chemother.* 52(11): 3933-3940.

Fournier, E.; Giraud, T.; Albertini, C. & Brygoo, Y. (2005). Partition of the *Botrytis cinerea* complex in France using multiple gene genealogies. *Mycologia*. 97(6): 1251-1267.

Fraaije, B.A.; Butters, J.A.; Coelho, J.M.; Jones, D.R. & Hollomon, D.W. (2002). Following the dynamics of strobilurin resistance in *Blumeria graminis* f. sp. *tritici* using quantitative allele-specific real-time PCR measurements with the fluorescent dye SYBR Green I. *Plant pathology*. 51(1): 45–54.

Fraaije, B.A.; Cools, H.J.; Kim, S.H.; Motteram, J.; Clark, W.S. & Lucas, J.A. (2007). A novel substitution I381V in the sterol 14alpha-demethylase (CYP51) of *Mycosphaerella graminicola* is differentially selected by azole fungicides. *Molecular Plant Pathology*. 8(3): 245–254.

Furuya, S.; Suzuki, S.; Kobayashi, H.; Saito, S. & Takayanagi, T. (2009). Rapid method for detecting resistance to a QoI fungicide in *Plasmopara viticola* populations. *Pest Manag Sci* 65:840–843

Gachotte, D.; Sen, S.E.; Eckstein, J.; Barbuch, R.; Krieger, M.; Ray, B.D. & Bard, M. (1999). Characterization of the *Saccharomyces cerevisiae* ERG27 gene encoding the 3-keto reductase involved in C-4 sterol demethylation. *Proc. Natl. Acad. Sci. U.S.A.* 96(22): 12655-12660.

Gilbert, R.J.C. (2010). Cholesterol-dependent cytolysins. *Adv. Exp. Med. Biol.* 677: 56-66.

Hac-Wydro, K.; Wydro, P.; Jagoda, A. & Kapusta, J. (2007). The study on the interaction between phytosterols and phospholipids in model membranes. *Chem. Phys. Lipids*. 150(1): 22-34.

Higashi, Y.; Kutchan, TM.; Smith, T.J. (2011). Atomic structure of salutaridine reductase from the Opium Poppy (*Papaver somniferum*). *Journal of Biological Chemistry* 286: 6532-6541.

Ishii, H. (2010). Fungicide resistance in rice. *Modern Fungicides and Antifungal Compounds VI*, Proceedings 16th *International Reinhardsbrunn Symposium* 2010 Eds H.W. Dehne, H.B. Deising, U. Gisi, K. H. Kuck, P.E. Russell, H. Lyr (Eds.), *in press*.

Kelly DE, Krasevec N, Mullins J, Nelson DR (2009) The CYPome (Cytochrome P450 complement) of Aspergillus nidulans. *Fungal Genetics and Biology* 46: S53-S61

Kianianmomeni, A.; Schwarz, G.; Felsenstein, F.G. & Wenzel, G. (2007). Validation of a real-time PCR for the quantitative estimation of a G143A mutation in the cytochrome bc1 gene of *Pyrenophora teres*. *Pest Management Science*. 63(3): 219–224.

Koenraadt, H.; Somerville, S.C. & Jones, A.L. (1992). Characterization of mutations in the beta-tubulin gene of benomyl-resistant field strains of *Venturia inaequalis* and other plant pathogenic fungi. *Phytopathology*. 82(11): 1348–1354.

Kretschmer, M.; Leroch, M.; Mosbach, M.; Walker, A.S.; Fillinger, S.; Mernke, D.; Schoonbeek, H.J.; Pradier, J.M.; Leroux, P.; De Waard, M.A. & Hahn, M. (2009). Fungicide-driven evolution and molecular basis of multidrug resistance in field populations of the Grey Mould Fungus *Botrytis cinerea*. *PLoS Pathogen* 5(12).

Leroux, P.; Debieu, D.; Albertini, C.; Arnold, A.; Bach, J.; Chapeland, F.; Fournier, E.; Fritz, R.; Gredt, M.; Giraud, T.; Hugon, M.; Lanen, C.; Malosse, C. & Thebaud. G. (2002). The Hydroxyanilide Botryticide Fenhexamid/ Mode of Action and Mechanism of Resistance. *Modern Fungicides and Antifungal Compounds III*. Proceedings 13th *International Reinhardsbrunn Symposium* 2001: 29-40. Eds H.W. Dehne, H.B. Deising, U. Gisi, K. H. Kuck, P.E. Russell, H. Lyr (Eds.), AgroConcept GmbH, Bonn, Germany.

Ma, Z. & Michailides, T.J. (2004). Characterization of iprodione-resistant *Alternaria* isolates from pistachio in California. *Pesticide Biochemistry and Physiology* 80: 75 – 84.

Ma Z, Michailides TJ (2005) Advances in understanding molecular mechanisms of fungicide resistance and molecular detection of resistant genotypes in phytopathogenic fungi. *Crop Protection 24: 853-863*

McKay, G.J.; Egan, D.; Morris, E. & Brown, A.E. (1998). Identification of benzimidazole resistance in *Cladobotryum dendroides* using a PCR-based method. *Mycological Research* 102: 671 – 676.

Mouches, C. (2005). Les mutations responsables de résistances aux insecticides. *Enjeux phytosanitaires pour l'agriculture et l'environnement. C. Regnaud R.* édition TEC DOC, Lavoisier, Paris. 207-224.

Moyano, C.; Gomez, V.; Melgarejo, P. (2004) Resistance to pyrimethanil and other fungicides in *Botrytis cinerea* populations collected on vegetable crops in Spain. *J Phytopathol* 152: 484-490.

Paila, Y.D. & Chattopadhyay, A. (2010). Membrane cholesterol in the function and organization of G-protein coupled receptors. *Subcell. Biochem.* 51: 439-466.

Pasay, C.; Arlian, L.; Morgan, M.; Vyszenski-Moher, D.; Rose, A.; Holt, D.; Walton, S.; McCarthy, .J (2008). High-resolution melt analysis for the detection of a mutation associated with permethrin resistance in a population of scabies mites. *Medical and Veterinary Entomology* 22: 82-88.

Petit, A.N.; Vaillant-Gaveau, N.; Walker, A.S.; Leroux, P.; Baillieul, F.; Panon, M.; Clément, C. & Fontaine, F. (2010). Determinants of fenhexamid effectiveness against grey mould on grapevine: Respective role of spray timing, fungicide resistance and plant defences. *Crop Protection.* 29(10): 1162-1167.

Rosslenbroich, H.J. & Stuebler, D. (2000). *Botrytis cinerea* - history of chemical control and novel fungicides for its management. *Crop Protection* 19: 557-561.

Sanglard D, Coste A, Ferrari S (2009) Antifungal drug resistance mechanisms in fungal pathogens from the perspective of transcriptional gene regulation. *FEMS Yeast Research* 9: 1029-1050

Saito, S.; Furuya, S.; Takayanagi, T. & Suzuki, S. (2010). Phenotypic analyses of fenhexamid resistant *Botrytis cinerea* mutants. Fungicides, ed. by Odile Carisse, InTech Publisher, pp. 247-260.

Sheng, C.; Miao, Z.; Ji, H.; Yao, J.; Wang, W.; Che, X.; Dong, G.; Lu, J.; Guo, W. & Zhang, W. (2009). Three-dimensional model of lanosterol 14α-demethylase from *Cryptococcus neoformans:* active-site characterization and insights into azole binding. *Antimicrobial agents and chemotherapy.* 53(8): 3487.

Sierotzki, H.; Parisi, S.; Steinfeld, U.; Tenzer, I.; Poirey, S. & Gisi, U. (2000). Mode of resistance to respiration inhibitors at the cytochrome bc1 enzyme complex of *Mycosphaerella fijiensis* field isolates. *Pest Management Science* 56: 833-841.

Suty, A.; Pontzen, R. & Stenzel, K. (1999). Fenhexamid-sensitivity of *Botrytis cinerea*: determination of baseline sensitivity and assessment of the risk of resistance. *Pflanzenschutz-Nachrichten Bayer* 52: 145-157.

Staats, M.; Van Baarlen, P.; & Van Kan, J.A.L. (2005). Molecular phylogeny of the plant pathogenic genus *Botrytis* and the evolution of host specificity. Molecular Biology and Evolution. 22(2): 333 -346.

Uesugi, Y. (2001). Fungal choline biosynthesis – a target for controlling rice blast. *Pesticide Outlook*, February.

Uesugi, Y. & Takenata, M. (1992). The mechanisms of action of phosphorothiolates fungicides. In Proceeding of the 10th International Symposium on Systemic Fungicides and Antifungal Compounds, eds H. Lyr and C. Polter, 159-164, German Phytomedical Society Series, Stuttgart, Germany.

Walker, A.S., Gautier, A.; Confais, C.; Martinho, D.; Viaud, M.; Le Pêcheur, P.; Dupont, J.; Fournier, J. (2011). *Botrytis pseudocinerea*, a new cryptic species causing grey mould in French vineyards in sympatry with *Botrytis cinerea*. *Phytopathology*, in press

Wheeler, I.; Kendall, S.; Butters, J. & Hollomon, D. (1995). Detection of benzimidazole resistance in *Rhynchosporium secalis* using allele-specific oligonucleotide probe. *EPPO Bulletin* 25, Issue 1-2, pages 113–116,

Yeagle, P.L. (1990). Frontiers of membrane research: lipid-protein complexes in membranes; membrane fusion. *Prog. Clin. Biol. Res.* 343: 15-28.

Zhang, C.Q.; Zhu, G.N.; Ma, Z.H. & Zhou, M. G. (2006). Isolation, characterization and preliminary genetic analysis of laboratory tricyclazole-resistant mutants of the Rice Blast Fungus, *Magnaporthe grisea*. *J. Phytopathology* 154, 392–397

Ziogas, B.N.; Markoglou, A.N. and Malandrakis, A.A. (2003). Studies on the inherent resistance risk to fenhexamid in *Botrytis cinerea*. *Eur J Plant Pathol* 109: 311–317.

5

Impact of Fungicide Timing on the Composition of the *Fusarium* Head Blight Disease Complex and the Presence of Deoxynivalenol (DON) in Wheat

Kris Audenaert[1,2], Sofie Landschoot[1,2], Adriaan Vanheule[1,2],
Willem Waegeman[3], Bernard De Baets[3] and Geert Haesaert[1,2]
[1]Associated Faculty of Applied Bioscience Engineering, Ghent University College, Ghent
[2]Laboratory of Phytopathology, Faculty of Bioscience Engineering, Ghent University, Ghent
[3]KERMIT, Department of Mathematical Modelling, Statistics and Bioinformatics,
Ghent University, Ghent
Belgium

1. Introduction

1.1 *Fusarium* head blight: A multi-faceted agricultural problem

Fusarium Head Blight (FHB) is one of the most important diseases in wheat, caused by a complex of up to 17 *Fusarium* species. The main causal agents of FHB in Europe are *Fusarium graminearum*, *Fusarium culmorum*, *Fusarium avenaceum*, *Fusarium poae* and *Microdochium nivale* (Audenaert *et al.* 2009; Brennan *et al.* 2003; Leonard & Bushnell 2003; Mudge *et al.* 2006; Parry *et al.* 1995). There is extensive work on the effect of FHB on grain yields of cereals. For example, in breeding programs aiming to generate resistant cultivars, yield losses have been observed ranging from 6 up to 74% (Snijders 1990; Snijders & Perkowski 1990). Symptoms of *Fusarium* occur just after anthesis. The partly white and partly green heads are diagnostic for the disease in wheat (Figure 1C). The fungus also may infect the peduncle immediately below the head, causing a brown/purplish discoloration of the stem tissue. Additional indications of FHB infection are pink to salmon-orange spore masses of the fungus often seen on the infected spikelets and glumes. Infected kernels are shriveled, lightweight and dull grayish or pinkish. These kernels sometimes are called "tomb-stones" because of their chalky, lifeless appearance. If infection occurs late in kernel development, *Fusarium*-infected kernels may be normal in size, but have a dull appearance or a pink discoloration.

Although FHB may cause wheat yield losses, the interest in FHB is primarily fuelled by the ability of *Fusarium* species to produce mycotoxins. FHB pathogens produce a diversified spectrum of mycotoxins depending on the species (Bennett & Klich 2003). Trichothecenes, zearalenon, moniliformin and fumonisins are the most important mycotoxins produced by *Fusarium* fungi. Among the trichothecenes, deoxynivalenol (DON) is the predominant mycotoxin throughout Europe and is mainly produced by *F. graminearum*. These secondary fungal metabolites can accumulate to significant doses and as such cause a serious impediment for human and animal health. Moreover, the European concern for several *Fusarium* mycotoxins has been concretized in regulations for maximum levels for human and animal consumption. These regulations provide an extra economic motive for farmers to prevent FHB infection and mycotoxin accumulation in small grain cereals such as wheat.

The prevention of DON and *Fusarium* in wheat is not easy, since the disease is primarily associated with weather conditions during anthesis of the crop. It is generally accepted that rainfall just before and during anthesis, which is situated in the month of June, favours the FHB pathogens and can cause serious yield losses. Conidia present on crop residues reach the ears by splashed rain droplets. A recent study by Landschoot *et al.* (2011a+b) fine-tuned the influence of weather conditions. These authors demonstrated nicely that also weather conditions during the vegetative growth of the crop in winter and spring are important parameters determining the disease incidence and DON level (Table 1). This conclusion is remarkable since infection with *Fusarium* starts in the months of June and July. A possible explanation for this remarkable findings may come from the influence of weather conditions during winter on the survival of the primary inoculums in soil, weeds and crop residues. In cold winters, survival of *Fusarium* conidia is poor, resulting in a lower primary inoculums pressure in June.

Until now, no absolute FHB resistance encoded by single dominant resistance genes has been characterized in wheat. Consequently, it is difficult to implement *Fusarium* resistance into breeding programs. Two major sources for resistance have been characterized. Type I resistance stops the pathogen at the level of penetration while type II resistance is involved the inhibition of fungal spread within the infected ear (Ban & Suenaga 2000; Singh *et al.* 1995). However, the implementation of quantitative trait loci associated with resistance into commercial wheat varieties is not for tomorrow because of economic drawbacks.

A. Month	Variable	negative	Variable	positive
November	Days with frost	-0.24	75%P Air pressure	0.47
December	Days with frost	-0.53	Average Dew point	0.68
January	90%P Air pressure	-0.28	75%P Dew point	0.69
February	Days with frost	-0.53	Median Dew point	0.74
March	Days with rainfall	-0.28	Average Dew point	0.62
April	Days with rainfall	-0.65	75%P Dew point	0.51
May	Average Air pressure	-0.58	Median wind speed	0.34
June	25%P Air pressure	-0.69	Days with RH >80%	0.66
July	75%P Temperature	-0.58	Total Rainfall	0.63
B. Month	Variable	negative	Variable	positive
November	25%P RH	-0.24	75%P Air pressure	0.23
December	Median RH	-0.19	90%P Temperature	0.33
January	Days with frost	-0.28	75%P Temperature	0.35
February	Days with frost	-0.34	75%P Temperature	0.48
March	90%P Dew point	-0.23	10%P Temperature	0.39
April	Days with rainfall	-0.31	90%P Air pressure	0.36
May	Average Air pressure	-0.32	25%P Temperature	0.34
June	Average Air pressure	-0.42	25%P RH	0.34
July	10%P Temperature	-0.31	10%P RH	0.28

Table 1. A. Highest pearson correlation coefficients for weather variables (Days with Frost, Air Pressure, Total Rainfall, Relative humidity (RH) Temperature, Dew point and wind speed) and the DI, 10%P, 25%P, 75%P, 90%P, respectively mean 10%, 25%, 75%, 90% percentiles.
B. Highest pearson correlation coefficients for weather variables (Days with Frost, Air Pressure, Total Rainfall, Relative humidity (RH) Temperature, Dew point and wind speed) and DON content in grain, 10%P, 25%P, 75%P, 90%P, respectively mean 10%, 25%, 75%, 90% percentiles.

Although good agricultural practices certainly help to reduce the risk for *Fusarium* epidemics, the application of fungicides remains the most important control measure to reduce *Fusarium* symptoms. Although there are a limited number of active ingredients with good control activity for FHB, the chemical control of this pathogenic disease complex remains a serious issue. The short vulnerable period of the pathogen, the fact that it is an ear pathogen, ands the fact that it mainly infects under wet conditions all hamper an efficient control of the FHB complex.

1.2 The Belgian situation

In order to get a better view on the FHB problem in Flanders, a region situated in the North of Belgium, an intensive survey started in 2002. Pursuing a combined approach of symptom evaluation, DON measurement and genetic characterization of the population, a comprehensive dataset was obtained. This dataset comprised data of ten growing seasons, on at least ten locations throughout Flanders. On each location 12 cultivars of wheat were sown in a complete randomized block design with four replications. An overview of the obtained results have previously been published (Audenaert *et al.*, 2009; Landschoot *et al.*, 2011a+b) and are presented in Figure 1. From this extensive survey, several solid conclusions could be drawn.

First, it was clear that the FHB population in Flanders is very dynamic evolving from a *F. graminearum/F. culmorum* dominated population in 2002 and 2003 towards a *F. poae/F. graminearum* population in 2008 and 2009 (Figure 1A). In addition, a correlative study on all variables elucidated some clear population characteristics. First, *F. poae* was shown to be a pathogen that is often occurring in association with other members of the disease complex. This is illustrated in the heat map presented in Figure 2A where *F. poae* clearly clusters with other species such as *F. avenaceum* and *M. nivale*.

A second layer of complexity is the link between DON level and the DON-producing species *F. graminearum* and *F. culmorum*. As illustrated in Figure 1B, the presence of DON was not really correlated with the presence of DON-producing species since it clustered separately in a different branch of the tree. In addition, the presence of *F. graminearum* and *F. culmorum* was rather linked to low disease classes such as Dc1 and Dc2 while the presence of the other species was linked with the higher disease classes Dc2, Dc3 and Dc4. Finally, the presence of DON was also associated with the higher disease classes. Nevertheless, although this link was apparent, no clear linear correlation was observed between quantitative DON presence and disease symptoms (Audenaert *et al.*, 2009).

2. Fungicides to control FHB and associated mycotoxins

2.1 Fungicides to control fungal growth

Several active ingredients such as triazoles and strobilurins have been reported for their efficiency against several species of the *Fusarium* complex. Triazoles are known inhibitors of the ergosterol biosynthesis in fungi while strobilurin fungicides inhibit mitochondrial electron transport by binding on the Qo site of the cytochrome BC1 complex. Where the effectiveness of triazole fungicides against *Fusarium* spp. is a certainty, the activity of strobilurins against *Fusarium* spp. is doubtable. A considerable amount of evidence shows that strobilurins are mainly active against *M. nivale*. Laboratories around the globe have devoted considerable efforts to develop a coherent view of the activity of fungicides against the FHB causing species. A comprehensive overview is illustrated in Table 2.

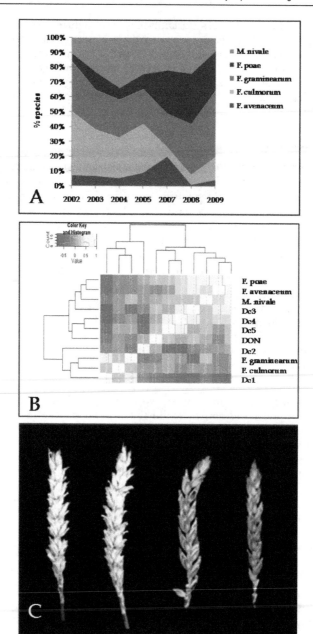

Fig. 1. Results of an extensive field survey in Flanders wheat fields on the FHB population, DON content and disease symptoms. A. shows the evolution of the population from 2002-2009. B shows the links between disease symptoms (Dc1 = disease class 1; Dc2=disease class 2; Dc3= disease class 3; Dc4 = disease class 4; Dc5 = disease class 5). C shows typical FHB symptoms at anthesis (two left ears) compared to asymptomatic ears (two right ears).

Impact of Fungicide Timing on the Composition of the Fusarium Head Blight Disease Complex and the
Presence of Deoxynivalenol (DON) in Wheat

83

Some exceptions notwithstanding no real contradictory reports are mentioned although Zhang
et al. (2009b) and Pirgozliev et al. (2002) obtained different results on the effect of azoxystrobin
to control F. graminearum and F. culmorum. Possibly, these differences originate from different
environmental conditions under which experiments were carried out. In line with this
assumption Magan et al. (2002) clearly highlighted the importance of the a_w value in the
efficiency of fungicides. Similarly, other researchers showed the importance of wheat cultivar
and isolate aggressiveness for control of Fusarium using fungicides (Mesterhazy et al. 2003).

2.2 Fungicides to control mycotoxin production

Where the effect of fungicides on fungal outgrowth is quite straightforward, reports on the
effect of fungicides on the production of mycotoxins is rather contradictory and information
is fragmentary. Indeed, to date, no studies are available describing the effect of fungicides to
the broad array of mycotoxins that can be produced by Fusarium. Most studies focus on just
one or two mycotoxins.

For tebuconazole it is generally accepted that it causes a reduction in the biosynthesis or
DON level (Edwards et al. 2001; Haidukowski et al. 2005; Ioos et al. 2005; Paul et al. 2008;
Simpson et al. 2001; Zhang et al. 2009a) and the trichothecene nivalenol (NIV) (Ioos et al.
2005). Information on another triazole fungicide propiconazole is contradictionary.
Application of propiconazole resulted in decreased DON levels in a study by Paul et al.
(2008), while other studies reported increased levels of DON upon propiconazole
application (Magan et al. 2002).

Application of the triazole metconazole generally results in decreased DON levels in grain
samples. This observation was corroborated by several scientific reports (Edwards et al. 2001;
Paul et al. 2008; Pirgozliev et al. 2002). Finally for prothioconazole Paul et al. (2008) mentioned
decreased DON levels. Consonant with this observation, Audenaert et al. (2010) described
reduced DON levels upon application of field doses of prothioconazole in an in vitro assay.
However, these authors added another layer of complexity in developing a coherent view on
the effect of prothioconazole on DON biosynthesis. Sub lethal application of prothioconazole
resulted in increased DON levels (Audenaert et al. 2010). This induction was shown to be
orchestrated through a reactive oxygen mediated pathway. Indeed, using an in vitro approach
the former authors succeeded to demonstrate that sub lethal application of prothioconazole
results in the prompt induction of H_2O_2 which preceded the DON accumulation. In addition,
elimination of H_2O_2 using catalase inhibited the production of DON.

The effect of the strobilurin fungicide azoxystrobin on DON varies from a proliferated DON
biosynthesis (Zhang et al. 2009; Magan et al. 2002; Simpson et al. 2001; Gaurilcikiene et al.
2011) towards reduced DON levels (Pirgozliev et al. 2002). It is tempting to speculate on this
observation. The fact that strobilurins often result in increased DON levels might be
explained by the pathogen spectrum of strobilurin which mainly targets M. nivale while
being less effective against F. graminearum. It is not unlikely that the niches that are not
longer occupied by M. nivale are taken by F. graminearum which consequently lead to
increased DON levels.

Although this kind of research is mainly carried out on the mycotoxin DON, the focus is also
shifted to other mycotoxins. Gaurilcikiene et al. (2011) demonstrated increased T-2 levels upon
azoxystrobin application. A similar result was obtained for NIV (Ioos et al. 2005).

TRIAZOLES					
	species	effect on species	mycotoxin	effect on mycotoxin	reference
Tebuconazole	*F. graminearum*	↓	DON	↓	Zhang *et al*. (2008)
	Fusarium spp.	↓	DON	↓	Edwards *et al*. (2001)
	F. culmorum	↓	DON	↓	Simpson *et al*. (2001)
	F. avenaceum	↓	ND	ND	Simpson *et al*. (2001)
	M. nivale	—	NR	NR	Paul *et al*. (2008)
	Fusarium spp.	↓	DON	↓	Haidukowski *et al*. (2005)
	F. culmorum	↓	NIV	—	Ioos *et al*. (2005)
Propiconazole	*F. culmorum*	↓*	DON	↑	Magan *et al*. (2002)
	Fusarium spp.	↓	DON	↓	Paul *et al*. (2008)
Metconazole	*F. culmorum*	↓	DON	↓	Pirgozliev *et al*. (2002)
	F. graminearum	↓	DON	↓	Pirgozliev *et al*. (2002)
	Fusarium spp.	↓	DON	↓	Edwards *et al*. (2001)
	Fusarium spp.	↓	DON	↓	Paul *et al*. (2008)
Prothioconazole	*Fusarium spp.*	↓	DON	↓	Paul *et al*. (2008)
	F. graminearum	↓	DON	↑	Audenaert *et al*. (2010)
STROBILURINS					
Azoxystrobin	*F. graminearum*	—	DON	↑	Zhang *et al*. (2008)
	F. culmorum	—	DON	↑	Magan *et al*. (2002)
	F. culmorum	↓	DON	↓	Pirgozliev *et al*. (2002)
	F. graminearum	↓	DON	↓	Pirgozliev *et al*. (2002)
	M. nivale	↓	NR	NR	Simpson *et al*. (2001)
	Fusarium spp.	↑	DON	↑	Simpson *et al*. (2001)
	Fusarium spp.	—	DON	—	Edwards *et al*. (2001)
	F. poae	↑	T2	↑	Gaurilcikiene *et al*. (2011)
	F. culmorum	↑	DON	↑	Gaurilcikiene *et al*. (2011)
	F. poae	↓	T2	↑	This work
	F. culmorum	—	NIV	—/↑	Ioos *et al*. (2005)
OTHERS					
Carbendazim	*F. graminearum*	—	DON	—	Zhang *et al*. (2008)
Thiram	*F. graminearum*	—	DON	—	Zhang *et al*. (2008)
Quintozene	*F. verticillioides*	↓	Fum	↓	Falcao *et al*. (2011)
Fludioxynil+metalaxyl-N	*F. verticillioides*	↓	Fum	↓	Falcao *et al*. (2011)

Table 2. Effect of several fungicides on *Fusarium* spp. and corresponding mycotoxin production. ↑: proliferated growth/production; ↓: reduced growth/production; —: no effect; NR: not relevant; ND: not detected;*: effect dependent on the a_w value.

Finally, some other fungicides namely carbendazim and thiram were tested for their efficiency to reduce DON in grain samples. However, no clear effect was observed (Zhang *et al.* 2009). A nice study with *F. verticiloides* showed decreased fumonisin levels upon application of respectively quintozene and fludioxynil+metalaxyl-N.

Although the above mentioned examples are not meant to provide a complete and extensive literature review on the use of fungicides against FHB, they clearly demonstrate that the infield control of FHB symptoms does not completely cover control of myctoxin production. We can conclude that when fungicides are not sprayed optimally, conditions which might be conducive for mycotoxin production might be created in the field. This conclusion will hopefully encourage further research in this scientific field.

3. Effect of fungicides on the fungal metabolome

Recently, interest in the effect of fungicides on the fungal metabolome has increased. Primarily fueling this interest in the interaction between sub lethal fungical concentrations and the fungus is that in practice, fungicidal treatments cannot always be carried out under optimal conditions. Consequently, the fungicide concentrations encountered by the pathogen are often lower than one would expect.

3.1 Short term effects

When *Fusarium* encounters fungicide concentrations that are not lethal, a complex spectrum of metabolic changes occurs. The full range of these metabolic changes is still not well dissected although the first steps have been taken to use genome wide approaches to disentangle transcriptional changes in *Fusarium* upon fungicide treatments (Liu *et al.* 2010).

Although the majority of these metabolic changes remain elusive, a fast growing number of papers focus on the oxidative stress induced by fungicides. In an *in vitro* approach it was demonstrated that exposing *F. graminearum* to sub lethal doses of prothioconazole resulted in proliferated production of DON. In addition, an increase in H_2O_2 which preceded the DON accumulation was observed. Addition of catalase, an H_2O_2 scavenger, resulted in loss of DON production. Similar results were obtained in a study using *F. graminearum* and *M. nivale*. This study provided evidence that H_2O_2 was produced by *F. graminearum* and *M. nivale* upon azoxystrobin application (Kaneko & Ishii 2009). However, this phenomenon is possibly isolate or species dependent. In a study by Covarelli *et al.* (2004), tebuconazole was shown to have a negative effect on the expression of the Tri5 gene, an indication for DON bioynthesis in *F. culmorum*.

3.2 Long term effects

The ability of fungi to adapt to stress is pivotal to their survival in the environment, and this adaptation ability is one of the key factors leading to mutations or adaptations that can give rise to more aggressive crop pathogens in an agricultural setting. In a recent study, evidence was brought forward showing that the initial efficacy of the triazole epoxiconazole eroded resulting in increasing EC_{50} values with a factor of approximately 1.4 (Klix *et al.* 2007). An interesting study illustrated that mutations in a β-tubulin conferred resistance of *F. graminearum* to benzimidazole fungicides (Chen *et al.* 2009). In addition, a benzimidazole binding site on the β-tubulin gene was suggested to be mutated conferring strains resistant to benzimidazole fungicides such as carbendazim (Qiu *et al.* 2011). An even more interesting observation in carbendazim resistant *F. graminearum* strains was that a proliferated production of DON was observed (Zhang *et al.* 2009b).

Typically for triazole fungicides, a slowly evolving fungicide resistance has been observed. Decreases in azole sensitivity can be caused by (i) point mutations in the target gene, (ii) overexpression of the target gene, (iii) alterations in ergosterol biosynthesis, (iv) enhanced efflux of toxic compounds, and (v) increased copy numbers of target genes or genes for efflux pumps (Becher *et al.* 2010). Similarly as in the carbendazim story, isolates displaying increased resistance to tebuconazole showed increased mycotoxin production.

Finally, for azoxystrobin, at least 27 fungal species are listed to be resistant. The majority of the resistance types are correlated with the G143A substitution in the quinol oxidation site of cytochrome b, the target for strobilurins. Also for members of the FHB complex, this type and other types of resistance towards strobilurins have been described in respectively *M. nivale* (Walker *et al.* 2009) and *F. graminearum* (Dubos *et al.* 2011). The results and examples given in the previous paragraphs clearly peeled away several layers of complexity in the chemical control of FHB. The divergence of fungicide effectiveness both at species and mycotoxin level hamper a simple control of this disease complex. Still, a better insight into the effect of fungicide application on *Fusarium* in the field is needed. European legislation for several *Fusarium* mycotoxins has been established. This legislation provides an extra economic motive for farmers to prevent FHB infection and mycotoxin accumulation in small grain cereals such as wheat.

A detailed study on the effect of fungicides at a population level will certainly contribute to new insights in the adaptive dynamics of a *Fusarium* population upon fungicide application. In addition, shifts in the population might have its consequences for the mycotoxin profiles present in these fields. In the present study, results from fungicide field trials from 2002-2010 are presented with regard to the effect of triazole and strobilurin fungicides on symptoms, population composition and the presence of the trichothecene mycotoxin DON in the field. These data provide new insights into the effect of fungicides on FHB both at a species- and mycotoxin level.

4. Experimental setup of this study

4.1 Experimental field trials

From 2002 to 2010, different field trials of winter wheat throughout Belgium were followed up for at least ten locations that were located in the most important wheat regions characterized by different growth conditions and crop husbandry measurements. The winter wheat area of Belgium is situated in the centre of the wheat growing region in Europe. Each year at each location, commercial wheat varieties were sown in a complete randomized block design with four replications. For all locations the normal crop husbandry measures were taken. Depending on the experiment, several fungicides and fungicide combinations were used and were applied at various Zadoks growth stages (GS39, GS55 up to GS65) of the crop. In this way both the effect of the active ingredient and the time of application was monitored. The wheat cultivars were sown in common crop rotation systems which lead to different previous crops, both host crops (maize or wheat) as well as non-host crops for *Fusarium* spp. (beans, sugar beets, onions or chicory).

Impact of Fungicide Timing on the Composition of the Fusarium Head Blight Disease Complex and the
Presence of Deoxynivalenol (DON) in Wheat

87

From GS71 to GS75 the experiments were evaluated for the presence of *Fusarium* symptoms. Both the FHB incidence and the FHB severity (disease classes 1-5 with 0, 25, 50, 75 or 100 % bleached ear surface, respectively) were scored. To take into account both assessments for 100 randomly chosen ears per plot, the disease index (DI) was computed as follows: DI = (0n1 + 1n2 + 2n3 + 3n4 + 4n5)/4n x 100%; with "n" the number of evaluated ears and "ni" the number of ears in disease class i.

In order to assess the composition of the FHB population, wheat ears were plated on PDA medium (potato dextrose agar, Oxoid, Belgium) for further species identification. Seeds were surface-sterilized for 1 minute in 1% NaOCl, washed for 1 min with 70% EtOH, washed with distilled sterile water, dried for 5 min and subsequently put on PDA plates. After five days of incubation at 20°C, outgrowing mycelium was transferred to a new PDA plate. For species determination, five mycelium plugs randomly taken from the fully grown PDA plates were transferred to liquid GPY-broth (10 g glucose, 1 g yeast extract and 1 g peptone, Oxoid, Belgium) and incubated for five days at 20°C. After five days, mycelium was transferred to eppendorf tubes, centrifuged for 10 min at 12,000 rpm and then freeze-dried for 6 h at -10°C and 4 h at -50°C (Christ Alpha 1-2 LD Plus, Osterode, Deutschland). DNA extraction was performed as described by Audenaert *et al.* (2009), based on the CTAB (hexadecyl trimethyl ammonium bromide) method (Saghai-Maroof *et al.*, 1984). PCR for single species detection was performed in a 25 µl reaction mixture (Demeke *et al.*, 2005).

DNA amplification was performed in an Applied Biosystems GeneAmp PCR system 97000 PCR. Amplicons were separated on 1.5% (wt/vol) agarose gels stained with 0.1 µl ethidium bromide. PCR was validated by including reference strains obtained from the MUCL/BCCM collection in each PCR run: *F. graminearum* MUCL 42841; *F. culmorum* MUCL 555; *F. poae* MUCL 6114; *M. nivale* MUCL 15949; *F. avenaceum* MUCL 6130 (Audenaert *et al.*, 2009, Isebaert *et al.*, 2009).

At harvest, DON levels was analyzed by enzyme-linked immunosorbent assay (ELISA) (Veratox DON 5/5 kit Biognost - Neogen). A subsample was taken from each field out of the randomized block design. All DON results are the average of at least four DON measurements per treatment.

4.2 *In vitro* trials

In the present study, fluoxastrobin+prothioconazole was tested for its efficiency to control several field isolates of *F. poae*. The field dose of the fungicide was the point of departure for the *in vitro* assay. The field dose mounted to 0.5 g/l + 0.5 g/l for fluoxastrobin+prothioconazole, A dilution series of the fungicide was prepared to obtain a final concentration of 1 mg/l, 5 mg/l, 10 mg/l and 50 mg/l in the 24-well plates in which the assay was executed. In these wells, 250 µl of conidial suspension was added and amended with 250 µl of the fungicide. The final concentration of the microconidia was 10^6 conidia per ml. These wells were incubated at 22°C. Two repetitions were done per dilution and the experiment was repeated two times independently in time. Control treatments consisted of 250 µl of spore suspension and 250 µl of distilled water. T-2 production kinetics were monitored using an ELISA (Veratox T-2 kit, Biognost-Neogen).

At each time point (4 h, 24 h, 48 h) after inoculation, the percentage of germinated conidia were counted. At each time point, three repetitions per treatment were counted.

5. Results

5.1 *In vitro* effect of fluoxastrobin+prothioconazole on *F. poae*

From previous work it is known that fluoxastrobin+prothioconazole provides good protection against *F. graminearum*. For several isolates it was shown that a dose of 50 mg/l of this fungicide resulted in a reduction in germination rate of 95% (Audenaert *et al.* 2010). Based on these results, we wanted to focus on the sensitivity of *F. poae* to this fungicide. *F. poae* was an underestimated species for long time since it was described being a weak pathogen. However, throughout Europe a steadily increase of this species has been observed. Yet, the basis for this increased importance remains elusive. To date research on this pathogen remains limited although research groups around the globe tend to initiate research initiatives with regard to this pathogens (Stenglein, 2009). From Figure 2 it is clear that a diversified spectrum of susceptibility can be observed in *F. poae* depending on the isolate. Most importantly, some of the isolated strains showed residual germination levels ranging from 20% up to 80% at fungicide concentrations three times the field dose (data not shown). In addition, among isolates very diverse reaction patterns were observed in the dilution series. This result highlights the high diversity of *F. poae* with regard to fungicide resistance.

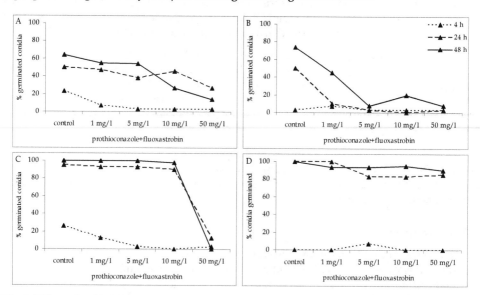

Fig. 2. Effect of a dilution series of fluoxastrobin+prothioconazole on the conidia germination of four different *F. poae* isolates (A,B,C, and D). Each panel shows the germination of a different isolate.

The results obtained in this work are rather contradictory with what has been described in literature. Generally it is accepted that *F. graminearum* is more resistant to fungicides than *F. poae*. Using several fungicides such as diphenoconazole, tebuconazole, iprodione,... it was demonstrated that the sensitivity of *F. graminearum* was lower compared to *F. poae* (Hudec, 2007; Mullenborn *et al.* 2008). Till now, we have no clear explanation for this discrepancy, however, possibly the isolate, the incubation temperature, the culture conditions might influence the sensitivity response in both species.

Impact of Fungicide Timing on the Composition of the Fusarium Head Blight Disease Complex and the
Presence of Deoxynivalenol (DON) in Wheat

89

In a second step, we wanted to investigate the interaction between stress induced by prothioconazole+fluoxastrobin and the toxin production by the *F. poae* isolates. Surprisingly, using an LC-MSMS approach, the *F. poae* isolates were characterized as T-2 chemotype. In addition, several other mycotoxins were produced such as diacetylscirpenol and nivalenol (data not shown). The ability of *F. poae* to produce T-2 is rather exceptional. Indeed, the ability of *Fusarium* to produce T-2 toxin has been described to be mainly restricted to *F. langsethiae* and *F. sporotrichoides*. A reason for this apparent discrepancy originates from the fact that no structural genetic information is available regarding the toxic metabolome of *F. poae*. Therefore, majority of the studies use artificial media in search for toxins of *F. poae*. However, recent research in our laboratory clearly illustrates that *F. poae* does not produce toxins on all media. There is some evidence that the nitrogen source and eventual amino acids or polyamines could play a role in the induction of T-2 production by *F. poae*. In Figure 2, the production kinetics of T-2 upon prothioconazole+fluoxastrobin is shown.

Similar to the results on the conidia germination (Figure 3), the effect of the fungicide prothioconazole+fluoxastrobin differed clearly depending on the isolate. Some isolates did not show a consistent proliferated T-2 production upon fungicide stress (Figure 3 A and C) others did (Figure 3B). Remarkably, the isolate that was extremely resistant to the fungicide prothioconazole+fluoxastrobin also showed extremely high basal levels of T-2 production which even increased upon fungicide application. Although these results are very preliminary, they pinpoint T-2 production as a possible protective mechanism upon fungicide application. Similar results were previously obtained with DON. Using a tri5 knockout mutant, it was demonstrated that DON-negative mutants of *F. graminearum* became hypersusceptible to fungicide application (Audenaert *et al.*, 2011).

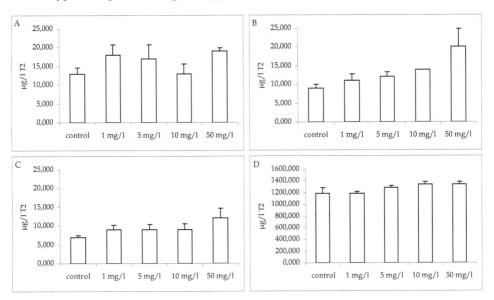

Fig. 3. Effect of sub lethal prothioconazole+fluoxastrobin concentrations on T-2 production on four different isolates (A,B,C and D).

5.2 Control of FHB by fungicide application at growth stage GS55

5.2.1 Effect of fungicides on DON content

In order to peel away the layers of complexity regarding control of *Fusarium* and corresponding DON contamination, a fungicide trial using prothioconazole, epoxyconazole+metconazole and prothioconazole+tebuconazole was set up during the growing seasons of 2007-2008, 2008-2009 and 2009-2010. Pursuing a combined approach of symptom scoring and DON measurement, we aimed to disentangle the effect of triazole fungicides on *Fusarium* development and mycotoxin production. For all treatments, fungicide applications resulted in a clear reduction in symptom development which was the same for all tested active ingredients (data not shown).

More interesting is the effect of these treatments on the DON level. In Figure 4, the effect of triazole application on the DON levels is displayed. By these results, evidence is brought forward demonstrating that the efficiency of triazole fungicides to reduce DON levels is depending on the background level of DON observed in the control treatments. Under conditions of low DON levels, fungicides do not result in decreases in DON level. On the contrary, several fungicide treatments resulted in increased DON levels compared to the control. This detrimental effect of fungicide treatments with respect to DON content has previously been described by other authors (Magan *et al.*, 2002).

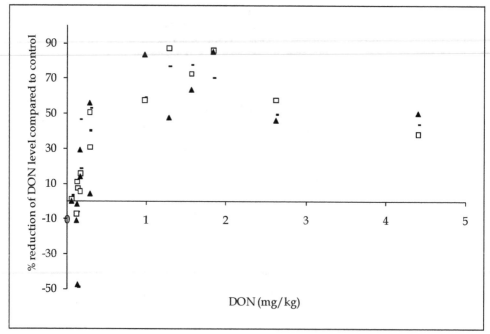

Fig. 4. Levels of DON after application of prothioconazole (▲), epoxyconazole+metconazole (□) and prothioconazole+tebuconazole (—) in function of the DON levels present in the untreated control fields. All data points are the result of four independent repetitions.

Surprisingly, our experimental field trials show similar efficiencies in function of the DON concentration for the three triazoles: prothioconazole, epoxyconazole+metconazole and prothioconazole+tebuconazole applied at GS55 were all the most efficient at DON concentrations between 1 mg/kg up to 2 mg/kg while the efficiency reduced to about 50 % for higher DON concentrations.

These results suggest that although fungicides have been described to be very effective to control *Fusarium* symptoms in the field, their efficiency to reduce DON seems to be limited. In addition, Figure 4 illustrates the usefulness of fungicides for fields with DON levels that are situated around the DON threshold values set by the European Commision: with reductions of the DON level from 50% to 90% this implies that samples that would exceed the European threshold limits drop below these limits when triazole fungicides are applied.

5.2.2 Effect of fungicides on the population structure

Besides the effect of fungicide application on FHB symptoms and DON levels, the impact of fungicides on the population constitution was monitored as well. The five predominant species in Flanders wheat fields were monitored i.e. *F. graminearum, F. poae, F.culmorum, F. avenaceum* and *M. nivale*. Results from these field trials clearly subscribe that fungicide application clearly influences the species distribution within the population (Figure 5).

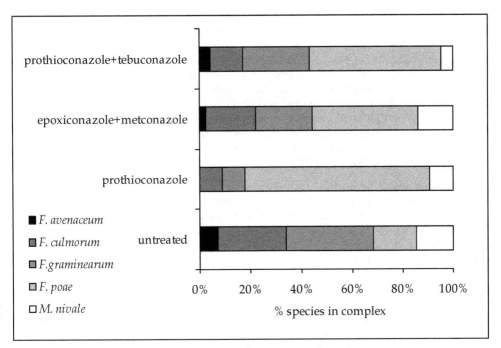

Fig. 5. Effect of prothioconazole, epoxiconazole+metconazole and prothioconazole+tebuconazole on the population of *Fusarium* species.

In fields treated with prothioconazole at GS55, *F. poae* became the predominant species whereas in the untreated fields, the population was initially dominated by *F. culmorum* and *F. graminearum*. This phenomenon was to a lesser extend also observed in the other treatments with prothioconazole+tebuconazole and epoxiconazole+metconazole. This result is in concordance with the *in vitro* experiments shown in Figure 1 which already suggested that a considerable portion of the *F. poae* isolates possesses a considerable level of resistance towards triazole fungicides. Finally, it was consistently surprising that this shift within the population towards an increase of *F. poae* was mainly at the expense of *F. culmorum*. Surprisingly, application of triazole fungicides did not result in a consistent increase of *M. nivale* in the population. *M. nivale* has previously been described for its resistance towards triazole fungicides but this could not be pinpointed in the present field study. No clear explanation for this observation can be found.

5.3 Fungicide application timing and control of FHB

5.3.1 Effect of fungicide timing on FHB symptoms and DON content

Another layer of complexity in developing a coherent view of the effects of fungicide application versus *Fusarium* and its mycotoxins is the timing of the application. In order to investigate the effect of timings for chemical control of *Fusarium*, several triazole fungicides were applied at different growth stages during wheat growth. One series of trials involved a *Fusarium* treatment at GS55 and GS65. Figure 5A clearly illustrates that application of triazole fungicides at GS65 does not result in reduced *Fusarium* symptoms. At the other hand, application of triazole fungicides at GS55 clearly reduces the impact of *Fusarium* at the level of the symptoms. The results for the concomitant DON levels are slightly different. Although chemical control of *Fusarium* at GS65 is inefficient at the level of symptom development, a consistent effect was observed at the level of DON (Figure 6C). This result came as a surprise since several authors report that suboptimal fungicide application can result in proliferated DON production (Audenaert *et al.*, 2010; Magan *et al.*, 2002). We assume that this late fungicide application at GS65 and at the normal application rate comes too late to avoid symptom development, however this application is still in time to decrease DON levels to some extent. It has been described by several authors that DON is a crucial virulence/pathogenicity factor at later stages of infection to facilitate migration of the pathogen in the ear (Mudge *et al.* 2006).

In a second series of time trials, application of prothioconazole was performed at GS39, GS65 and a combination at GS39+GS65. Results are presented in Figure 6B+D. Similar as in Figure 6A, application of prothioconazole at GS65 results in a small decreased number of symptoms, although this reduction was not significant. Remarkably, application of prothioconazole at GS39 resulted in a significant decrease in *Fusarium* symptoms. In addition, a combined application at GS39+GS65 had a synergistic effect at the level of symptoms. For DON, results were quite similar. Application of prothioconazole at GS39 resulted in reduced DON levels compared to untreated plots and plots treated at GS65. A combined application of prothioconazole did not result in further reduction of the DON level. This result confirms that the ability of fungicides to reduce DON levels in *Fusarium* infected fields is limited to reductions of about 50%.

Scientific research on the effects of fungicide timing with regard to *Fusarium* symptom development and DON levels are scarce. Previously, Wiersma and Motteberg (2005) reported

Impact of Fungicide Timing on the Composition of the Fusarium Head Blight Disease Complex and the
Presence of Deoxynivalenol (DON) in Wheat

93

GS60 as ideal for control of FHB. In addition, using tebuconazole, these authors checked the efficiency to control FHB and DON levels but regarding *Fusarium* symptoms, they could not come up with solid conclusions regarding the timing of application. For the efficiency of tebuconazole to reduce DON levels, applications at GS39 and GS60 performed equally.

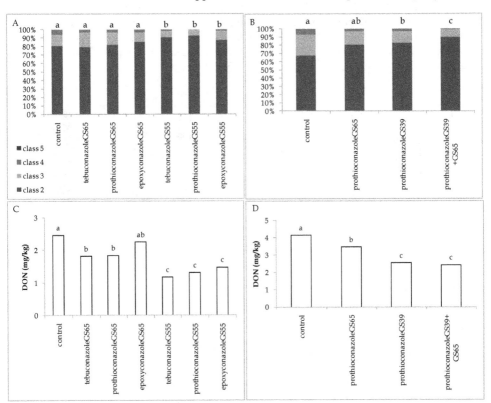

Fig. 6. Effect of timing of fungicide application on *Fusarium* symptom development and DON content. A+C: effect of prothioconazole, tebuconazole and epoxyconazole applied at GS65 and GS55 on symptoms (A) and DON level (C). B+D: effect of prothioconazole applied at GS39, GS65 and GS39+GS65 on *Fusarium* symptom development (B) and DON level (D).

5.3.2 Fungicide timing and population structure

The timing of fungicide application to control *Fusarium* clearly had an effect on the composition of the population. When we compare the population in the untreated control of the experimental field trial presented in Figure 7 with the untreated control in Figure 4, the huge differences in population composition is obvious. This enormous elasticity of the FHB population depending on field location has previously been described by Audenaert *et al.* (2009).

Application of prothioconazole at GS55 clearly favoured *F. poae* which was not present in the control fields but which popped up in the prothioconazole treatment (Figure 7A).

For the other triazole fungicides, a consistent reduction of *F. avenaecum* was observed. In addition, the niches seemed to solely colonized by *F. graminearum*. In the other field trial, where prothioconazole was applied at GS39, GS65 and GS39+GS65, major shifts in the population were observed although no consistent changes were observed between the treatments. At first sight this might be unexpected, however both application of GS39 and GS65 can be considered to be apsecific for FHB. Therefore, the population causing disease symptoms has not yet been established (at GS39) or is already fading (GS65). This might explain the less consistent results in this experiments compared to the results obtained in the experiment where fungicides were applied at GS55 (Figure 5).

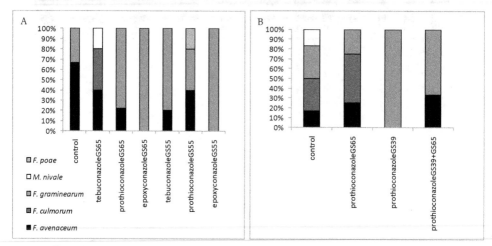

Fig. 7. Effect of timing of fungicide application on the FHB population. A: effect of prothioconazole, tebuconazole and epoxyconazole applied at GS65 and GS55. B: effect of prothioconazole applied at GS39, GS65 and GS39+GS65.

It was remarkable that the increased portion of *F. graminearum* in the fungicide treatments did not result is increased DON levels. This result provides indirect evidence that the presence of DON-producing chemotypes within the population does not necessarily result in a proliferated DON level in the field. This finding underscores the work by Landschoot *et al.* (2011a+b).

6. Conclusions

The aim of this study was to disentangle the effect of fungicide application on the FHB disease complex. Pursuing a combined approach of *in vitro* and *in vivo* field trials, some very interesting conclusions could be drawn. First, it is important to state that fungicides applied at normal rate under optimal conditions are effective in controlling FHB disease. However, working in the field often implies working under suboptimal conditions, and then, problems regarding FHB presence of concomitant mycotoxins might occur. First, it is important to highlight that not all species comprised in the FHB disease show the same susceptibility towards fungicide application. The data presented in the present work provide a compelling amount of evidence illustrating that *F. poae* is more resistant to triazoles than the other members of the disease complex. This was illustrated in *in vitro* trials which showed that a considerable amount of isolates were able to grow at field doses of fungicide applications. Consonant with these *in vitro* trials field trials demonstrated that triazole application generally resulted in a shift in the FHB population in favour of *F. poae.*

At the level of symptom development, all tested triazole fungicides when optimally applied at GS55 resulted in similar disease symptoms reduction. When varying the timing of application, it was obvious that triazole application at GS65 came too late to efficiently reduce FHB symptoms. On the contrary, fungicide spraying at GS39 resulted in a significant reduction of disease symptoms. We suggest that these treatments reduce the primary inoculums present in the vegetative crop. A combined application at GS39+GS65 resulted in a synergistic effect of the treatments.

With regard to DON, several lines of evidence corroborate a role for timing of fungicide application. The results obtained in the present study demonstrate nicely that where complete control of FHB can be obtained at the level of disease symptoms, there seems to be some sort of threshold efficiencies that cannot be exceeded when DON levels are higher than 2 mg/kg. In all the field trials comprised in the present study, a maximum DON reduction of 50% was obtained compared to the untreated control fields. In addition, where application of triazoles was not affecting disease symptoms when applied at GS65, a minor reduction of DON levels was observed. Contrary to observations on the symptom level, no additional effect was observed when performing a combined application of triazoles at GS39+GS65 compared to single applications at GS39 or GS65.

In conclusion, this study peeled away several layers of complexity in the chemical control of FHB in wheat. We are convinced that fungicide use is an important hurdle that can be included in crop management systems to prevent FHB and concomitant mycotoxin present. However, it is clearly an oversimplification to pretend that fungicide use is the "holy grail" in the control of FHB disease. The disease is far too complex and too multifaceted to draw this conclusion. It will take all branches of scientific research to keep this problem under control. The use of wheat varieties with high levels of resistance, the use of good culture practices such as broad crop rotations and intelligent soil tillage will certainly contribute

7. Acknowledgment

Kris Audenaert is a postdoctoral fellow of the Ghent University Research Fund. Sofie Landschoot and Adriaan Van Heule are PhD students financed by the Ghent University College research fund. We greatly acknowledge the help of Bernard De Baets and Willem Waegeman from Ghent University, research group KERMIT, Belgium for the nice collaboration in the data analysis. Secondly, our gratitude goes to Lies Willaert, Melvin Berten and Daniël Wittouck from the Landbouw Centrum Granen for helping to establish all field trials troughout Flanders during the last 10 years. Sarah De Saeger and Sofie Monbaliu, from the faculty of Pharmaceutical Sciences, Ghent University helped with the mycotoxin analysis using LC-MSMS.

Part of this work was financially supported by the Flemish Institute for the Stimulation of Scientific - Technological Research in Industry project 70575 (IWT, Brussels, Belgium).

8. References

Audenaert, K., Callewaert, E., Hofte, M., De Saeger, S., & Haesaert, G. (2010). Hydrogen peroxide induced by the fungicide prothioconazole triggers deoxynivalenol (DON) production by *Fusarium graminearum*. *BMC Microbiology*, vol.10, pp.1-10.

Audenaert, K., De Schuyffeleer, N., Maene, P., Monbaliu, S., Vekeman, F., De Saeger, S., Haesaert, G., & Eeckhout, M. (2011). Efficacy of neutral electrolyzed water to reduce *Fusarium* spp and deoxynivalenol (DON) production in vitro and on wheat kernels. Submitted to *Food control*.

Audenaert, K., Van Broeck, R., Bekaert, B., De Witte, F., Heremans, B., Messens, K., Hofte, M., & Haesaert, G. (2009). *Fusarium* head blight (FHB) in Flanders: population diversity, inter-species associations and DON contamination in commercial winter wheat varieties. *European Journal of Plant Pathology*, vol.125, pp. 445-458.

Ban, T., & Suenaga, K. (2000). Genetic analysis of resistance to *Fusarium* head blight caused by *Fusarium graminearum* in Chinese wheat cultivar Sumai 3 and the Japanese cultivar Saikai 165. *Euphytica*, vol.113, pp. 87-99.

Becher, R., Hettwer, U., Karlovsky, P., Deising, H.B., & Wirsel, S.G.R. (2010). Adaptation of *Fusarium* graminearum to tebuconazole yielded descendants diverging for levels of fitness, fungicide resistance, virulence, and mycotoxin production. *Phytopathology*, vol.100, pp. 444-453.

Bennett, J.W., & Klich, M. (2003) Mycotoxins. *Clinical Microbiology Reviews*, vol.16, pp. 497-512.

Brennan, J.M., Fagan, B., van Maanen, A., Cooke, B.M., & Doohan, F.M. (2003). Studies on in vitro growth and pathogenicity of European *Fusarium* fungi. *European Journal of Plant Pathology*, vol.109, pp. 577-587.

Chen, C.J., Yu, J.J., Bi, C.W., Zhang, Y.N., Xu, J.Q., Wang, J.X., & Zhou, M.G. (2009). Mutations in a beta-Tubulin Confer Resistance of *Gibberella zeae* to Benzimidazole Fungicides. *Phytopathology*, vol.99, pp. 1403-1411.

Covarelli, L., Turner, A.S., & Nicholson, P. (2004). Repression of deoxynivalenol accumulation and expression of Tri genes in *Fusarium culmorum* by fungicides *in vitro*. *Plant Pathology* vol.53, pp. 22-28.

Demeke, T., Clear, R.M., Patrick, S.K., & Gaba, D. (2005). Species-specific PCR-based assays for the detection of *Fusarium* species and a comparison with the whole seed agar plate method and trichothecene analysis. *International Journal of Food Microbiology*, vol.103, pp 271-284.

Dubos, T., Pasquali, M., Pogoda, F., Hoffmann, L., & Beyer, M. (2011). Evidence for natural resistance towards trifloxystrobin in *Fusarium* graminearum. *European Journal of Plant Pathology*, vol.130, pp. 239-248.

Edwards, S.G., Pirgozliev, S.R., Hare, M.C., & Jenkinson, P. (2001). Quantification of trichothecene-producing *Fusarium* species in harvested grain by competitive PCR to determine efficacies of fungicides against *Fusarium* head blight of winter wheat. *Applied and Environmental Microbiology*, vol.67, pp. 1575-1580.

Falcao, V.C., Ono, M.A., Vizoni, E., de Avila Miguel, T., Hirooka, E. Y., & Ono, E.Y. (2011). *Fusarium verticillioides*: evaluation of fumonisin production and effect of fungicides on in vitro inhibition of mycelial growth. *Mycopathologia*, vol.171, pp. 77-84.

Gaurilcikiene, I., Mankeviciene, A., & Suproniene, S. (2011). The effect of fungicides on rye and triticale grain contamination with *Fusarium* fungi and mycotoxins. *Zemdirbyste*, vol.98, pp. 19-26.

Haidukowski, M., Pascale, M., Perrone, G., Pancaldi, D., Campagna, C., & Visconti, A. (2005). Effect of fungicides on the development of *Fusarium* head blight, yield and deoxynivalenol accumulation in wheat inoculated under field conditions with *Fusarium graminearum* and *Fusarium culmorum*. *Journal of the Science of Food and Agriculture*, vol.85, pp. 191-198.

Hudec, K. (2007). Pathogenicity of fungi associated with wheat and barley seedling emergence and fungicide efficacy of seed treatment. *Biologia*, vol.62, pp. 287-291.

Ioos, R., Belhadj, A., Menez, M., & Faure, A. (2005). The effects of fungicides on *Fusarium* spp. and *Microdochium nivale* and their associated trichothecene mycotoxins in French naturally-infected cereal grains. *Crop Protection*, vol. 24, pp. 894-902.

Isebaert, S., De Saeger, S., Devreese, R., Verhoeven, R., Maene, P., Heremans, B., & Haesaert, G. (2009). Mycotoxin-producing *Fusarium* species occurring in winter wheat in Belgium (Flanders) during 2002-2005. *Journal of Phytopathology*, vol.157, pp. 108-116.

Kaneko, I., & Ishii, H. (2009). Effect of azoxystrobin on activities of antioxidant enzymes and alternative oxidase in wheat head blight pathogens *Fusarium graminearum* and *Microdochium nivale*. *Journal of General Plant Pathology*, vol.75, pp. 388-398.

Klix, M.B., Verreet, J.A., & Beyer, M. (2007). Comparison of the declining triazole sensitivity of *Gibberella zeae* and increased sensitivity achieved by advances in triazole fungicide development. *Crop Protection*, vol.26, pp. 683-690.

Landschoot, S., Audenaert, K., Waegeman, W., Pycke, B., Bekaert, B., De Baets, B. & Haesaert, G. (2011). Connection between the primary inoculum on weeds, soil and crop residue and the *Fusarium* population on wheat plants. *Crop protection*, vol.30, pp. 1297-1305

Landschoot, S., Waegeman, W., Audenaert, K., Vandepitte, J., Baetens, J., Haesaert, G., & De Baets, B. (2011). An empirical analysis of explanatory variables affecting *Fusarium* infection and deoxynivalenol production in wheat (2010). *Submitted to Journal of Plant Pathology*.

Leonard, K., & Bushnell, W. (2003). *Fusarium* head blight of wheat and barley. APS Press.

Lima, P.; Bonarini, A. & Mataric, M. (2004). *Application of Machine Learning,* InTech, ISBN 978-953-7619-34-3, Vienna, Austria

Liu, X., Jiang, J.H., Shao, J.F., Yin, Y.N., & Ma, Z.H. (2010). Gene transcription profiling of *Fusarium graminearum* treated with an azole fungicide tebuconazole. *Applied Microbiology and Biotechnology*, vol.85, pp. 1105-1114.

Magan, N., Hope, R., Colleate, A., & Baxter, E.S. (2002). Relationship between growth and mycotoxin production by *Fusarium* species, biocides and environment. *European Journal of Plant Pathology*, vol.108, pp. 685-690.

Mesterhazy, A., Bartok, T., & Lamper, C. (2003). Influence of wheat cultivar, species of *Fusarium*, and isolate aggressiveness on the efficacy of fungicides for control of *Fusarium* head blight. *Plant Disease*, vol.87, pp. 1107-1115.

Mudge, A.M., Dill-Macky, R., Dong, Y.H., Gardiner, D.M., White, R.G., & Manners, J.M. (2006). A role for the mycotoxin deoxynivalenol in stem colonisation during crown rot disease of wheat caused by *Fusarium graminearum* and *Fusarium pseudograminearum*. *Physiological and Molecular Plant Pathology*, vol.69, pp. 73-85.

Mullenborn, C., Steiner, U., Ludwig, M., & Oerke, E.C. (2008). Effect of fungicides on the complex of *Fusarium* species and saprophytic fungi colonizing wheat kernels. *European Journal of Plant Pathology*, vol.120, pp. 157-166.

Parry, D.W., Jenkinson, P., & McLeod, L. (1995). *Fusarium* ear blight (scab) in small grain cereals. A review. *Plant Pathology*, vol.44, pp. 207-238.

Paul, P.A., Lipps, P.E., Hershman, D.E., McMullen, M.P., Draper, M.A., & Madden, L.V. (2008). Efficacy of triazole-based fungicides for *Fusarium* head blight and deoxynivalenol control in wheat: A multivariate meta-analysis. *Phytopathology,* vol.98, pp. 999-1011.

Pirgozliev, S.R., Edwards, S.G., Hare, M.C., & Jenkinson, P. (2002). Effect of dose rate of azoxystrobin and metconazole on the development of *Fusarium* head blight and the accumulation of deoxynivalenol (DON) in wheat grain. *European Journal of Plant Pathology,* vol.108, pp. 469-478.

Qiu, J.B., Xu, J.Q., Yu, J.J., Bi, C.W., Chen, C.J., & Zhou, M.G. (2011). Localisation of the benzimidazole fungicide binding site of *Gibberella zeae* beta(2)-tubulin studied by site-directed mutagenesis. *Pest Management Science,* vol.67, pp. 191-198.

Saghai-Maroof, M.A., Soliman, K.M., Jorgensen, R.A., & Allard, R.W. (1984). Ribosomal DNA spacer-length polymorphisms in barley: Mendelian inheritance, chromosomal location, and population dynamics. *Proceedings of the National Academic of Sciences of the USA,* vol.81, pp. 8014-8018.

Simpson, D.R., Weston, G.E., Turner, J.A., Jennings, P., & Nicholson, P. (2001). Differential control of head blight pathogens of wheat by fungicides and consequences for mycotoxin contamination of grain. *European Journal of Plant Pathology,* vol.107, pp. 421-431.

Singh, R.P., Ma, H., & Rajaram, S. (1995). Genetic analysis of resistance to scab in spring wheat cultivar Frontana. *Plant Disease,* vol.79, pp. 238-240.

Snijders, C.H.A. (1990). *Fusarium* head blight and mycotoxin contamination of wheat, a review. *Netherlands Journal of Plant Pathology,* vol.96, pp. 187-198.

Snijders, C.H.A., & Perkowski, J. (1990). Effects of head blight caused by *Fusarium culmorum* on toxin content and weight of wheat kernels. *Phytopathology,* vol.80, pp. 566-570.

Stenglein, S.A. (2009). *Fusarium poae*: a pathogen that needs more attention. *Journal of Plant Pathology,* vol. 91, pp. 25-36.

Walker, A.S., Auclair, C., Gredt, M., & Leroux, P. (2009). First occurrence of resistance to strobilurin fungicides in *Microdochium nivale* and *Microdochium majus* from French naturally infected wheat grains. *Pest Management Science,* vol.65, pp. 906-915.

Wiersma, J.J., & Motteberg, C.D. (2005). Evaluation of five fungicide timings for the control of leaf - spot diseases and *Fusarium* head blight in hard red spring wheat. *Canadian Journal of Plant Pathology,* vol. 27, pp. 25-37.

Zhang, Y.J., Fan, P.S., Zhang, X. , Chen, C.J., & Zhou, M.G. Quantification of *Fusarium graminearum* in harvested grain by Real-Time Polymerase Chain Reaction to assess efficacies of fungicides on Fusarium head blight, deoxynivalenol contamination, and yield of winter wheat.(2008.) *Phytopathology,* vol.99, pp.95-100.

Zhang, Y.J., Fan, P.S., Zhang, X., Chen, C.J., & Zhou, M.G. (2009a). Quantification of *Fusarium graminearum* in Harvested Grain by Real-Time Polymerase Chain Reaction to Assess Efficacies of Fungicides on *Fusarium* Head Blight, Deoxynivalenol Contamination, and Yield of Winter Wheat. *Phytopathology,* vol.99, pp. 95-100.

Zhang, Y.J., Yu, J.J., Zhang, Y.N., Zhang, X., Cheng, C.J., Wang, J.X., Hollomon, D.W., Fan, P.S., Zhou, M.G. (2009b). Effect of Carbendazim Resistance on Trichothecene Production and Aggressiveness of *Fusarium graminearum*. *Molecular Plant-Microbe Interactions,* vol.22, pp. 1143-1150.

6

State of the Art and Future Prospects of Alternative Control Means Against Postharvest Blue Mould of Apple: Exploiting the Induction of Resistance

Simona Marianna Sanzani and Antonio Ippolito
Department of Environmental and Agro-Forestry Biology and Chemistry,
University of Bari "Aldo Moro"
Italy

1. Introduction

Fresh fruit and vegetables supply essential nutrients, such as vitamins and minerals, and are a major source of complex carbohydrates, antioxidants, and anticarcinogenic substances which are important to human health and well being (Arul, 1994). Being aware of the advantages potentially coming from their use, the consumers' demand for fresh products has greatly increased during recent years. On the other end, the possible presence of chemical residues, mycotoxins and other contaminants of fruit and vegetables (Paster et al., 1995), creates great concern among consumers for safety issues.

Therefore, aim of an adequate storage is to help the harvested fruit and vegetables to arrive at their destination fresh, disease-free, and safe for consumers, despite the complexity of treatments they have to undergo prior to or during storage, and the long period between harvest and consumption. All the means and methods with the power to aid in preserving the quality of the harvested products and in protecting them from decay agents during storage and shelf-life, are aimed at this objective.

2. Apple: Origin, distribution and economic importance

Apples (*Malus domestica* Borkh.) belongs to the *Pomoideae* subfamily *Rosaceae*, along with pear (*Pyrus communis* L.), apricot (*Prunus armeniaca* L.), prune (*Prunus domesticus* L.), cherry (*Prunus avium* L.) and peach (*Prunus persica* L.). Actually there is a debate about whether *M. domestica* originated from hybridization among various wild species or from a single species, still growing on the Northern slopes of the Tien Shan mountains at the border between northwest China and the former Soviet Republic of Kazakhstan (Phipps *et al.*, 1990; Juniper & Mabberley, 2006). Apples were consumed by human beings since the Stone Age, as documented by numerous Neolithic finds discovered in Northern Europe. In the medieval times monasteries were responsible for selection, propagation, and perpetuation of hundred of different cultivar types. In the Nineteenth Century these types became the ideal stock for horticulturists to develop techniques to cross desirable selections. From that time *M. domestica* cultivars spread throughout the world, particularly in North America, Russia, Australia, and Japan. Over 7500 apple cultivars are known, but even though nowadays breeders worldwide create new selections annually, only very few of them are widely produced (Janick, 1996).

In 2008 apple constituted the third fruit most cultivated in the world with and area of 4,696,472 hectares and a production of 69,304,442 tonnes (Food and Agriculture Organization of the United Nations [FAO], 2011). Moreover, Italy represented in the world the sixth highest producing country (more than 2 million tonnes), and the first exporting one (683 thousand tonnes); whereas, apple consumption was estimated to be 47.78 g/day/capita (FAO, 2011). The apple production is concentrated in Northern Italy: 80% of the crop, in fact, comes from Trentino-Alto Adige (46%), Emilia-Romagna (17%), and Veneto (14%) regions. Golden Delicious and Granny Smith are two of the most popular apple cultivars among consumers. In Italy apples are mainly consumed as fresh commodity, whereas in Northern Europe and America they are used for juices and cider production.

Apples are commonly harvested from the end of August till mid-October, although, because of the postharvest practices (pre-refrigeration, controlled atmosphere, etc.), they can be stored for long times and so are available on commerce all through the year.

3. Apple pathogens and diseases

The surface of fruit or vegetables is covered by fungal and bacterial propagules that they have acquired from the air during their development on the parent plant, or with which they have come in contact during picking or any of the subsequent stages of handling. However, most fungal and bacterial propagules that reach the harvested product do not cause decay, even when conditions suitable for penetration and development are present.

Harvested fruit and vegetables are naturally attacked by a relatively small group of pathogens: approximately forty species. However, each fruit or vegetable has its own typical pathogens out of this particular group. Eckert & Ogawa (1988) divided the major postharvest pathogens of pome fruits into two groups: (a) those that cause quiescent infections of lenticels, including *Phlyctema vagabunda* Desm. (syn. *Gloeosporium album*), *G. perennans* Zeller & Childs (syn. *Cryptosporiopsis curvispora*) and *Nectria galligena* Bresad; and (b) those that preferably enter through wounds after harvest, including *Penicillium expansum* Link, *Botrytis cinerea* Pers. ex Fr., *Monilinia* spp., *Mucor* spp., *Rhizopus* spp., *Alternaria alternata* (Fr.) Keissl, *Stemphylium botryosum* Wallr, and *Cladosporium herbarum* (Pers.) Link. The rots in the lenticels are initiated in the orchard in the late summer, and are a major problem for apples grown in areas with late summer rainfall such as the United Kingdom and Northern European Countries (Edney, 1983). In drier apple production areas, main problems are caused by wound pathogens that invade the fruit after harvest through injuries sustained during harvesting and handling and via puncture wounds, bruised lenticels, etc. Other important pathogens of pome fruits are species of *Phytophthora*, that may become a serious problem during rainy seasons for fruit from orchards with heavy soils (Edney, 1978), and *Colletotrichum gloeosporioides* Penz (syn. *Gloeosporium fructigenum)*, the bitter rot fungus, which is capable of direct penetration of the intact skin (Brook, 1977). *Botryosphaeria* spp., the black and white rot fungi, are of importance in several areas of the USA (Snowdon, 1990). Finally, pathogens of minor importance that may occasionally be found on harvested apples and pears include *Trichothecium roseum* Link, and species of *Phomopsis, Nigrospora, Fusarium, Epicoccum, Aspergillus,* and *Trichoderma* (Snowdon, 1990).

3.1 *Penicillium expansum*

The blue mould rot, caused by *P. expansum*, is one of the most common and destructive rots of harvested apples and pears, but it can also be found on sweet cherries and other commodities such as apricots, grapes, blueberries, peaches, strawberries, walnuts, pecans, hazelnuts, and acorns (Andersen *et al.*, 2004; Murphy *et al.*, 2006). Blue mould is a worldwide severe disease, even in production areas where the most advanced storage technologies are available. In Northern Europe the estimated incidence of this disease varies between seasons and cultivars, ranging from 5 to 20% (Mari *et al.*, 2002). *P. expansum* is primarily a necrotroph and a wound parasite, most frequently gaining entrance through fresh mechanical injuries such as stem punctures, bruises and insect injuries, finger-nail scratches by pickers and necrotic tissues of diverse origin, for instance due to infections by other pathogens, such as *Gloeosporium* spp., *Phytophthora* spp., and *Mucor* spp. (Snowdon, 1990). Thus, resistance of the epidermis to breakage may be an important factor in the resistance of apple cultivars to decay. Studying the force required to break the epidermis of several cultivars, as a criterion for resistance to wound pathogens, Spotts *et al.* (1999) found that the epidermis of Golden Delicious and Jonagold was more easily broken than that of other cultivars, while the epidermal tissues of Fuji and Granny Smith were the most resistant to puncture. Sometimes infections may occur through normal stems, open calyx canals or lenticels, especially when they are damaged by cracking after a sudden abundant supply of water following a period of dryness, or after bruising late in the storage season when fruit have been weakened by ripening and aging (Janisiewicz *et al.*, 1991).

The fungus produces pale brown to brown soft-watery spots that enlarge rapidly under shelf life conditions. Under humid conditions, conidia-bearing conidiophores group to form coremia on the surface of the lesion. As the conidia mature, they turned from nearly snow white to blue-green and form masses which give the decay its typical colour. Another characteristics, important in the recognition of *P. expansum*, is the earthy, musty odour.

Since decay development is favoured by high humidity, the blue mould is a particularly serious problem on fruits stored or shipped in plastic film liners (Hall & Scott, 1989). Decay can progress, albeit slowly, during cold storage; rapid development begins when the fruits are transferred to warmer conditions. The fungus can spread during the postharvest handling, since the blue mold spores are long-lived and may easily survive from season to season on contaminated bins, where the fungus can grow and produce copious amounts of spores. Contamination with these spores may come from various other sources including orchard soil present on bins, decaying fruit or air. Inoculation of the fruit going into storage is believed to occur mainly from the diphenylamine (DPA) drenching solution used for protection against superficial scald, where the spore concentration increases with each successively drenched bin and may reach high levels if solutions are not changed regularly. Inoculation can also occur during fruit handling in water contaminated with the fungus in packing houses. A single decayed fruit may contain enough spores to contaminate water of the entire packing line (Janisiewicz *et al.*, 1991). At present few chemicals are permitted in Europe to be used in the postharvest phase: for instance, only thiabendazole and pyrimethanilare are allowed in Italy for postharvest control of blue mold. Moreover, during the past few decades thiabendazole has lost efficacy due to the establishment of resistant pathogen populations (Baraldi *et al.*, 2003).

Finally, it should be considered that contamination of fruits by *P. expansum* not only results in economic losses during storage and shelf-life, but it also has a potential public health significance, since some strains of *P. expansum* produces the mycotoxin patulin.

3.2 Patulin

Thousands of mycotoxins exist, but only a few represent significant food safety challenges. The most important mycotoxins impacting food production and manufacturing are: patulin, produced by *Penicillium* species (mainly from *P. expansum*); ochratoxin, which is present in a large variety of foods as it is produced by several fungal strains of *Penicillium* and *Aspergillus* species; zearalenone, found prevalently in grains infected by *Fusarium graminearum* Rank; aflatoxins, mainly produced by *Aspergillus flavus* Link and *A. parasiticus* Speare; trichothecene, whose production is known for several *Fusarium* species; and, finally, fumonisins mainly produced by the maize pathogens *Fusarium verticillioides* (formerly *F. moniliforme*) (Sacc.) Nirenberg and *F. proliferatum* (Matsush.) Nirenberg ex Gerlach and Nirenberg.

Of special interest to postharvest pathology is the production of patulin by *P. expansum* in pome fruits. Patulin {4-hydroxy-4H-furo[3,2-c]pyran-2(6H)-one} is a polar compound transferred in fruit juices if rotten fruits, especially apples, are not picked up during fruit juice processing. There is no clear evidence that patulin is carcinogenic, however, it has been shown to cause immunotoxic effects (Pacoud *et al.*, 1990) and to be neurotoxic in animals (Deveraj *et al.*, 1982), so that the European Commission in 2006 established the maximum levels of patulin permitted in foodstuffs (European Commission, 2006). The amount of patulin produced by *P. expansum* may vary greatly according to the strain involved. Sommer *et al.* (1974) found that patulin production in different strains of *P. expansum* in Golden Delicious apples ranged from 2 to 100 µg gram^{-1} of tissue. On the other hand, other constitutive and environmental parameters, such as the cultivar and the storage temperature may influence patulin accumulation. Paster *et al.* (1995) demonstrated that while more patulin was produced in Starking apples than in Spadona pears held at 0-17°C, higher toxin levels were produced in pears than in apples held at 25°C. Furthermore, a consistent reduction in patulin production was observed when inoculated apples were held in a $3\%CO_2/2\%O_2$ atmosphere. Thus, it can be concluded that the ability to produce patulin and the amounts of patulin produced depend on fungal strain, fruit cultivar, storage temperature, and composition of storage atmosphere (Lovett *et al.*, 1975; Paster *et al.*, 1995). Finally, it has to be considered that, patulin, as many fungal secondary metabolites, has an antibiotic activity. This ability is particularly useful because the secondary products may represent competitive weapons in nature (Sutton, 1996).

4. Alternative control means

The development of resistance, together with the increasing concern about possible adverse effects on human health and environment caused by fungicides, have contributed to arouse interest in the development of alternative means for controlling plant pathogens, capable of integrating, if not totally replacing, synthetic fungicides. Substantial progress has been made in finding alternatives to synthetic fungicides for the control of postharvest diseases of fruit and vegetables (Ippolito *et al.*, 2004; Palou *et al.*, 2008; Sanzani *et al.*, 2009a; Schena *et al.*, 2007; Sharma *et al.*, 2009; Zhang *et al.*, 2009).

Emerging postharvest biocontrol technology employs different approaches, such as use of physical means, natural biocides, antagonistic microorganisms and their products, or the intensification of natural defense mechanisms. Treatments that have been evaluated for effectiveness against *P. expansum* on apples include chlorine dioxide administration in aqueous environment (Okull *et al.*, 2006); surface application of cinnamon oil or potassium sorbate (Ryu

& Holt, 1993); fumigation with acetic acid vapor (Sholberg *et al.*, 2000); immersion in electrolyzed oxidizing water (Okull & LaBorde, 2004). A wash treatment with ≥ 2% acetic acid for more than 1 min proved to be effective in completely inhibiting *P. expansum* growth and relevant patulin production on apples destined for cider (Chen *et al.*, 2004). Hot water immersion inhibited decay development in *P. expansum*-inoculated apples (Fallik, 2010).

Several examples of success in preventing *P. expansum* rots during postharvest phase of fruits by using yeasts antagonists (Ippolito *et al.*, 2000; Zhang *et al.*, 2009), bacteria (Morales *et al.*, 2008) and biologically active natural products (Mari *et al.*, 2002), have been reported. The antifungal effectiveness of an antagonist can be increased by addition of substances, as in the case of sodium bicarbonate in combination with a strain of *Metschnikowia pulcherrima* (Spadaro *et al.*, 2004). Finally the controlled atmosphere (CA), used for the long-term storage of apples, may also influence the efficacy of biocontrol agents by affecting their vitality or by altering the physiological status of the treated fruits (Chalutz & Droby, 1997).

4.1 Host protection and defence mechanisms

Induced disease resistance has been adopted as a general term and defined as 'the process of active resistance, dependent on the physical or chemical barriers of the host plant, activated by biotic or abiotic agents (inducing agents)' (Kloepper *et al.*, 1992). In compatible plant-fungus interactions resistance mechanisms may be activated too slowly to be effective or be suppressed by the invading pathogen. So the level of basic resistance may simply not be sufficient to halt infection and prevent extensive tissue colonization and symptom development. Whereas, in induced tissues the balance may be shifted in favour of the plant, by an earlier and quicker response, that can be effective in limiting tissue colonization to various extents, depending on the specific plant-pathogen relationship.

Plants respond to invasion by pathogens with an array of biochemical and genetic changes, including the production of reactive oxygen species (ROS), antimicrobial compounds, antioxidants, and signalling molecules such as salicylic acid (SA), ethylene and jasmonic acid (JA) (Mahalingam *et al.*, 2003). They also respond by the localized activation of a cell-death program, designated "hypersensitive response (HR)", and by the systemic activation of cellular and molecular defences, termed "systemic acquired resistance (SAR)" (Ryals *et al.*, 1996). There is evidence for commonalities between plant responses to pathogens (referred to as defence responses) and environmental stresses (referred to as stress responses). However, a plant response to each environmental challenge is unique and tailored to increasing the plant ability to survive the inciting stress.

Most plant antimicrobial natural products have relatively broad spectrum activity, and specificity is often determined by whether or not a pathogen has the enzymatic machinery to detoxify a particular host product. Accumulation of inducible antimicrobial compounds is often orchestrated through signal-transduction pathways linked to perception of the pathogen by host receptors. The simplest functional definitions recognize phytoalexins as compounds that are synthesized *de novo* and phytoanticipins as pre-formed infectional inhibitors. However, the distinction between phytoalexin and phytoanticipin is not always obvious, as some compounds may be phytoalexins in one species and phytoanticipins in others (Dixon, 2001).

Phytoalexins are low-molecular-weight toxic compounds mainly produced in the host tissue in response to initial infection by microorganisms (Harborne, 1999). In other words, in order to overcome an attack by the pathogen, the host is induced to produce antifungal compounds that would prevent pathogen development. However, the accumulation of phytoalexins does not depend on infection only. Such compounds may be elicited by microbial metabolites, mechanical damage, plant constituents released after injury, a wide diversity of chemical compounds, or by low temperature, irradiation, and other stress conditions. Phytoalexins are, thus, considered to be general stress-response compounds, produced after biotic or abiotic stress. The most available evidence on the role of phytoalexins shows that disruption of cell membranes is a central factor in their toxicity (Smith, 1996), and that the mechanism is consistent with the lipophilic properties of most phytoalexins (Arnoldi & Merlini, 1990).

Earlier studies by Müller & Borger (1940) already provided strong evidence that resistance of potato to *Phytophthora infestans* (Mont.) de Bary is based on the production of fungitoxic compounds by the host. A terpenoid compound, rishitin, produced in potato tubers following infection by *P. infestans*, was first isolated by Tomiyama *et al.* (1968) from resistant potatoes inoculated with the fungus. It accumulates rapidly in the tuber and reaches levels much higher than those required to prevent fungal development. The relationship between the accumulation of rishitin in the tuber and its resistance to late blight may point to its role in resistance development (Kuc, 1976). Other sesquiterpenoids that have been found in potatoes may also play a role in tuber disease resistance; they include rishitinol (Katsui *et al.*, 1972), lubimin (Katsui *et al.*, 1974), oxylubimin (Katsui *et al.*, 1974), solavetivone (Coxon *et al.*, 1974), and others. These sesquiterpenoids proved also to be effective in suppressing mycelial growth of the potato pathogen *P. infestans* on a defined medium (Engstrom *et al.*, 1999). Moreover, several phytoalexins, such as umbelliferone, scopoletin, and esculetin, are produced in sweet potato roots infected by the fungus *Ceratocystis fimbriata* Ellis and Halsted; it was noted that these compounds accumulate more rapidly in roots resistant to this fungus than in sensitive roots (Minamikawa *et al.*, 1963).

Further studies with celery (Afek *et al.*, 1995a) indicate that (+) marmesin, the precursor of linear furanocoumarins in this crop, is the major compound involved in celery resistance to pathogens, indeed increased susceptibility of stored celery to pathogens is accompanied by a decrease in (+) marmesin concentration. Indeed treatment of celery prior to storage with gibberellic acid (GA3), a naturally occurring phytohormone in juvenile plant tissue, resulted in decay suppression during 1 month of storage at 2°C, although GA3 does not have any effect on fungal growth *in vitro* (Barkai-Golan & Aharoni, 1976). It was suggested that the phytohormone retards celery decay during storage by slowing down the conversion of (+) marmesin to psoralens, thereby maintaining high level of (+) marmesin and low levels of psoralens and, thus increasing celery resistance to storage pathogens (Afek *et al.*, 1995b). Another phytoalexin found in celery tissue is columbianetin, which probably also plays a more important role than psoralens in celery resistance to decay (Afek *et al.*, 1995c). This hypothesis is derived from the following facts: (a) columbianetin exhibits strong activity against *A. alternata, B. cinerea*, and *Sclerotinia sclerotiorum* (Lib.) de Bary, the main postharvest pathogens of celery; (b) the concentration of columbianetin in the tissue is close to that required for their suppression; (c) increased sensitivity of celery to pathogens during storage occurred in parallel with the decrease in the concentration of columbianetin.

The phytoalexin capsidiol is a sesquiterpenoid compound produced by pepper fruits in response to infection by a range of fungi (Stoessl *et al.*, 1972). Pepper fruits inoculated with

B. cinerea and *Phytophthora capsici* Leon contain only small quantities of capsidiol, whereas fruits inoculated with saprophytic species or with weak pathogens may produce higher concentrations of the phytoalexin, which inhibit spore germination and mycelial growth. Inoculating peppers with *Fusarium* species results in increased capsidiol concentration from 6 to 12 h after inoculation. In such cases, the capsidiol accumulation in the tissue is rapid, whereas in other cases it is quickly oxidized to capsenone, which is characterized by a much weaker toxic effect. When unripe pepper fruits were inoculated with *Glomerella cingulata* Stonem, the causal agent of anthracnose, a phytoalexin was readily identified in tissue extracts (Adikaram *et al.*, 1982). This compound, possibly related to capsidiol but much less water soluble, has been named capsicannol (Swinburne, 1983).

The resistance of unripe banana to anthracnose incited by *Colletotrichum musae* Berk and Curt has been attributed to the accumulation of five fungitoxic phytoalexin compounds that were not present in healthy tissue. As the fruit ripened these compounds diminished and, at a progressive stage of disease development, no phytoalexins were detected (Brown & Swinburne, 1980). Elicitors composed of a glucan-like fraction of the cell walls of hyphae and conidia of C. *musae* elicited both necrosis and the accumulation of the two major phytoalexins found in naturally infected tissues.

The apple main phytoalexins are phloretin, naringenin, quercetin, (+)-catechin and benzoic acid (Burse *et al.*, 2004; Gottstein *et al.*, 1992; Iwashina, 2003). Benzoic acid is produced in apples as a result of infection by *Nectria galligena* and other pathogens. Fruit resistance to this pathogen at the beginning of a long storage period was attributed to the formation of this phytoalexin (Swinburne, 1973). *Nectria* penetrates apples via wounds or lenticels prior to picking, but its development in the fruit is very limited. Benzoic acid is the compound isolated from the limited infected area. The elicitor of benzoic acid synthesis was found to be a protease produced by the pathogen (Swinburne, 1975). This protease is a non-specific elicitor and a number of proteases from several sources may elicit the same response. On the other hand, *P. expansum, B. cinerea, Sclerotinia fructigena* Pers., and *Aspergillus niger* van Tieghem, which do not produce protease in the infected tissue and do not induce the accumulation of benzoic acid, can rot immature fruit (Swinburne, 1975).

The *in vivo* levels of the principal phenolic compounds found in olive plants infected by *Phytophthora megasperma* Drechsler and *Cylindrocarpon destructans* (Zinssm.) Scholten differed from the levels observed in non-infected plants. When the antifungal activity of these compounds against both fungi was studied *in vitro*, the most active were quercetin and luteolin aglycons, followed by rutin, oleuropein, *p*-coumaric acid, luteolin-7-glucoside, tyrosol, and catechin (Baidez *et al.*, 2006). Moreover, it has been reported that the antifungal activity of aqueous neem leaf extract against *P. expansum* was related to the presence of highly bioactive compounds including quercetin (Allameh *et al.*, 2001; Mossini *et al.*, 2004).

Inoculating citrus fruits with their specific pathogens *Penicillium digitatum* Pers. and *Penicillium italicum* Wehm results in the accumulation of the phytoalexins scopoletin, scoparone, and umbelliferone. The induced compounds have a greater toxic effect than that of the preformed antifungal compounds naturally found in the fruit tissue, such as citral and limetin, as indicated by the inhibition of *P. digitatum* spore germination (Ben-Yehoshua *et al.*, 1992). The antifungal activity of both scoparone and scopoletin against and *P. digitatum* was observed in UV-C irradiated grapefruits (D'hallewin *et al.*, 2000; Rodov *et al.*, 1992). Indeed, a correlation has been drawn between the level of phytoalexin accumulated in the flavedo of irradiated fruits and its increased resistance. In particular, decay reduction was achieved when irradiation was applied to the fruit prior to its inoculation, and therefore without any direct exposure of the pathogen to the radiation; this finding led to speculate that disease inhibition stems from increased resistance of the fruit to infection and not from the direct fungicidal effect of UV on the pathogen (Rodov *et al.*, 1992). Moreover, Kim *et al.* (1991) reported that the increased concentration of scoparone in heat-treated lemon fruits was in good correlation with their increasing resistance to *P. digitatum* and enhanced antifungal activity of the fruits extract. Similarly, Afek *et al.* (1999) reported that umbelliferone accumulated in the albedo of pathogen-challenged grapefruit played a role in defence mechanisms of immature grapefruit against wound pathogens such as *P. digitatum*.

Biosynthesis of toxic compounds as a result of wounding or other stress conditions, is a ubiquitous phenomenon in various plant tissues. An example of such a synthesis is the production of the toxic compound 6-methoxymellein in carrot roots in response to wounding or to ethylene application (Chalutz *et al.*, 1969; Coxon *et al.*, 1973); the application of *B. cinerea* conidia and other fungal spores to the wounded area was found to stimulate the formation of this compound (Coxon *et al.*, 1973). This 6-methoxymellein probably has an important role in the resistance of fresh carrots to infection. Carrots that have been stored for a long period at a low temperature lose the ability to produce this compound and, in parallel, their susceptibility to pathogens increases.

Resveratrol is a phenolic substance present in both grape skin and wines in response to various fungal infections, UV radiation, or chemicals (Adrian *et al.*, 1997; Jeandet *et al.*, 1995; Langcake, 1981) and it is involved in grey mould resistance (Celotti *et al.*, 1996; Gonzalez Ureña *et al.*, 2003).

The garlic and strawberry phytoalexin esculetin showed a strong activity against fungal strains, especially *Trichophyton mentagrophytes* Malmsten and *Rhizoctonia solani* J.G. Kühn (Cespedes *et al.*, 2006), whereas a moderate antifungal activity against *Fusarium* spp. was reported for ferulic acid (Walker *et al.*, 2003).

4.2 A case study: The flavonoid quercetin

Enhanced protection of host plant tissue during periods of susceptibility through induced/acquired resistance is considered a preferred strategy for achieving integrated pest management (Kuć, 2000). Luckey (1980) reported that natural disease resistance may be induced by low or sub-lethal doses of an elicitor/agent, such as a chemical inducer or a physical stress. For instance, chemical activators could act by modifying the plant-pathogen interaction so that it resembles an incompatible interaction with defence-related mechanisms induced prior to or after challenge (Sticher *et al.*, 1997).

Flavonoids is the general name of compounds that have a fifteen-carbon skeleton, which consists of two phenyl rings connected by a three-carbon bridge. They are potent dietary antioxidants that are found in several plant materials. They are also thought to improve human health and this effect seems related, at least partially, to their antioxidant effect (Nijveldt *et al.*, 2001). Ingham *et al.* (1972) have reviewed the role of flavonoid phytoalexins and other natural products as factors in plant disease resistance. Among these compounds, quercetin was considered the most prominent (Bock, 2003). It consists of 3 rings and 5 hydroxyl groups and occurs in food (i.e. apple, tea, onion, nuts, berries, cauliflower and cabbage) as the aglycone (attached to a sugar molecule) of many plant glycosides. Quercetin can scavenge superoxide and hydroxyl radicals and reduce lipid peroxidation. In addition, it has been reported that quercetin reduces the biosynthesis of heat shock proteins, by reducing the heat shock factor which is the transcriptional factor contributing to their expression (Ishida *et al.*, 2005). However, nothing is known about its possible mode of action in reducing blue mould incidence and severity in apple fruit.

In recent publications the flavonoid quercetin, commonly available on commerce, even as dietary supplements, was tested both *in vitro* and *in vivo* against blue mould and relevant patulin accumulation on apples (Sanzani *et al.*, 2009b). The *in vitro* trials on amended and non-amended PDA plates revealed that although quercetin, at the tested concentrations (100 μg/dish), exerted a slight reduction on fungal growth, it proved to be effective in reducing patulin accumulation, in a dose-dependent way. These results were not surprising, as the antitoxigenic properties of phenolic antioxidants with any interference on fungal growth has already been reported (Kim *et al.*, 2008). However, since quercetin proved to be much more stable in acidic conditions (S.M. Sanzani, unpublished data), as those typically present in apple tissues, and considering that the *in vitro* evaluation of the antifungal activity of a compound is just the first step to test its suitability in preventing the growth and development of a phytopathogenic microorganism, quercetin was further tested *in vivo* by adding it to wounds (100 μg/wound) of Golden Delicious and Granny Smith apples. The compound not only confirmed the suppressive activity on toxin accumulation, but proved to be effective in controlling blue mould incidence and severity. In particular, the best results were observed on Golden Delicious, the apple cultivar considered more resistant to diseases, and on disease severity control, i.e. practically slowing down disease development. The chance of an application by dipping was investigated; however, results were weaker since the wider surface to be treated was not taken into account.

Considering that, although slightly active on *in vitro* fungal growth, quercetin was effective in consistently controlling *in vivo* blue mould, it was hypothesized that quercetin could act by enhancing host natural defence response. A further clue came from a paper by Sanzani *et al.* (2010) in which quercetin proved to exert its activity even in lack of direct contact with the pathogen, particularly when *P. expansum* was inoculated 24-48 h after quercetin application. Thus, to try to gain much insight into the mode of action of quercetin, a molecular technique, called Suppression Subtractive Hybridization (SSH), was applied to identify apple genes putatively induced by its application. Results revealed that a substantial number of enzymes or proteins, which have a function in the adaptation process to oxidative or more general stresses, seemed to be up-regulated after quercetin application. In particular, the antifungal activity seemed to be mainly associated to pathogenesis-related proteins family 10, expressed also in response to challenge with *Venturia inaequalis*, and to PhzC/PhzF proteins. These proteins are tightly linked to structural genes and enzymes from the shikimic acid and tryptophan biosynthetic pathway, known to be involved in the biosynthesis of all the plant phenolic compounds, including phytoalexins and lignin. As a consequence of this response, the fruit should be prepared for a successful defence against pathogens.

Concerning quercetin effect on patulin production, the good results obtained independently from the effect on the fungal growth, suggested that the compound might act directly on toxin biosynthetic pathway, which is mainly associated with fungal secondary metabolism. In a study on the topic Sanzani *et al.* (2009c) proved that the expression level of selected genes, known to code enzymes involved in patulin biosynthesis (i.e. the 6-methylsalicilic acid synthase, the isoepoxydon dehydrogenase, an ABC transporter and two cytocrome P450 monoxygenases), was determined by quantitative real-time PCR. Results evidenced that quercetin down-regulated the expression of two P450 monooxygenases involved in oxygen activation. These results are consistent with quercetin antioxidant properties and with a similar study in which it is reported that the application of the antioxidant phenolic compound caffeic acid reduced aflatoxin production and down-regulated P450 monooxygenases (Kim *et al.*, 2008). Therefore, the initial hypothesis about a presumed direct activity of quercetin on patulin biosynthetic pathway, particularly at the transcriptional level, seems to be confirmed.

5. Conclusions

From this review it appears that some significant progress has been made toward alternative control of postharvest diseases on fruit. Some biofungicides are already on the market in a few countries, and will probably become more widely available as they are registered in more areas. For instance, quercetin might represent an interesting alternative to synthetic fungicides to be applied in the postharvest phase of apples against blue mould and relevant patulin accumulation. Indeed, on the basis of new available information on the mode of action, further studies might lead to the determination of proper conditions to improve their applicability on a commercial scale.

Under field conditions, many alternative control agents have provided limited success, which is often attributed to "uncontrollable" environmental conditions. However, the likelihood of success greatly increases during the postharvest phase due to better environmental control. Moreover, it is often easier to effectively apply alternative control agents while commodities are being processed after harvest.

The success of alternative control greatly depends on influencing the consumer to prefer inner quality to outward appearance. Indeed, alternative control means might represent an important if not essential component of an integrated disease management scheme aimed to reduce economic losses and risks for consumers' health.

6. References

Adikaram, N.K.B., Brown, A.E., & Swinburne, T.R. 1982. Phytoalexin involvement in the latent infection of *Capsicum annuum* L. fruit by *Glomerella cingulata* (Stonem.). *Physiological Plant Pathology*, 21, pp. 161-170, ISSN 0048-4059

Adrian, M., Jeandet, P, Veneau, J., Weston, L.A., & Bessis, R. 1997. Biological activity of resveratrol, a stilbenic compound from grapevines, against *Botrytis cinerea*, the causal agent for gray mould. *Journal of Chemical Ecology*, 23, pp. 1689-1702, ISSN 0098-0331

Afek, U., Aharoni, N., & Carmeli, S. 1995a. The involvement of marmesin in celery resistance to the pathogens during storage and the effect of temperature on its concentration. *Phytopathology*, 85, pp. 1033-1103, ISSN 0031949X

Afek, U., Aharoni, N., & Carmeli, S. 1995b. Increasing celery resistance or pathogens during storage and reducing high-risk psoralen concentration by treatment with GA3. *American Society for Horticultural Science*, 120, pp. 562-565, ISSN 0003-1062

Afek, U., Carmeli, S., & Aharoni, N. 1995c. Columbianetin, a phytoalexin associated with celery resistance to pathogens during storage. *Phytochemistry*, 39, pp. 1347-1350, ISSN 0031-9422

Afek, U., Orenstein, J., Carmeli, S., Rodovc, V., & Joseph, M.B. 1999. Umbelliferone, a phytoalexin associated with resistance of immature Marsh grapefruit to *Penicillium digitatum*. *Phytochemistry*, 49, pp. 1129-1132, ISSN 0031-9422

Allameh, A., Razzaghi, A.M., Shams, M., Rezaee, M.B., & Jaimand, K. 2001. Effects of neem leaf extract on production of aflatoxins and activities of fatty acid synthetase, isocitrate dehydrogenase and glutathione S-transferase in *Aspergillus parasiticus*. *Mycopathologia*, 154, pp. 79–84, ISSN 0301-486X

Andersen, B., Smedsgaard, J., & Frisvad, J.C. 2004. *Penicillium expansum*: consistent production of patulin, chaetoglobosins, and other secondary metabolites in culture and their natural occurrence in fruit products. *Journal of Agricultural and Food Chemistry*, 52, pp. 2421-2428, ISSN 0021-8561

Arnoldi, A., & Merlini, L. 1990. Lipophilicity-antifungal relationships for some isoflavonoid phytoalexins. *Journal of Agricultural and Food Chemistry*, 38, pp. 834-838, ISSN 0021-8561

Arul, J. 1994. Emerging technologies for the control of postharvest diseases of fresh fruits and vegetables. In: *Biological Control of Postharvest Diseases - Theory and Practice*. Wilson, C.L., & Wisniewski, M.E., pp. 1-10, CRC Press, Boca Raton, FL, USA

Báidez, A.G., Gómez, P., Del Río J.A., & Ortuño A. 2006Antifungal capacity of major phenolic compounds of *Olea europaea* L. against *Phytophthora megasperma* Drechsler and *Cylindrocarpon destructans* (Zinssm.) Scholten. *Physiological and Molecular Plant Pathology*, 69, 4-6, pp. 224-229, ISSN 0885-5765

Baraldi, E., Mari, M., Chierici, E., Pondrelli, M., Bertolini, P., & Pratella, G.C. 2003. Studies on thiabendazole resistance of *Penicillium expansum* of pears: pathogenic fitness and genetic characterization. *Plant Pathology*, 52, pp. 362-370, ISSN 1365-3059

Barkai-Golan, R., & Aharoni, Y. 1976. The sensitivity of food spoilage yeasts to acetaldehyde vapors. *Journal of Food Science*, 41, pp. 717-718, ISSN 1750-3841

Ben-Yehoshua, S., Rodov, V., Kim, J.J., & Carmeli, S. 1992. Preformed and induced antifungal materials of citrus fruits in relation to the enhancement of decay resistance by heat and ultraviolet treatments. *Journal of Agricultural and Food Chemistry*, 40, pp. 1217-1221, ISSN 0021-8561

Bock, K.W. 2003. Vertebrate UDP-glucuronosyltransferases: functional and evolutionary aspects. *Biochemical Pharmacology*, 66, pp. 691–696, ISSN 0006-2952

Brook, P.J. 1977. *Glomerella cingulata* and bitter rot of apples. *New Zealand Journal of Agricultural Research*, 20, pp. 547-555, ISSN ISSN 1175-8775

Brown, A.E., & Swinburne, T.R. 1980. The resistance of immature banana fruits to anthracnose *Colletotrichum musae* (Berk. & Curt. Arx). *Journal of Phytopathology*, 99, pp. 70-80, ISSN 1439-0434

Burse, A., Weingart, H., & Ullrich, M.S. 2004. The phytoalexin-inducible multidrug efflux pump AcrAB contributes to virulence of fire blight pathogen *Erwinia amylovora*. *Molecular Plant Microbe Interactions*, 17, 1, pp. 43-54, ISSN 0894-0282

Celotti, E., Ferrarini, R., Zironi, R., & Conte, L.S. 1996. Resveratrol content of some wines obtained from dried Valpolicella grapes: Recioto and Amarone. *Journal of Chromatography A.*, 730, 1-2, pp. 47-52, ISSN 0021-9673.

Céspedes, C.L., Avila, J.G., Martínez, A., Serrato, B., Calderón-Mugica, J.C., & Salgado-Garciglia, R. 2006. Antifungal and Antibacterial Activities of Mexican Tarragon (*Tagetes lucida*). *Journal of Agricultural and Food Chemistry*, 54, 10, pp. 3521-3527, ISSN 0021-8561

Chalutz, E., Devay, J.E., & Maxie, E.C. 1969. Ethylene induced isocoumarin formation in carrot root tissue. *Plant Physiology*, 44, pp. 235-241, ISSN 1532-2548

Chalutz, E., & Droby, S. 1997. Biological control of postharvest disease. In: *Plant–microbe interactions and biological control*. Boland, G.J., & Kuykendall, L.D., pp. 157–170, Marcel Dekker Inc., New York, USA

Chen, L., Ingham B.H., & Ingham, S.C. 2004. Survival of *Penicillium expansum* and patulin production on stored apples after wash treatments. *Journal of Food Science*, 69, pp. C669-C675, ISSN 1750-3841

Coxon, D.T., Curtis, R.F., Price, K.R., & Levett, G. 1973. Abnormal metabolites produced by *Daucus carota* roots stored under conditions of stress. *Phytochemistry*, 12, pp. 1881-1885, ISSN 0031-9422

Coxon, D.T., Pwuce, K.R., Howard, B., Osman, S.F., Kaan, E.B., Zacharius, R.M. 1974 Two new vetispirane derivatives: stress metabolites from potato (*Solanum tuberosum*) tubers. *Tetrahedron Letters*, 34, pp. 2921-2924, ISSN 0040-4039

D'hallewin, G., Schirra, M., Pala, M., & Ben-Yehoshua, S. 2000. Ultraviolet C irradiation at 0.5 kJâm^{-2} reduces decay without causing damage or affecting postharvest quality of Star Ruby grapefruit (*C. paradisi* Macf.). *Journal of Agricultural and Food Chemistry*, 48, pp. 4571-4575, ISSN 0021-8561

Deveraj, H., Shanmugasundaram, K.R., & Shanmugasundaram, E.R.B. 1982. Neurotoxic effect of patulin. *Indian journal of experimental biology*, 20, pp. 230-231, ISSN:0019-5189

Dixon, R.A. 2001. Natural products and plant disease resistance. *Nature*, 411, pp. 843-847, ISSN 0028-0836

Eckert, J.W., & Ogawa, J.M. 1988. The chemical control of postharvest diseases: deciduous fruits, berries, vegetables and root/tuber crops. *Annual Review of Phytopathology*, 26, pp. 433-469, ISSN 0066-4286

Edney, K.L. 1978. The infection of apples by *Phytophthora syringae*. *Annals of Applied Biology*, 88, pp. 31-36, ISSN 0003-4746

Edney, K.L. 1983. Top fruit. In: *Post-Harvest Pathology of Fruits and Vegetables*, Dennis, C., pp. 43-71, Academic Press, London, UK

Engstrom, K., Widmark, A.K., Brishammar, S., & Helmersson, S. 1999. Antifungal activity to *Phytophthora infestants* of sesquiterpenoids from infected potato tubers. *Potato Research*, 42, pp. 43-50, ISSN 0014-3065

European Commission, 2006. Commission Regulation (EC) no. 1881/2006 setting maximum levels for certain contaminants in foodstuffs. *Official Journal of the European Union*, L, 364, pp.5-24, ISSN 1725-2555

Fallik, E. 2010. Hot Water Treatments of Fruits and Vegetables for Postharvest Storage. In: *Horticultural Reviews* (volume 38), Janick, J., John Wiley & Sons, Inc., Hoboken, NJ, USA, doi: 10.1002/9780470872376.ch5

FAO Statistical Database (FAOSTAT), 2011. Statistics on agriculture, nutrition, fisheries, forestry, food aid, land use and population. Retrieved from: http://faostat.fao.org/

Gonzalez Ureña A., Orea, J.M., Montero, C., & Jimenez, J.B. 2003. Improving postharvest resistance in fruits by external application of trans-resveratrol. *Journal of Agricultural and Food Chemistry*, 51, pp. 82-89, ISSN 0021-8561

Gottstein, D., & Gross, D. 1992. Phytoalexins of woody plants. *Trees*, 6, pp. 55-68, ISSN 1432-2285

Hall, E.G., & Scott, K.J. 1989. Pome fruit. In: *Temperate Fruit*, Beattie, BB., McGlasson, W.B., & Wade, N.L., pp. 7-35, CSIRO Pub., Melbourne, Australia.

Harborne, J.B. 1999. The comparative biochemistry of phytoalexin induction in plants. *Biochemical Systematics and Ecology*, 27, pp. 335-337, ISSN 0305-1978

Ingham, J.L. 1972 Phytoalexins and other natural products as factors in plant disease resistance. *Botanical Review*, 38, pp. 343-424, ISSN 0006-8101

Ippolito, A., El Ghaouth, A., Wilson, C.L., & Wisniewski, M. 2000. Control of postharvest decay of apple fruit by *Aureobasidium pullulans* and induction of defense responses. *Postharvest Biology and Technology*, 19, pp. 265-272, ISSN 0925-5214

Ippolito A., Nigro, F., & Schena, L. 2004. Control of postharvest diseases of fresh fruits and vegetables by preharvest application of antagonistic microorganisms. In: *Crop Management and Postharvest Handling of Horticultural Products, Volume 4: Diseases and Disorders of Fruits and Vegetables*. Dris, R., Niskanen, R., & Jain, S.M., pp. 1- 29, Science Publishers, Inc. Enfield, NH, USA

Ishida, T., Naito, E., Mutoh, J., Takeda, S., Ishii, Y., & Yamada, H. 2005. The plant flavonoid, quercetin, reduces some forms of dioxin toxicity by mechanisms distinct from aryl hydrocarbon receptor activation, heat shock protein induction and quenching oxidative stress. *Journal of Health Science*, 51, pp. 410-417, ISSN 1916-9744

Iwashina, T. 2003. Flavonoid Function and Activity to Plants and Other Organisms. *Biological Sciences in Space*, 17, pp. 24-44, ISSN 1349-967X

Janick, J., Cummings, J.N., & Hemmat, M. 1996. Apples. In: *Fruit Breed, vol I: Tree and tropical Fruit*, Janick, J., & Moore, J.M., Wiley & Sons Inc, Hoboken, NJ, USA, ISBN 978-0-471-31014-3

Janisiewicz, W., Yourman, L., Roitman, J., & Mahoney, N. 1991. Postharvest control of blue mold and gray mold of apples and pears by dip treatment with pyrrolnitrin, a metabolite of *Pseudomonas cepacia*. *Plant Disease*, 75, pp. 490-494, ISSN 0191-2917

Jeandet, P., Bessis, R., Sbaghi, M., & Meunier, P. 1995. Production of the phytoalexin resveratrol by grapes as a response to Botrytis attack under natural conditions. *Journal of Phytopathology*, 143, pp. 135–139, ISSN 1439-0434

Juniper, B.E., & Mabberley, D.J. 2006. *The Story of the Apple*. Timber Press, OR, USA, ISBN 0-88192-784-8

Katsui, N., Matsunaga, A., & Masamune, T. 1974. The structure of lubimin and oxylubimin, antifungal metabolites from diseased potato tubers. *Tetrahedron Letters*, 51/52, pp. 4483-4486, ISSN 0040-4039

Katsui, N., Matsunaga, A., Imaizumi, K., Masamune, T., & Tomiyama, K. 1972. Structure and synthesis of rishitinol, a new sesquiterpene alcohol from diseased potato tubers. *Bulletin of the Chemical Society of Japan*, 45, 2871-2877, ISSN 1348-0634

Kim, J.J., Ben-Yehoshua, S., Shapiro, B., Henis, Y., & Carmeli, S. 1991. Accumulation of scoparone in heat-treated lemon fruit inoculated with *Penicillium digitatum* Sacc. *Plant Physiology*, 97, pp. 880-885, ISSN 1532-2548

Kim, J.H., Yu, J., Mahoney, N., Chan, K.L., Molyneux, R.J., Varga, J., Bhatnagar, D., Cleveland, T.E., Nierman, W.C., & Campbell, B.C. 2008. Elucidation of the functional genomics of antioxidant-based inhibition of aflatoxin biosynthesis. *International Journal of Food Microbiology*, 122, pp. 49–60, ISSN 0168-1605

Kloepper, J.W., Tuzun, S., & Kuc, J.A., 1992. Proposed definitions related to induced disease resistance. *Biocontrol Science and Technology*, 2, pp. 349–351, ISSN 1360-0478

Kuc, J. 1976. Phytoalexins. In: *Encyclopedia of Plant Physiology, New Sen Vol. 4, Physiological Plant Pathology*. Heitefuss, R., & Williams, P.H., pp. 632-652, Springer-Verlag Berlin Heidelberg, New York, USA

Kuc, J. 2000. Development and future direction of induced systemic acquired resistance in plants. *Crop Protection*, 19, pp. 859–861, ISSN 0261-2194

Langcake, P. 1981. Disease resistance of *Vitis* spp. and the production of the stress metabolites resveratrol, ε-viniferin, α-viniferin and pterostilbene. *Physiological and Molecular Plant Pathology*, 18, pp. 213–226, ISSN 0885-5765

Lovett, J., Thompson, R.G., & Boutin B.K. 1975. Patulin production in apples stored in a controlled atmosphere. *Journal of the Association of Official Analytical Chemists*, 58, pp. 912-914, ISSN 0004-5756

Luckey, T.D. 1980. *Hormesis with Ionizing Radiation*. CRC Press, Boca Raton, FL, USA

Mahalingam, R., Gomez-Buitrago, A.M., Eckardt, N., Shah, N., Guevara-Garcia, A., Day, P., Raina, R., & Fedoroff, N.V. 2003, Characterizing the stress/defense transcriptome of Arabidopsis. *Genome Biology*, 4, pp. R20, ISSN 1465-6914

Mari, M., Leoni, O., Iori, R., & Cembali, T. 2002. Antifungal vapour-phase activity of ally-isothiocyanate against *Penicillium expansum* on pears. *Plant Pathology*, 51, 231-236, ISSN 1365-3059

Minamikawa, T., Akazawa, T., & Utitani, I. 1963. Analytical study of umbelliferone and scopoletin synthesis in sweet potato roots infected by *Ceratocystis fimbriata*. *Plant Physiology*, 38, pp. 493-497, ISSN 1532-2548

Morales, H., Sanchis, V., Usall, J., Ramos, A.J., & Marin, S. 2008. Effect of biocontrol agents *Candida sake* and *Pantoea agglomerans* on *Penicillium expansum* growth and patulin accumulation in apples. *International Journal of Food Microbiology*, 122, 1-2, pp. 61-67, ISSN 0168-1605

Mossini, S.A.G., De Oliveira K.P., & Kemmelmeier, K. 2004. Inhibition of patulin production by *Penicillium expansum* cultured with neem (*Azadirachta indica*) leaf extracts. *Journal of Basic Microbiology*, 44, pp. 106-113, ISSN 1521-4028

Muller, K., & Borger, H. 1940. Experimentelle Untersuchungen uber die *Phytophthora* Resistenz der Kartoffel. *Arb Biol Reichsanst Land Forstwirtsch*, 23, pp. 189-231, ISSN 0077-961

Murphy, P.A., Hendrich, S., Landgren, C., & Bryant C.M. 2006. Food Mycotoxins: An Update. *Journal of Food Science* 71 (5), pp. R51-R65, ISSN 0022-1147

Nijveldt, R.J., van Nood, E., van Hoorn, D.E.C., Boelens, P.G., van Norren, K., & van Leeuwen P.A.M. 2001. Flavonoids: a review of probable mechanisms of action and potential applications. *American Journal of Clinical Nutrition*, 74, pp.418–425, ISSN 1938-3207

Okull, D.O., & LaBorde, L.F. 2004. Activity of electrolyzed oxidizing water against *Penicillium expansum* in suspension and on wounded apples *Journal of Food Science*, 69, 23-27, ISSN 0022-1147

Okull, D.O., Demirci, A., Rosenberger, D., LaBorde, L.F. 2006. Susceptibility of *Penicillium expansum* spores to sodium hypochlorite, electrolyzed oxidizing water, and chlorine dioxide solutions modified with nonionic surfactants. *Journal of Food Protection*, 69, 8, pp. 1944-1948, ISSN 0362-028X

Palou, L., Smilanick, J.L., & Droby, S. 2008. Alternatives to conventional fungicides for the control of citrus postharvest green and blue moulds. *Stewart Postharvest Review*, 4, 2, pp. 1-16, ISSN 1745-9656

Paster, N., Huppert, D., & Barkai-Golan R. 1995. Production of patulin by different strains of *Penicillium expansum* in pear and apple cultivars stored at different temperatures and modified atmospheres. Food Additives and Contaminants, 12, 51-58, ISSN 1464-5122.

Paucod, J.C., Krivobok, S., & Vidal, D. 1990. Immunotoxicity testing of mycotoxins T-2 and patulin on Balb/c mice. *Acta Microbiologica et Immunologica Hungarica*, 37, 143-146, ISSN 1588-2640.

Phipps, J.B., Robertson, K.R., Smith, P.G., & Rohrer, J.R. 1990. A checklist of the subfamily Maloideae (Rosaceae). *Canadian Journal of Botany*, 68, pp. 2209–2269, ISSN 1480-3305

Rodov, V., Ben-Yehoshua, S., Kim, J.J., Shapiro, B., & Ittah, Y. 1992. Ultraviolet illumination induces scoparone production in kumquat and orange fruit and improves decay resistance. *Journal of the American Society for Horticultural Science*, 117, pp. 788-792, ISSN 0003-1062

Ryals, J.A., Neuenschwander, U.H., Willits, M.G., Molina, A., Steiner H.Y., & Hunt, M.D.. 1996. Systemic acquired resistance. *The Plant Cell*, 8, pp. 1809–1819, ISSN 1532-298X

Ryu, D., & Holt, D.L. 1993. Growth inhibition of *Penicillium expansum* by several commonly used food ingredients. *Journal of Food Protection*, 56, pp. 862–867, ISSN 0362-028X

Sanzani, S.M., Nigro, F., Mari, M., & Ippolito, A., 2009a. Innovations in the control of postharvest diseases of fresh fruit and vegetables (short paper). *Arab Journal of Plant Protection. Arab Journal of Plant Protection*, 27, 2, pp. 240-244, ISSN 0255-983X

Sanzani, S.M., De Girolamo, A., Schena, L., Solfrizzo, M., Ippolito, A., &Visconti, A. 2009b. Control of *Penicillium expansum* and patulin accumulation on apples by quercetin and umbelliferone. *European Food Research and Technology*, 228(3), pp. 381–389, ISSN 1438-2385

Sanzani, S.M., Schena, L., De Girolamo, A., Nigro, F., & Ippolito, A. 2009c. Effect of quercetin and umbelliferone on the transcript level of *Penicillium expansum* genes involved in patulin biosynthesis. *European Journal of Plant Pathology*, 125, 2, pp. 223-233, ISSN 1573-8469

Sanzani, S.M., Schena, L., De Girolamo, A., Ippolito, A., & González-Candelas, L., 2010. Characterization of genes associated to induced resistance against *Penicillium expansum* in apple fruit treated with quercetin. *Postharvest Biology and Technology*, 56, pp. 1-11, 0925-5214, ISSN 0925-5214

Schena, L., Nigro, F., & Ippolito, A. 2007. Natural antimicrobials to improve storage and shelf-life of fresh fruit, vegetables and cut flowers. In: *Microbial Biotechnology in Horticulture, Vol. 2.*, Ray, R.C., & Ward, O.P., pp. 259-303, Oxford & IBH Publishing Co., New Delhi, India, ISBN 1578084172

Sharma, R.R., Singh, D., Singh, R., 2009. Biological control of postharvest diseases of fruits and vegetables by microbial antagonists: A review. *Biological Control*, 50, 3, pp. 205-221, ISSN 1049-9644

Sholberg, P., Haag, P., Hocking, R., & Bedford, K. 2000. The use of vinegar vapor to reduce postharvest decay of harvested fruit. Horticultural Science, 35, pp. 898–903, ISSN 0862-867X.

Smith, C.J. 1996. Accumulation of phytoalexins: defence mechanism and stimulus response system. New phytologist, 132, pp. 1-45, ISSN 0028-646X

Snowdon, A.L. 1990. Post-Harvest Diseases and Disorders of Fruits and Vegetables, Vol. 1. General Introduction and Fruits, CRC Press, Inc., Boca Raton, FL, USA.

Sommer, N.F., Buchanan, J.R., & Fortlage, R.J. 1974. Production of patulin by Penicillium expansum. Applied and Environmental Microbiology, 28, pp. 589-593, ISSN 1098-5336.

Spadaro, D., Garibaldi, A., & Gullino, M.L. 2004. Control of Penicillium expansum and Botrytis cinerea on apple combining a biocontrol agent with hot water dipping and acibenzolar-S-methyl, baking soda, or ethanol application. Postharvest Biology and Technology, 33, 2, pp. 141-151, ISSN 0925-5214

Spotts, R.A., Cervantes, L.A., & Mielke, E.A. 1999. Variability in postharvest decay among apple cultivars. Plant Disease, 83, pp. 1051-1054, ISSN 0191-2917

Sticher, L., Mauch-Mani, B., & Métraux J.P. 1997. Systemic acquired resistance. Annual Review of Phytopathology, 35, pp. 235–270, ISSN 0066-4286

Stoessl, A., Unwin, C.H., & Ward, E.W.B. 1972. Post-infectional inhibitor from plants. I. Capsidiol, an antifungal compound from Capsicum frutescens. Phytopathology Z., 74, pp. 141-152, ISSN 0031-9481

Sutton, B.C. 1996. A century of mycology. British Mycological Society Century Symposium. Ed. Cambridge University Press, UK, ISBN 9780521050197

Swinburne, T.R. 1973. The resistance of immature Bramley's seedling apples to rotting by Nectria galligena Bres. In: Fungal Pathogenicity and the Plant's Response. Byrde, R.J.W., & Cutting, C.V., pp. 365-382, Academic Press, London, UK, ISBN 0121488500

Swinburne, T.R. 1975. Microbial proteases as elicitors of benzoic acid accumulation in apples. Phytopathology Z., 82, pp.152-162, ISSN 0031-9481

Swinburne, T.R. 1983. Quiescent infections in postharvest diseases. In: Postharvest Pathology of Fruits and Vegetables, Dennis, C., pp. 1-21, Academic Press, London, UK

Tomiyama, K., Sakuma, T., Ishizaka, N., Katsui, N., Takasugi, M., & Masamune, T. 1968. A new antifungal substance isolated from resistant potato tuber tissue infected by pathogens. Phytopathology, 58, pp. 115-116, ISSN 0031-949X

Walker, T.S., Bais, H.P., Halligan, K.M., Stermitz, F.R., & Vivanco, J.M. 2003. Metabolic Profiling of Root Exudates of Arabidopsis. Journal of Agricultural and Food Chemistry, 51, pp. 2548-2554, ISSN 0021-8561

Zhang, H., Wang, L., Ma, L., Dong, Y., Jiang, S., Xu, B., & Zheng, X., 2009. Biocontrol of major postharvest pathogens on apple using Rhodotorula glutinis and its effects on postharvest quality parameters. Biological Control, 48, 1, pp. 79-83, ISSN 1049-9644

7

Inoculation of Sugar Beet Seed with Bacteria *P. fluorescens*, *B. subtilis* and *B. megaterium* – Chemical Fungicides Alternative

Suzana Kristek, Andrija Kristek and Dragana Kocevski
University of J. J. Strossmayer, Faculty of Agriculture
Croatia

1. Introduction

During the process of sugar beet cultivation, depending on weather conditions, the degree of soil infection with pathogenic microorganisms, and the application of soil management, the presence of sugar beet root rot is frequently observed. Pathogens are numerous, inflicting serious damage.

According to the occurrence time of root rot symptoms, the pathogens can be divided into three major groups: pathogens occurring during the emergence phase, pathogens occurring in the growing season, and ones that cause sugar beet root rot during the period of storage.

The most frequent decay pathogens on sugar beet seedlings are the fungi *Aphanomyces cochlioides*, *Pythium ultimum* and *Pythium debarianum*, as well as *Rhizoctonia solani*. In the growing season plant decay is most frequently caused by the fungi *Rhizoctonia solani*, *Fusarium* spp., and by the bacteria *Erwinia* spp. From the root rot pathogens most damage is inflicted by *Rhizoctonia solani*. Besides weather conditions that influence this type of root decay, the inadequate soil management – irregular soil cultivation resulting in damaged soil structure and rise in soil acidity – is a contributory factor.

Apart from the root pathogens occurring during cultivation in the field, certain pathogens cause sugar beet decay in the period of storage, and before processing. Storage conditions, as well as plant damages caused when digging are contributory factors in the occurrence of fungal diseases.

The most frequent pathogenic microorganisms observed are fungi, such as *Rhizoctonia solani*, *Botrytis cinerea*, *Penicillium* spp., *Aspergillus* spp. ... The pathogenic fungus *Rhizoctonia solani*, if compared with all other sugar beet root rot pathogens, is able to infect the plant from the emergence phase, during growing season, until the process of sugar beet storage.

Fungicidal treatment of seeds, use of tolerant hybrids, or proper soil management does not ensure full plant protection against this pathogen. Moreover, heavily infected soils and weather conditions favourable for development of the pathogen, contribute to the infection to higher or lower degree.

Agronomists are faced with a growing problem of evolving resistance of soil pathogens to chemical fungicides. Also, there is an increasing sensitivity to the importance of health food production and rising awareness for environmental protection. Namely, all pesticides which are not biodegraded in the soil are being washed out in deeper layers causing eutrophication of underground waters. Therefore, natural resources and processes in the soil are widely used as an alternative to chemical fungicides.

Seed inoculation with the bacteria showing antagonistic activity against pathogenic fungi such as *Rhizoctonia solani, Pythium ultimum, P. debarianum, Phoma betae* and *Aphanomyces cochlioides* account for an alternative to chemical fungicides and an option of solving problem of disease control, not only in sugar beet, but in other, mainly vegetable crops. Many authors (Koch et al., 2002; Sorensen et al., 2001; Thrane et al., 2000; Whipps, 2001) have studied control of these pathogens by the beneficial bacterium *Pseudomonas fluorescens* which shows pronounced antagonistic activity against the fungi. Similarly, results of their studies proved positive effect of the bacterium *Bacillus megaterium* against the same root decay agents of sugar beet (Asaka & Shoda, 1996; Choudhary & Johri, 2009; Thrane et al., 2000). However, some authors proved antagonistic effect of certain beneficial fungi and yeast (*Trichoderma harzianum, Candida valida, Rhodotula glutinis, ...*) against these pathogenic fungi (Abada, 1994; El-Tarabily & Sivasithamparam, 2006).

Pseudomonas fluorescens belongs to rod-shaped, asporogenous, gram-negative bacteria that as saprophytes prevalent in soil and water belong to a group of soil microorganisms crucial to the process of soil denitrification.

Due to the production of antimicrobial agents, these bacteria also show a distinguished antibiosis against pathogens causing diseases on arable crops, by inactivating their growth and reproduction (Whipps, 2001). For the capacity to synthesize toxic cyclic lipopeptides (Andersen et al., 2003; Koch et al., 2002; Thrane et al., 2000) they are used as effective biological control agents (Sorensen et al., 2001). Many authors have reported that these purified lipopeptides show an antagonistic activity against certain fungi pathogenic on sugar beet roots such as *Rhizoctonia solani* (Andersen et al., 2003; Nielsen et al., 2000, 2002), *Aphanomyces cochlioides* (Raaijmakers et al., 2010; Sorensen et al., 2001), *Pythium ultimum* and *Pythium debarianum* (Andersen et al., 2003; Gorlach-Lira & Stefaniak, 2009; Lee et al., 2000; Nielsen et al., 2000; Thrane et al., 2000). This suggests that bacteria producing lipopeptides could have a potential role in the biocontrol of fungal diseases, which was approved in both laboratory and field trials (Thrane et al., 2000, 2001). Lipopeptides may also function as biosurfactants (Desai et al., 1997) which can facilitate bacterial growth on water – insoluble carbon sources (Koch et al., 1991; Ron & Rosenberg, 2001) or their interaction with hydrophobic surfaces (Neu, 1996), e. g., surface motility (Engelhardt et al., 2009; Lindum et al., 1998).

Bacillus subtilis and *Bacillus megaterium* form symbiotic relationship with the root system of field crops, and like *P. fluorescens* show antagonistic activity against these pathogens. In this study we have investigated synergistic effect of the bacteria in order to achieve as better results as possible.

Bacillus is a genus of Gram-positive rod-shaped bacteria and a member of the division *Firmicutes*. *Bacillus* species can be obligate aerobes or facultative anaerobes. The species *Bacillus subtilis* and *Bacillus megaterium* participate in nitrogen cycle, and in the soil processes of organic matter humification and mineralization of humus, especially in mineralization of phosphorus and potassium converting them into plant-accessible forms. These bacteria belong to biological control agents, i. e. they produce antimicrobial substances that exhibit antagonistic activity against soilborne pathogens, agents of diseases in arable crops (Sorensen et al., 2001).

Therefore, seed inoculation with bacteria showing antagonism against pathogenic fungi is an acceptable alternative to chemical pesticide application (Andersen et al., 2003).

Moreover, considering heavily infected soils with the pathogen R. solani and the fact that beneficial bacteria *Pseudomonas fluorescens*, *B. subtilis* and *B. megaterium* are not sensitive to a low dose fungicide (Pedersen et al., 2002), it is possible to treat the seeds combining low doses of the chemical agents aiming to control growth and reproduction of the pathogenic fungi, which in consequence, produce positive effect on all the parameters of sugar beet yield and quality.

2. Materials and methods

The experiment was set up on two types of the soil: Mollic Gleysols (FAO, 1998) and Eutric Cambisols.

On Mollic Gleysols soil type in the period of 2005 – 2009 pathogenic fungus R. solani (FAO, 1998) was determined, whilst on Eutric Cambisols (Table 1) soil type in the same period P. debarianum was determined.

Investigated properties in a field	Type of soil	
Layer (0 – 0.3 m)	Mollic Gleysols	Eutric Cambisols
pH (H₂O)	7.58	7.33
pH (KCl)	6.90	6.62
Humus (%)	3.40	2.15
P (mg/ 100 g soil)	22.80	20.55
K (mg/ 100 g soil)	23.10	19.08

Table 1. Soil characteristics

In 2007, 2008 and 2009 the experiment was set up by completely randomised block design in 4 repetitions and 8 various seed treating variants: 1. control (untreated seed); 2. Thiram 42-S fungicide treated seed (48% Thiram, 600 ml/100 kg seed); 3. seed inoculated with P. fluorescens No 8569 (1.2 x 10^{10} bacteria/ha); 4. seed inoculated with B. subtilis No 2109 (1.2 x 10^{10} bacteria/ha); 5. seed inoculated with B. megaterium No 2894 (1.2 x 10^{10} bacteria/ha); 6. seed inoculated with B. subtilis No 2109 (0.6 x 10^{10} bacteria/ha) + seed inoculated with B. megaterium No 2894 (0.6 x 10^{10} bacteria/ha); 7. seed inoculated with P. fluorescens No 8569 (0.4 x 10^{10} bacteria/ha) + seed inoculated with B. subtilis No 2109 (0.4 x 10^{10} bacteria/ha) + seed inoculated with B. megaterium No 2894 (0.4 x 10^{10} bacteria/ha); 8. seed inoculated with P. fluorescens No 8569 (0.4 x 10^{10} bacteria/ha) + seed inoculated with B. subtilis No 2109 (0.4 x 10^{10} bacteria/ha) + seed inoculated with B. megaterium No 2894 (0.4 x 10^{10} bacteria/ha) + Thiram 42-S fungicide treated seed (48% Thiram, 200 ml/100 kg seed).

Bacteriological cultures applied in the experiment were from German Collection of Microorganisms and Cell Cultures (DSMZ). *P. fluorescens* No 8569 were transferred from lyophilised into vegetative form, and multiplied on Caso agar (Merck 105458) having following composition: peptone from casein 15.0 g, peptone from soymeal 5.0 g, NaCl 5.0 g, agar 15.0 g, distilled water 1000.0 ml. For *B. megaterium* No 2894 and *B. subtilis* No 2109 nutrient agar of the following composition was used: peptone 5.0 g, meat extract 3.0 g, agar 15.0 g, $MnSO_4$ x H_2O 10.0 mg, distilled water 1000.0 ml.

Month	Precipitation amounts					Air temperatures				
	Necessity	1901-91	2007	2008	2009	Necessity	1901-91	2007	2008	2009
April	40	56	3	50	19	-	11.2	13.3	12.5	12.4
May	50	63	56	67	39	14.2	16.8	18.3	18.1	16.5
June	50	88	33	76	63	18.0	19.4	22.2	21.4	20.3
July	80	66	27	79	12	18.5	21.2	23.9	21.8	23.2
August	65	61	46	46	61	18.2	20.4	22.2	21.8	21.7
September	35	46	65	86	10	14.0	16.8	14.5	15.6	15.6
October	40	56	94	30	55	-	11.1	10.3	13.0	9.1
Total	*360*	*436*	*324*	*434*	*259*	-	-	-	-	-
Average	-	-	-	-	-	-	*16.7*	*17.8*	*17.7*	*17.0*

Table 2. Weather conditions in experimental years and years – long average in Osijek. (Meteorological and hydrological service of Croatia (DHMZ). Agrometeorological bureau in Osijek)

Hybrid of Belinda (KWS) sugar beet was used in the sowing. In the three trial years the sowing was conducted in the second decade of March. The row spacing was 50 cm, with 20 cm within row. Adequate plant stand was obtained with no corrections necessary.

Percentage of the plants infected with pathogenic fungi *R. solani* and *P. debarianum* as well as percentage of decayed plants was stipulated in 2 - 4 true leaves phase. The sugar beet digging conducted in the middle of October was followed by determination of root yield (t/ha), sugar content (%), sugar in molasses (%) and sugar yield (t/ha).

Weather conditions appeared to affect sugar beet growth. Data obtained from the Agrometeorological bureau in Osijek, Croatia (Table 2) were used in weather analysis. Weather conditions in 2007 were characterized by the increase in air temperature of more than 1⁰C above multiyear average. Precipitation rates in the growing season measured 74% out of the average, with rainfall deficiency in June and July, and excessive rainfall in September and October which was unfavourable for sugar beet growth. 2008 and 2009 were also characterized by higher air temperatures in comparison with multiyear average, which was unfavourable for sugar beet growth due to the values significantly (2.5⁰C in 2008; and 3.0⁰C in 2009) above the optimum. Precipitation rates in the growing season 2008 were level with the multiyear average, with rainfall deficiency in August and excessive rainfall in September. Considering precipitation amount in the growing season (April – September), 2009 was dry. Only 59 % out of the average precipitation rates were measured, with rainfall deficiency throughout the year except for June. Such weather conditions were favourable for sugar beet growth until July, and then followed by the unfavourable conditions until the end of the growing season.

Results were processed by modern statistical methods (ANOVA) using the computer program StatSoft Inc. (2001) STATISTICA (data analysis software system), version 6.

3. Results and discussion

3.1 Percentage of infected and decayed plants

During the three - year research into the influence of beneficial bacteria and chemical fungicide Thiram 42-S on the intensity of infection and decay of sugar beet plants caused by the parasitic fungi *Rhizoctonia solani* on Mollic Gleysols, and by the parasitic fungi *Pythium debarianum* on Eutric Cambisols significant influence of the bacteria and the chemical fungicide was determined in all tested variants when compared with the control (Table 3).

On Mollic Gleysols in the three trial years the best results were recorded in the variant 7 - seed inoculated with *P. fluorescens* No 8569 (0.4 x 10^{10} bacteria/ha) + seed inoculated with *B. subtilis* No 2109 (0.4 x 10^{10} bacteria/ha) + seed inoculated with *B. megaterium* No 2894 (0.4 x 10^{10} bacteria/ha). All other variants of treated seeds obtained highly significant larger number (p < 0.01) of infected and decayed plants due to the attack of the parasitic fungi *Rhizoctonia solani* if compared with all other tested variants. Among the variants 2 - Thiram 42-S fungicide treated seed (48% Thiram, 600 ml/100 kg seed), 3 - seed inoculated with *P. fluorescens* No 8569 (1.2 x 10^{10} bacteria/ha) and 8 - seed inoculated with *P. fluorescens* No 8569 (0.4 x 10^{10} bacteria/ha) + seed inoculated with *B. subtilis* No 2109 (0.4 x 10^{10} bacteria/ha) + seed inoculated with *B. megaterium* No 2894 (0.4 x 10^{10} bacteria/ha) + Thiram 42-S fungicide treated seed (48% Thiram, 200 ml/100 kg seed) no significant differences were determined (p > 0.01), though the smallest number of the infected and decayed plants was recorded in the variant 8 in the three trial years. Average number of infected plants in treated variants in the three trial years was 10.86% or 58.17% lower than the three-year average in the control variants which was 25.96%. The difference in decayed plants was 74.70%.

The results obtained are in agreement with those of Whipps (2001) who reported that the plants inoculated with bacteria *P. fluorescens* were characterized with rapid initial growth that enabled fast undergoing through the most vulnerable phase when the damage of the pathogen affecting was the most pronounced. Due to the pronounced antagonism of the bacterium to the root rot agent of sugar beet – fungi *R. solani* high survival percentage of inoculated plants was obtained in comparison with the non-inoculated ones and decrease in the damage of infected plants, which consequently, influenced the sugar beet root yield. Kristek et al. (2007) in their research results of the effects of the beneficial bacterium *P. fluorescens* on the pathogenic fungus *R. solani* recorded obvious decrease in the number of infected and decayed plants in variants treated with the bacterium *P. fluorescens* when compared with the control variants. They also reported significant difference in the number of infected and decayed plants between tolerant hybrids compared with the hybrids sensitive to the pathogenic fungi. With tolerant hybrids in treated variants 4.10 % smaller number of infected plants was obtained than in the same variants of sensitive hybrids. The difference in the number of decayed plants was 3.86 %.

On Eutric Cambisols in the three trial years the best results were recorded in the variant 8 - seed inoculated with *P. fluorescens* No 8569 (0.4 x 10^{10} bacteria/ha) + seed inoculated with *B. subtilis* No 2109 (0.4 x 10^{10} bacteria/ha) + seed inoculated with *B. megaterium* No 2894

(0.4 x 10^{10} bacteria/ha) + Thiram 42-S fungicide treated seed (48% Thiram, 200 ml/100 kg seed), though no significant difference (p > 0.01) was determined between the above variant and variant 2 - Thiram 42-S fungicide treated seed (48% Thiram, 600 ml/100 kg seed). All other variants obtained highly significant (p < 0.01) larger number of infected and decayed plants due to the attack of the parasitic fungi *P. debarianum*. Average number of infected plants in the three trial years in treated variants was 11.70% or 56.16% lower than in the three-year average in the control variants which was 26.69%. The difference in decayed plants was 70.97%.

Investigated parameter	Variants	Mollic Gleysols (*Rhizoctonia solani*)				Eutric Cambisols (*Pythium debarianum*)			
		Year			*Average*	Year			*Average*
		2007	2008	2009		2007	2008	2009	
Infected plants (%)	1	29.46	20.78	27.65	*25.96*	28.46	23.51	28.10	*26.69*
	2	11.39	9.59	10.43	*10.47*	11.29	10.55	10.96	*10.93*
	3	11.35	9.66	10.39	*10.47*	11.71	10.96	11.44	*11.37*
	4	12.97	11.59	12.21	*12.26*	13.18	12.68	12.73	*12.86*
	5	12.85	11.53	12.29	*12.22*	12.99	12.43	12.65	*12.69*
	6	11.14	10.27	10.94	*10.78*	12.35	11.55	11.97	*11.96*
	7	10.25	8.90	9.66	*9.60*	11.64	10.89	11.39	*11.31*
	8	10.99	9.46	10.27	*10.24*	11.13	10.41	10.92	*10.82*
$LSD_{0.05}$		0.24	0.15	0.13	0.16	0.11	0.12	0.10	0.10
$LSD_{0.01}$		0.43	0.27	0.24	0.28	0.21	0.23	0.19	0.18
Decayed plants (%)	1	25.90	18.40	23.66	*22.65*	25.98	21.04	24.18	*23.73*
	2	5.61	5.27	5.28	*5.39*	6.70	5.99	6.24	*6.31*
	3	5.70	5.19	5.33	*5.41*	7.11	6.22	6.57	*6.63*
	4	7.12	6.33	6.90	*6.78*	7.98	7.08	7.85	*7.64*
	5	7.06	6.38	6.84	*6.76*	8.03	7.12	7.78	*7.64*
	6	6.19	5.65	5.92	*5.92*	7.65	6.79	7.20	*7.21*
	7	5.03	4.11	4.80	*4.65*	7.04	6.13	6.48	*6.55*
	8	5.48	4.95	5.11	*5.18*	6.56	5.96	6.16	*6.23*
$LSD_{0.05}$		0.14	0.18	0.13	0.15	0.09	0.08	0.06	0.07
$LSD_{0.01}$		0.26	0.34	0.25	0.27	0.17	0.15	0.13	0.14

1. control (untreated seed); 2. Thiram 42-S fungicide treated seed (48% Tiram, 600 ml/100 kg seed); 3. seed inoculated with *P. fluorescens* No 8569 (1.2 x 10^{10} bacteria/ha); 4. seed inoculated with *B. subtilis* No 2109 (1.2 x 10^{10} bacteria/ha); 5. seed inoculated with *B. megaterium* No 2894 (1.2 x 10^{10} bacteria/ha); 6. seed inoculated with *B. subtilis* No 2109 (0.6 x 10^{10} bacteria/ha) + seed inoculated with *B. megaterium* No 2894 (0.6 x 10^{10} bacteria/ha); 7. seed inoculated with *P. fluorescens* No 8569 (0.4 x 10^{10} bacteria/ha) + seed inoculated with *B. subtilis* No 2109 (0.4 x 10^{10} bacteria/ha) + seed inoculated with *B. megaterium* No 2894 (0.4 x 10^{10} bacteria/ha); 8. seed inoculated with *P. fluorescens* No 8569 (0.4 x 10^{10} bacteria/ha) + seed inoculated with *B. subtilis* No 2109 (0.4 x 10^{10} bacteria/ha) + seed inoculated with *B. megaterium* No 2894 (0.4 x 10^{10} bacteria/ha) + Thiram 42-S fungicide treated seed (48% Tiram, 200 ml/100 kg seed)

Table 3. Percentage of the infected and decayed plants as a consequence of *R. solani* (Mollic Gleysols), *P. debarianum* (Mollic Gleysols) infection in the 2–4 true leaves stage

Similar results in examining the effect of the bacterium *B. megaterium* (a constituent of the applied microbiological preparation) in the control of pathogenic fungi *P. ultimum* and *P. debarianum* were reported in the studies on the same parameters by Evačić et al. (2008).

It is evident that weather conditions in the growing season of sugar beet influenced intensity of the infection by pathogenic fungi *Rhizoctonia solani* and *Pythium debarianum*. Namely, in the second trial year precipitation rates and air temperatures were favourable for crop development which enabled faster undergoing of young sugar beet plants through the most vulnerable phase of the growth. Consequently, intensity of the infection and the number of decayed plants as the result of the pathogen attack were both decreased which was recorded in the controls. Furthermore, there was obvious difference in the number of infected and decayed plants on different types of the soil. As it did not concern the same pathogen, it could be higher sensitivity of the hybrid to *Pythium debarianum* than to *Rhizoctonia solani*. Nevertheless, pathogenic fungi *Rhizoctonia solani* usually cause more economic damage in the middle and eastern European countries than pathogenic fungi *Pythium debarianum*. Belinda hybrid was chosen due to the high sensitivity to the fungi *Rhizoctonia solani*. Therefore chemical and microbiological soil properties could be presumed to have great importance in plant development and resistance against the infection caused by the pathogenic fungi. Namely, Mollic Gleysols has chemical (Table 1) and microbiological (Figure 1, 2, 3, 4, 5) properties of better quality.

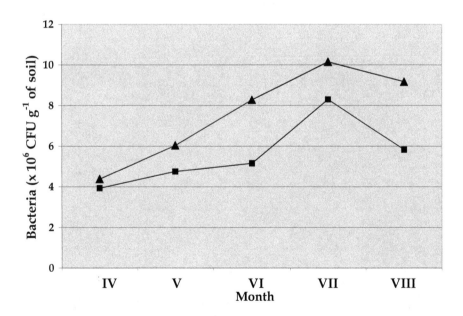

(▲ – Mollic Gleysols; ■ – Eutric Cambisols)

Fig. 1. Average number of total Bacteria in soil during three trial years

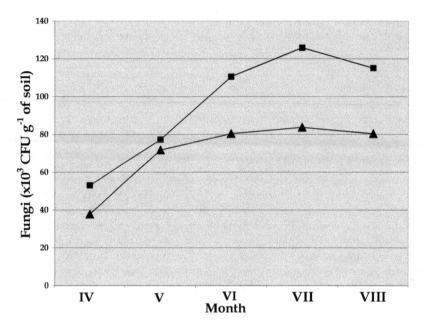

(▲ – Mollic Gleysols; ■ – Eutric Cambisols)

Fig. 2. Average number of total Fungi in soil during three trial years

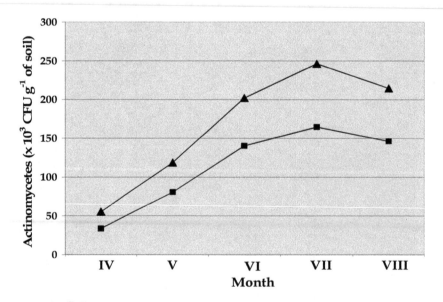

(▲ – Mollic Gleysols; ■ – Eutric Cambisols)

Fig. 3. Average number of total Actinomycetes in soil during three trial years

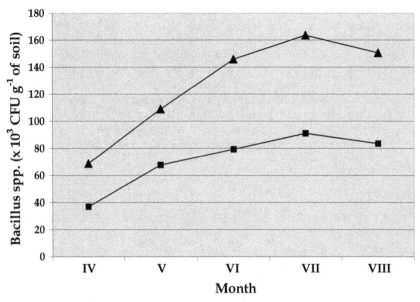

(▲ – Mollic Gleysols; ■ – Eutric Cambisols)

Fig. 4. Average number of *Bacillus* spp. in soil during three trial years

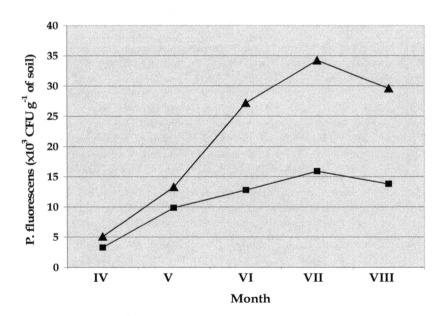

(▲ – Mollic Gleysols; ■ – Eutric Cambisols)

Fig. 5. Average number of *Pseudomonas fluorescens* in soil during three trial years

It is the soil of neutral reaction with high portion of organic matter and large supply of phosphorus and potassium. The soil is abundant in bacteria and actinomycetes that participate in the processes of soil organic matter humification and mineralization of humus. Large portion in microbial mass is comprised of non-symbiotic nitrogen fixers and cellulolytic microorganisms. The presence of the large number of the bacteria of the *Pseudomonas fluorescens* species, as well as *Bacillus* spp., showing antagonistic activity against the soilborne pathogenic fungi is a factor responsible for smaller number of infected plants in the controls if compared with the same on Eutric Cambisols, a soil of poorer chemical and microbiological properties. Average number of infected plants was in highly significant positive correlation with average number of decayed plants (r=0.965; p<0.01).

3.2 Root yield

Sugar beet root yield depended on soil types, or on chemical and microbiological soil properties, and on trial years.

On Mollic Gleysols in the three trial years the highest average sugar beet root yield was recorded in the variant 7 - seed inoculated with *P. fluorescens* No 8569 (0.4×10^{10} bacteria/ha) + seed inoculated with *B. subtilis* No 2109 (0.4×10^{10} bacteria/ha) + seed inoculated with *B. megaterium* No 2894 (0.4×10^{10} bacteria/ha). All other variants obtained statistically highly significant lower average root yield (Table 4). Average sugar beet root yield in treated variants was 80.63 t/ha, or 21.12% higher than average root yield in the control variant (66.57 t/ha). Since the variant with the strains of beneficial bacteria being applied together obtained significantly higher average root yield than the variant with the strains applied one at a time, it can be concluded that there was highly positive synergistic activity between them.

On Eutric Cambisols in the three trial years the highest average sugar beet root yield was recorded in the variant 8 - seed inoculated with *P. fluorescens* No 8569 (0.4×10^{10} bacteria/ha) + seed inoculated with *B. subtilis* No 2109 (0.4×10^{10} bacteria/ha) + seed inoculated with *B. megaterium* No 2894 (0.4×10^{10} bacteria/ha) + Thiram 42-S fungicide treated seed (48% Thiram, 200 ml/100 kg seed), though no significant difference (p > 0.01) was determined between this variant and variant 2 - Thiram 42-S fungicide treated seed (48% Thiram, 600 ml/100 kg seed). All other variants obtained highly significant (p < 0.01) lower sugar beet root yield. Average sugar beet root yield in treated variants was 69.45 t/ha or 21.08 % higher than the average root yield in the control variant (57.02 t/ha).

It is evident that on the soil type of poorer microbiological properties the best results were obtained with the application of chemical fungicides. It is also obvious that between variant 3 with application of the bacteria *P. fluorescens*, and variant 7 with application of

P. fluorescens, *B. subtilis* and *B. megaterium* no significant differences were determined. Therefore, it can be concluded that the bacteria *P. fluorescens* played a crucial role in the control of pathogenic fungi *P. debarianum*.

Significantly better results in sugar beet production by applying plant growth – promoting rhizobacteria *P. fluorescens* in the control of pathogenic fungi *R. solani* were reported by Esh & El-Kholi (2005), as well as by Kiewnick et al. (2001) in the control of fungi *Rhizoctonia crown* by combination of beneficial bacteria.

P. fluorescens shows distinguished biosurfactant properties (Raaijmakers et al., 2010; Ron & Rosenberg, 2001) and improves general plants condition resulting in arable crops yield and quality. As the results of their research studies some authors stated that *P. fluorescens* stimulate nutrients uptake (Duijff et al., 1993; Loper & Buyer, 1991), increases photosynthesis intensity (Zhang et al., 2002) as well as solubility of phosphorus inorganic forms. The aforesaid improves plants vigour and increases yield by 15-20% (Whipps, 2001).

3.3 Sugar content

This parameter was also influenced by weather conditions in the growing season of sugar beet. Namely, higher sugar content in sugar beet root (%) was recorded in the second trial year (Table 4).

On Mollic Gleysols in the three trial years the highest average sugar content in sugar beet was recorded in the variant 7 - seed inoculated with *P. fluorescens* No 8569 (0.4 x 10^{10} bacteria/ha) + seed inoculated with *B. subtilis* No 2109 (0.4 x 10^{10} bacteria/ha) + seed inoculated with *B. megaterium* No 2894 (0.4 x 10^{10} bacteria/ha). All other variants obtained statistically highly significant lower (p<0.01) average sugar content. Average sugar content in sugar beet in treated variants was 15.27% or 7.01% higher than sugar content in the control (14.27%).

No significant differences among the variants 2 - Thiram 42-S fungicide treated seed (48% Thiram, 600 ml/100 kg seed), 3 - seed inoculated with *P. fluorescens* No 8569 (1.2 x 10^{10} bacteria/ha) and 8 - seed inoculated with *P. fluorescens* No 8569 (0.4 x 10^{10} bacteria/ha) + seed inoculated with *B. subtilis* No 2109 (0.4 x 10^{10} bacteria/ha) + seed inoculated with *B. megaterium* No 2894 (0.4 x 10^{10} bacteria/ha) + Thiram 42-S fungicide treated seed (48% Thiram, 200 ml/100 kg seed) were determined. Due to the fact that variant 3, as well as variant 7- the one which obtained the best results, were not treated with a chemical fungicide, it can be left out from application in this case.

Fungicide treated seed on finishing is obviously neither sufficient nor safe measure since the fungi fast develop resistance to these pesticide active agents (Cooke et al., 1999; Durrant et al., 1998). Furthermore, fungicides are exposed to washing out, hence the possibility of underground waters eutrophication, i.e. direct danger for a human health and environment. For its physical characteristics, this beneficial bacterium represents an acceptable alternative to chemical fungicides application (Kristek et al., 2005, 2006; Thrane et al., 2000).

On Eutric Cambisols in the three trial years the highest average sugar content in sugar beet was recorded in the variant 8 - seed inoculated with *P. fluorescens* No 8569 (0.4 x 10^{10} bacteria/ha) + seed inoculated with *B. subtilis* No 2109 (0.4 x 10^{10} bacteria/ha) + seed inoculated with *B. megaterium* No 2894 (0.4 x 10^{10} bacteria/ha) + Thiram 42-S fungicide treated seed (48% Thiram, 200 ml/100 kg seed). Between this variant and variant 2 - Thiram 42-S fungicide treated seed (48% Thiram, 600 ml/100 kg seed) no significant difference was determined. High sugar content in sugar beet was obtained in variant 3 - seed inoculated with *P. fluorescens* No 8569 (1.2 x 10^{10} bacteria/ha) and variant 7 - seed inoculated with *P. fluorescens* No 8569 (0.4 x 10^{10} bacteria/ha) + seed inoculated with *B. subtilis* No 2109 (0.4 x 10^{10} bacteria/ha) + seed inoculated with *B. megaterium* No 2894 (0.4 x 10^{10} bacteria/ha) with no statistical significance determined. Average sugar content in sugar beet in treated variants was 14.44 % or 4.94 % higher than sugar content in the control (13.76 %).

126   Encyclopedia of Fungicides: Benefits and Drawbacks

Similar results were reported in the studies of Esh & El-Kholi (2005), Kristek et al. (2006, 2007) and Pedersen et al. (2002).

Investigated parameter	Variants	Mollic Gleysols (*Rhizoctonia solani*)				Eutric Cambisols (*Pythium debarianum*)			
		Year			*Average*	Year			*Average*
		2007	2008	2009		2007	2008	2009	
Root yield (t/ha)	1	63.40	70.21	66.10	*66.57*	50.65	68.30	52.10	*57.02*
	2	73.91	95.50	78.92	*82.77*	65.97	90.24	70.66	*75.62*
	3	72.96	94.63	79.45	*82.35*	58.79	87.10	64.98	*70.29*
	4	64.28	89.45	70.55	*74.76*	53.20	75.45	59.75	*62.80*
	5	65.96	88.16	69.18	*74.43*	54.88	76.21	60.30	*63.80*
	6	68.50	93.10	74.80	*78.80*	56.90	80.20	62.50	*66.53*
	7	78.10	98.35	86.54	*87.66*	60.25	86.47	65.10	*70.61*
	8	74.13	96.45	80.24	*83.61*	66.80	90.55	72.18	*76.51*
$LSD_{0.05}$		0.75	0.99	0.81	0.84	0.84	0.72	0.97	0.68
$LSD_{0.01}$		1.42	1.86	1.51	1.48	1.56	1.39	1.83	1.22
Sugar content (%)	1	14.08	14.53	14.21	*14.27*	13.46	14.28	13.55	*13.76*
	2	14.73	16.01	15.65	*15.46*	14.27	15.71	14.51	*14.83*
	3	14.68	15.84	15.59	*15.37*	13.99	15.29	14.09	*14.46*
	4	14.31	15.26	14.85	*14.81*	13.70	14.68	13.82	*14.07*
	5	14.37	15.32	14.62	*14.77*	13.69	14.70	13.90	*14.10*
	6	14.50	15.55	15.12	*15.06*	13.85	15.05	13.87	*14.26*
	7	15.26	16.45	16.08	*15.93*	13.82	15.36	14.11	*14.43*
	8	14.75	16.08	15.73	*15.52*	14.30	15.90	14.59	*14.93*
$LSD_{0.05}$		0.08	0.17	0.14	0.11	0.11	0.17	0.20	0.08
$LSD_{0.01}$		0.14	0.31	0.25	0.20	0.20	0.29	0.31	0.14

1. control (untreated seed); 2. Thiram 42-S fungicide treated seed (48% Tiram, 600 ml/100 kg seed); 3. seed inoculated with *P. fluorescens* No 8569 (1.2×10^{10} bacteria/ha); 4. seed inoculated with *B. subtilis* No 2109 (1.2×10^{10} bacteria/ha); 5. seed inoculated with *B. megaterium* No 2894 (1.2×10^{10} bacteria/ha); 6. seed inoculated with *B. subtilis* No 2109 (0.6×10^{10} bacteria/ha) + seed inoculated with *B. megaterium* No 2894 (0.6×10^{10} bacteria/ha); 7. seed inoculated with *P. fluorescens* No 8569 (0.4×10^{10} bacteria/ha) + seed inoculated with *B. subtilis* No 2109 (0.4×10^{10} bacteria/ha) + seed inoculated with *B. megaterium* No 2894 (0.4×10^{10} bacteria/ha); 8. seed inoculated with *P. fluorescens* No 8569 (0.4×10^{10} bacteria/ha) + seed inoculated with *B. subtilis* No 2109 (0.4×10^{10} bacteria/ha) + seed inoculated with *B. megaterium* No 2894 (0.4×10^{10} bacteria/ha) + Thiram 42-S fungicide treated seed (48% Tiram, 200 ml/100 kg seed)

Table 4. Root yield (t/ha) and sugar content (%) on two soil types during three trial years

3.4 Sugar in molasses

Sugar content in molasses depended, respectively, on seed treated variant, hybrid properties (tolerant, sensitive), soil type tested, and trial year.

The lowest average sugar content in molasses in the three year trial (Table 5) was recorded on Mollic Gleysols in the variant 7 - seed inoculated with *P. fluorescens* No 8569 (0.4 x 10^{10} bacteria/ha) + seed inoculated with *B. subtilis* No 2109 (0.4 x 10^{10} bacteria/ha) + seed inoculated with *B. megaterium* No 2894 (0.4 x 10^{10} bacteria/ha). Average sugar content in molasses in treated variants was 2.60% or 14.75% lower than the average sugar content in molasses in the control variants (3.05%).

Investigated parameter	Variants	Mollic Gleysols (*Rhizoctonia solani*)				Eutric Cambisols (*Pythium debarianum*)			
		Year			*Average*	Year			*Average*
		2007	2008	2009		2007	2008	2009	
Sugar in molassess (%)	1	3.15	2.95	3.04	*3.05*	4.26	3.58	4.06	*3.97*
	2	2.64	2.37	2.59	*2.53*	3.41	2.75	3.22	*3.13*
	3	2.62	2.38	2.61	*2.54*	3.69	2.99	3.44	*3.37*
	4	2.92	2.65	2.82	*2.80*	3.81	3.28	3.80	*3.63*
	5	2.85	2.61	2.88	*2.78*	3.80	3.36	3.80	*3.65*
	6	2.78	2.42	2.74	*2.65*	3.87	3.12	3.60	*3.53*
	7	2.54	2.26	2.38	*2.40*	3.66	2.94	3.38	*3.33*
	8	2.65	2.35	2.55	*2.52*	3.38	2.75	3.15	*3.09*
LSD$_{0.05}$		0.028	0.039	0.050	0.025	0.034	0.030	0.043	0.037
LSD$_{0.01}$		0.051	0.072	0.086	0.043	0.062	0.058	0.081	0.066
Sugar yield (t/ha)	1	6.92	8.13	7.38	*7.48*	4.66	7.30	4.94	*5.63*
	2	8.93	13.03	10.30	*10.75*	7.16	11.69	7.98	*8.94*
	3	8.79	12.73	10.31	*10.61*	6.06	10.71	6.92	*7.90*
	4	7.32	11.27	8.48	*9.02*	5.26	8.60	5.99	*6.62*
	5	7.59	11.20	8.28	*9.02*	5.43	8.64	6.09	*6.72*
	6	8.03	12.22	9.26	*9.84*	5.66	9.57	6.42	*7.22*
	7	9.93	13.95	11.86	*11.90*	6.12	10.74	6.99	*7.95*
	8	9.22	13.24	10.57	*11.01*	7.29	11.90	8.26	*9.15*
LSD$_{0.05}$		0.28	0.30	0.21	0.24	0.13	0.16	0.17	0.15
LSD$_{0.01}$		0.50	0.56	0.38	0.46	0.24	0.29	0.31	0.27

1. control (untreated seed); 2. Thiram 42-S fungicide treated seed (48% Tiram, 600 ml/100 kg seed); 3. seed inoculated with *P. fluorescens* No 8569 (1.2 x 10^{10} bacteria/ha); 4. seed inoculated with *B. subtilis* No 2109 (1.2 x 10^{10} bacteria/ha); 5. seed inoculated with *B. megaterium* No 2894 (1.2 x 10^{10} bacteria/ha); 6. seed inoculated with *B. subtilis* No 2109 (0.6 x 10^{10} bacteria/ha) + seed inoculated with *B. megaterium* No 2894 (0.6 x 10^{10} bacteria/ha); 7. seed inoculated with *P. fluorescens* No 8569 (0.4 x 10^{10} bacteria/ha) + seed inoculated with *B. subtilis* No 2109 (0.4 x 10^{10} bacteria/ha) + seed inoculated with *B. megaterium* No 2894 (0.4 x 10^{10} bacteria/ha); 8. seed inoculated with *P. fluorescens* No 8569 (0.4 x 10^{10} bacteria/ha) + seed inoculated with *B. subtilis* No 2109 (0.4 x 10^{10} bacteria/ha) + seed inoculated with *B. megaterium* No 2894 (0.4 x 10^{10} bacteria/ha) + Thiram 42-S fungicide treated seed (48% Tiram, 200 ml/100 kg seed)

Table 5. Sugar in molasses (%) and sugar yield (t/ha) on two soil types during three trial years

By drawing comparison among the variants with the strains of the beneficial bacteria applied one at a time, the best results in this parameter as well were recorded with the bacteria *P. fluorescens*. The variants comprising the bacteria *B. subtilis* and *B. megaterium*, being applied one at a time, or together, obtained highly significant (p<0.01) lower values in all tested parameters.

This is in agreement with the results of Lifshitz et al. (1987) and Whipps (2001) who stated that *P. fluorescens* as plant growth – promoting bacteria increased the solubility of phosphorus in inorganic forms around the active root zone. By mobilizing soil phosphorus plant vigour improves and yield increases by 15–20% (great root availability reduce negative nitrogen effect on sugar beet achieving balanced plant nutrition and reduced production of alpha - amino nitrogen, potassium and sodium i.e. decrease in the amount of sugar which cannot be isolated in the production process but turns into molasses).

The lowest average sugar content in molasses in the three year trial was recorded on Eutric Cambisols with the variant 8 - seed inoculated with *P. fluorescens* No 8569 (0.4 x 10^{10} bacteria/ha) + seed inoculated with *B. subtilis* No 2109 (0.4 x 10^{10} bacteria/ha) + seed inoculated with *B. megaterium* No 2894 (0.4 x 10^{10} bacteria/ha) + Thiram 42-S fungicide treated seed (48% Thiram, 200 ml/100 kg seed). Between this variant and variant 2 - Thiram 42-S fungicide treated seed (48% Thiram, 600 ml/100 kg seed) no statistical significance (p>0.05) was determined. All other variants obtained highly significant (p<0.01) higher values of sugar in molasses. Average sugar content in molasses in treated variants was 3.39% or 14.61% lower than average sugar content in molasses in the control variants (3.97%).

3.5 Sugar yield

On Mollic Gleysols the highest average sugar yield in the three trial years was recorded in the variant 7 - seed inoculated with *P. fluorescens* No 8569 (0.4 x 10^{10} bacteria/ha) + seed inoculated with *B. subtilis* No 2109 (0.4 x 10^{10} bacteria/ha) + seed inoculated with *B. megaterium* No 2894 (0.4 x 10^{10} bacteria/ha). All other variants obtained statistically highly significant lower average sugar yield. Average sugar yield in treated variants was 10.31 t/ha or 37.83% higher than average sugar yield in the control (7.48 t/ha). It is evident that in this parameter as well, the best results were obtained in the second year of the trial (Table 5).

On Eutric Cambisols the highest average sugar yield in the three trial years was recorded in the variant 8 - seed inoculated with *P. fluorescens* No 8569 (0.4 x 10^{10} bacteria/ha) + seed inoculated with *B. subtilis* No 2109 (0.4 x 10^{10} bacteria/ha) + seed inoculated with *B. megaterium* No 2894 (0.4 x 10^{10} bacteria/ha) + Thiram 42-S fungicide treated seed (48% Thiram, 200 ml/100 kg seed), though no statistical significant difference (p > 0.05) was determined between the above variant and variant 2 - Thiram 42-S fungicide treated seed (48% Thiram, 600 ml/100 kg seed). All other variants obtained highly significant (p < 0.01) larger number of infected and decayed plants. Average sugar yield in treated variants was 7.78 t/ha or 38.18% higher than average sugar yield in the control variant (5.63 t/ha).

Sugar yield was in very significant positive correlation with root yield (r=0.975; p<0.01) and sugar content (r=0.946; p<0.01).

4. Conclusion

Inoculation of sugar beet seed with the bacteria *Pseudomonas fluorescens, Bacillus subtilis* and *Bacillus megaterium* affected root decay agents *Rhizoctonia solani* and *Pythium debarianum* and showed highly significant influence on all the elements of sugar beet yield and quality in the three trial years.

The best results in all elements of the research on both soil types were recorded in the second trial year (2008) due to the more favourable weather conditions (air temperatures, precipitation values) in the growing season 2008 than in the dry 2007 and 2009.

Better results in all tested parameters in the three trial years were obtained on Mollic Gleysols. The reason is in chemical and microbiological properties of much better quality than on Eutric Cambisols. In Mollic Gleysols total number of the bacteria and actinomycetes is higher enabling more intensive humification of organic matter and mineralization of humus in the plant-accessible mineral compounds. Significantly higher number of beneficial bacteria in the soil (*P. fluorescens, Bacillus* spp.) that show antagonism against soilborne pathogens is evident.

On Mollic Gleysols the best results in all tested parameters were recorded in the variant treated with the beneficial bacteria (*P. fluorescens, B. subtilis, B. megaterium*), whereas all other variants obtained significantly (p<0.01) lower values. Due to the beneficial bacteria in the soil, population of pathogenic microorganisms was reduced, making application of the beneficial bacteria in the sowing period sufficient to reduce or control the infection with pathogenic fungi *R. solani*. Among the beneficial bacteria *P. fluorescens, B. subtilis* and *B. megaterium* highly positive synergism is evident.

On Eutric Cambisols the best results in all tested parameters were reached in the variant treated with chemical fungicide and the beneficial bacteria (*P. fluorescens, B. subtilis, B. megaterium*). Nevertheless, in most cases no statistically significant difference (p>0.05) was determined between this variant and the variant being treated solely with chemical fungicide. The reason is in poorer quality of chemical and microbiological soil properties which is favourable for reproduction of pathogenic fungi *P. debarianum* in the soil.

If necessary measures of improving chemical soil properties are to be carried out, neutralization of soil acidity in the first place, microbiological soil properties will be improved, and the beneficial soil microorganisms will increase in number. Due to the antagonism of beneficial microorganisms against soilborne plant pathogens the number of pathogenic microorganisms will be reduced as well as the infection of arable crops. In this case the application of beneficial microorganisms will reach satisfactory level without application of chemical pesticides. By introducing beneficial microorganisms microbiological soil properties will be consequently improved and number of pathogens in the forthcoming growing season will be decreasing. The reduction in the application of chemical pesticides is the matter of great importance from the both economic and ecological point of view.

5. Acknowledgment

I would like to show my gratitude to Ministry of Science, Education and Sports of the Republic of Croatia for financial support of our project research (079-0791843-1933). Results presented in this chapter were obtained during this project research.

6. References

Andersen, J.B.; Koch, B.; Nielsen, T.H.; Sørensen, D.; Hansen, M.; Nybroe, O.; Christophersen, C.; Sørensen, J.; Molin, S. & Giskov, M. (2003). Surface motility in *Pseudomonas sp.* DSS73 is required for efficient biological containment of the root – pathogenic microfungi *Rhizoctonia solani* and *Pythium ultimum*. *Microbiology*, Vol.149, No. 1, (January 2003), pp. 37-46, ISSN 1350-0872

Asaka, O. & Shoda, M. (1996). Biocontrol of *Rhizoctonia solani* damping-off of tomato with *Bacillus subtilis* RB14. *Applied and Environmental Microbiology*, Vol.62, No.11, (November 1996), pp. 4081-4085, ISSN 0099-2240

Choudhary, D.K. & Johri, B.N. (2009). Interactions of *Bacillus* spp. and plants-with special reference to induced systemic resistance (ISR). *Mirobiological research*, Vol.164, No.5, (September 2009), pp. 493-513, ISSN 0944-5013

Cooke, D.A.; Dewar, A.M. & Asher, M.J.C. (1989). Pests and diseases of sugar beet, In: *Pest and Disease Control Handbook*, N. Scopes & L. Stables, (Ed.), 241-259, British Crop Protection Council, ISBN 0-948404-28-0, Thornton Heath, UK

Desai, J.D. & Banat, I. M. (1997). Microbial production of surfactants and their commercial potential. *Micr. Mol. Biol. Rev.* Vol.61, No.1, (March 1997), pp. 47-64, ISSN 1092-2172

Duijff, B.J.; Meijer, J.W.; Bakker, P.A.H.M. & Schippers, B. (1993). Siderophore mediated competition for iron and induced disease resistance in the suppression of *Fusarium wilt* of carnation by fluorescent *Pseudomonas* spp. *Neth. J. Plant Pathol.*, Vol.99, No.5-6, (September 1993), pp. 277-289, ISSN 0028-2944

Durrant, M.J.; Payne, P.A.; Prince, J.W.F. & Fletcher, R. (1988). Thiram steep seed treatment to control *Phoma betae* and improve the establishment of the sugar beet plant stand. *Crop protection*, Vol.7, No.5, (October 1988), pp. 319-326, ISSN 0261-2194

El-Tarabily, K.A. & Sivasithamparam, K. (2006). Potential of yeasts as biocontrol agents of soil-borne fungal plant pathogens and as plant growth promoters. *Mycoscience*, Vol.47, No. 1, (February 2006), pp. 25-35, ISSN 1340-3540

Engelhardt, S.; Lee, J.; Gäbler, Y.; Kemmerling, B.; Haapalainen, M.L.; Li, C.M.; Wei, Z.; Keller, H.; Joosten, M. & Taira, S. (2009). Separable roles of the *Pseudomonas syringae* pv. *phaseolicola* accessory protein HrpZ1 in ion-conducting pore formation and activation of plant immunity. *Plant J.*, Vol.57, No.4, (February 2009), pp. 706-717, ISSN 0960-7412

Esh, A.M.H. & El-Kholi, M.M. (2005). Effect of *Pseudomonas fluorescens* extra cellular enzymes and secondary metabolites on *Rhizoctonia solani* the causal of sugar beet damping off disease. *Zagazig Journal of Agricultural Research*, Vol.32, pp. 1537-1557, ISSN 1110-0338

Evačić, M.; Kristek, S. & Stanisavljević, A. (2008). Use of the bacteria *Bacillus megaterium* in the control of sugar beet root decay agents – *Pythium ultimum* and *Pythium debarianum*. *Cereal Resear.Comm.*, Vol.36, Suppl.5, Part1, pp. 383-386, ISSN 0133-3720

FAO. World Reference Base for Soil Resources, (1998). FAO Roma

Gorlach-Lira, K. & Stefaniak, O. (2009): Antagonistic activity of bacteria isolated from crops cultivated in a rotation system and a monoculture against *Pythium debaryanum* and *Fusarium oxysporum*. *Folia microb.*, Vol.54, No.5, (July 2009), pp. 447-450, ISSN 0015-5632

Kiewnick, S.; Barry, J.J.; Braun-Kiewnick, A.; Eckhoff, L.A. & Bergman, J.W. (2001). Integrated control of *Rhizoctonia crown* and root rot of sugar beet with fungicides and antagonistic bacteria. *Plant Disease*, Vol.85, No.7, (July 2001), pp. 718-722, ISSN 0191-2917

Koch, A.K.; Kappeli, O.; Fiechter, A. & Reiser, J. (1991). Hydrocarbon assimilation and biosurfactant production in *Pseudomonas aeruginosa* mutants. *Journal of Bacteriology*, Vol.173, No.13, (July 1991), pp. 4212-4219, ISSN 0021-9193

Koch, B.; Nielsen, T.H.; Sørensen, D.; Andersen, J.B.; Christophersen, C.; Molin, S.; Giskov, M.; Sørensen, J. & Nybroe, O. (2002). Lipopeptide production in *Pseudomonas* sp. strain Dss73 is regulated by components of sugar beet seed exudates via the Gac two – component regulatory system. *Applied and Environmental Microbiology*, Vol.68, No.9, (September 2002), pp. 4509-4516, ISSN 0099-2240

Kristek, S. & Kristek, A. (2005). Inoculation of sugar beet seed by the bacterium *Pseudomonas fluorescens* - chemical fungicides alternative. *Listy Cukrovarnicke a Reparske*, Vol.121, No.4, pp. 49-53, ISSN 0024-4449

Kristek, S.; Kristek, A.; Guberac, V. & Stanisavljević, A. (2006). Effect of bacterium *Pseudomonas fluorescens* and low fungicide dose seed treatments on parasite fungus *Aphanomyces cochlioides* and sugar beet yield and quality. *Plant, soil and environment*, Vol.52, No.7, (July 2006), pp. 314-320, ISSN 1214-1178

Kristek, S.; Kristek, A.; Pospišil, M.; Glavaš Tokić, R. & Ćosić, J. (2007). Influence of bacterium *Pseudomonas fluorescens* on the pathogen of root rot *Rhizoctonia solani*, storage period and elements of sugarbeet yield and quality. *Zuckerindustrie*, Vol.132, No.7, (July 2007), pp. 568-575, ISSN 0344-8657

Lee, C.H.; Kempf, H.J.; Lim, Y. & Cho, H. (2000). Biocontrol activity of *Pseudomonas cepacia* AF2001 and anthelmintic activity of its novel metabolite, cepacidine A. *Journal of Microb. and Biotechnology*, Vol.10, No.4, (August 2000), pp. 568-571, ISSN 1017-7825

Lifshitz, R.; Kloepper, J.W. & Kozlowski, M. (1987). Growth promotion of canola (repeseed) seedlings by a strain of *Pseudomonas putida* under gnotobiotic conditions. *Can. J. Microbiol.*, Vol.33, No.5, (May 1987), pp. 390-395, ISSN 0008-4166

Lindum, P.W., Anthoni, U., Christophersen, C., Eberl, L., Molin, S. & Givskov, M. (1998). N – Acyl – L – homoserine lactone autoinducers control production of an extracellular lipopeptide biosurfactant required for swarming motility of *Serratia liquefaciens* MG1. *J. Bacteriol*, Vol. 180, No. 23, pp. 6384-6388, ISSN 0021-9193

Loper, J.E. & Buyer, J.S. (1991). Siderophores in microbial interactions on plant surfaces. *Mol. Plant - Microbe Interact.*, Vol.4, No.1, (Jan.-February 1991), pp. 5-13, ISSN 0894-0282

Neu, T.R. (1996). Significance of bacterial surface – active compounds in interaction of bacteria with interfaces. *Microbiol. Rev.*, Vol.60, No.1, (March 1996), pp. 151-166, ISSN 0146-0749

Nielsen, T.H.; Thrane, C.; Christophersen, C.; Anthoni, U. & Sørensen, J. (2000). Structure, production characteristics and fungal antagonism of tensin – a new antifungal cyclic lipopeptide from *Pseudomonas fluorescens* strain 96.578. *Journal of Applied Microbiology*, Vol.89, No.6, (December 2000), pp. 992-1001, ISSN 1364-5072

Nielsen, T.H.; Sørensen, D.; Tobiasen, C.; Andersen, J.B.; Christophersen, C.; Giskov, M. & Sørensen, J. (2002). Antibiotic and biosurfactant properties of cyclic lipopeptides produced by fluorescent *Pseudomonas* spp. from the sugar beet rhizosphere. *Applied and Envir. Microbiology*, Vol.68, No.7, (July 2002), pp. 3416-3423, ISSN 0099-2240

Pedersen, H.C.; Weiergang, I.; Pontoppidan, M.M.; Jørgensen, L. & Svingel, A. (2002). Danisco Seed, Seed Technology Dept., Hoejbygaardvej 31, DK – 4960 Holeby, Denmark

Raaijmakers, J.M.; De Brujin, I.; Nybroe, O. & Ongena, M. (2010). Natural functions of lipopeptides from *Bacillus* and *Pseudomonas*: more than surfactants and antibiotics. FEMS *Microb. Reviews*, Vol.34, No.6, pp.1037-1062, ISSN 1574-6976

Ron, E.Z. & Rosenberg, E. (2001). Natural roles of biosurfactants. *Environmental Microbiology*, Vol.3, No.4, (April 2001), pp. 229-236, ISSN 1462-2912

Sørensen, D.; Nielsen, T.H.; Christophersen, C.; Sørensen, J. & Gajhede, M. (2001). Cyclic lipoundecapeptide amphisin from *Pseudomonas* sp. Strain Dss73. *Acta Crystallographica Section C Crystal Structure Communications*, Vol.57, No.9, (September 2001), pp. 1123 – 1124, ISSN 0108-2701

Thrane, C.; Nielsen, T.H.; Nielsen, M.N.; Sørensen, J. & Olsson, S. (2000). Viscosinamide – producing *Pseudomonas fluorescens* DR54 exerts a biocontrol effect on *Pythium ultimum* in sugar beet rhizosphere. *FEMS Microbiology Ecology*, Vol.33, No.2, (August 2000), pp. 139-146, ISSN 0168-6496

Zhang, S.; Reddy, M.S. & Kloepper, J.W. (2002). Development of assays for assessing induced systemic resistance by plant growth-promoting rhizobacteria against blue mold of tobacco. *Biological Control*, Vol.23, No.1, (January 2002), pp. 79-86, ISSN 1049-9644

Whipps, J.M. (2001). Microbial interactions and biocontrol in the rhizosphere. *Journal of Experimental Botany*, Vol.52, Suppl.1, (March 2001), pp. 487-511, ISSN 0022-0957

www.statsoft.com

8

Combined Effects of Fungicides and Thermotherapy on Post-Harvest Quality of Horticultural Commodities

Maurizio Mulas
University of Sassari
Italy

1. Introduction

Production, market and consumption of fresh fruit and vegetables (Fig. 1) are continuously increasing in the world (FAOSTAT, 2011). This increase is not only linked to the growth of the world population and to the consequent demand of food but also to the continuous improving of the food quality in the emerging countries. Immediately after the satisfaction of the energy needs, in fact, more lipids are introduced in the diet with the aim of improving the organoleptic quality of the food. Subsequently, a wide variety of ingredients, including fresh fruit and vegetables are requested to integrate the daily diet. Cereals are absolutely fundamental as they provide energy, proteins, lipids and minerals. However, fibre, minerals, vitamins, sugars, acids, aminoacids, and other compounds, like polyphenols and other antioxidants of particular importance because of their nutritional and nutraceutical value are provided by fresh fruits and vegetables (Schreiner & Huyskens-Keil, 2006).

In spite of the large variability of horticultural commodities marketed today in the world, not all the regions of the globe showed possibilities to give high yields of many crops. Often geographical reasons made impossible the production of any commodity, like in the extreme latitude North and South, or in arid and desert areas. Technology advances may contribute to extend the growing area of some crops, but not always there is a real economic convenience in this effort. Furthermore, in other lands poverty and lack of professional abilities make difficult cultivation even when the natural environment is potentially favourable. For these reasons the international market of horticultural commodities is very big and diversified. The most developed countries manage the highest part of exchanges, while a part of producer countries showed an increasing activity with the aim of taking a part of the international market management due to quality of commodities and competitive prices. Some of these "emerging" countries have also an internal market absolutely wide, and an increasing demand which is the market base to export commodities of excellent and selected quality. This is the case of Brazil, South Africa, Mexico, India, and China. Some other advanced countries have in the export of horticultural commodities a point of excellence of their economy, like Spain, Italy, Turkey, Israel, and Greece. Moreover, other "emerging" countries are small producers and have a small internal market, but have found a good specialization in the export of high value horticultural commodities thus obtaining an interesting added value from this activity: like Cyprus, Chile, Cuba, Costa Rica and Morocco.

The big quantity of fresh horticultural commodities that are the object of import/export on the international markets, but also on the big internal markets, is characterized by quality and nutritional properties that should be maintained until the final consumer (Kader, 2003). Many difficulties operate against this objective. The first is the physiological weight loss of horticultural commodities determined by both water loss, due to transpiration of living plant tissues, and by dry matter loss due to product respiration. This weight and value loss may be comprised between 10 and 20%. Other losses are determined by some physiological disorders like the chilling injury which can occur during and after the cold storage. The cold storage is often a need in order to preserve the product quality by minimizing the physiological losses and the mould decay. The losses for chilling injury are comprised between 10 and 15%, mainly for tropical and subtropical fruits and vegetables. Finally, the biggest losses are caused by specific postharvest diseases. Fungi and other microorganisms may be the responsible of 15-25% decay of horticulture commodities during storage and transport (Barkai-Golan, 2001).

Since the first experiences of the fruit and vegetable postharvest export industry, fungicide treatments have been applied to horticultural commodities in order to reduce mould decay in the best effective way. However, approved fungicides for postharvest treatments have been always a small number of compounds because of the short period of tolerance between application and product consumption. Severity in the control of fungicides residues is also usual in technological advanced importing countries. Moreover, the amounts of fungicides applied to horticultural commodities during postharvest process are small respect to the applied on the field, and the cost for the authorization of a new product is very high, thus determining low interest in the development of new fungicides by the specialized chemical industries (Narayanasamy, 2006).

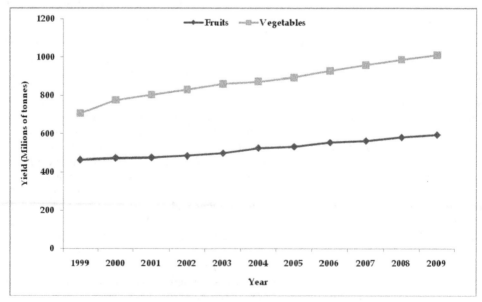

Fig. 1. World yield of fruits and vegetables in the last decade (data source: FAOSTAT 2011).

As a consequence of the repeated and exclusive use of the few admitted fungicides for postharvest treatments of fresh horticultural commodities in a very limited space like the processing and storage packing houses, many resistant strains for these fungicides of the main pathogens have been widely detected in the last decades.

The urgency to find alternative technologies of control against postharvest pathogens of horticultural commodities is clearly evidenced and sustained by the following arguments:

- the growth of the horticultural commodity quantity marketed and submitted to import/export movements;
- the increased risk to spread dangerous pests and pathogens with the fruits and vegetables shipped all over the world;
- the detection and isolation every day more frequent of strains of the main postharvest pathogens resistant to the commonly used fungicides;
- the claimed exigency of consumers and of importing countries to have foods free from any chemical compound dangerous for the human safety;
- the need to have more fungicides as possible alternatives in postharvest treatments;
- the possibility to increase fungicide effectiveness by combination with thermotherapy and with other friendly chemical or physical means generally recognized as safe (GRAS).

2. The thermotherapy of horticultural commodities

The use of the heat to control some postharvest citrus pathogens was experimented in the first decades of the past century (Mulas & Schirra, 2007). The first experiment cited in the literature was the postharvest use of hot water and hot sodium hydroxide solutions to control citrus moulds. Subsequently, the use of the thermotherapy was also widely applied against parasites of horticultural commodities (Barkai-Golan & Phillips, 1991; Lurie, 1998; Schirra et al., 2000a; Shellie & Robert, 2000; Fallik, 2004; Fallik & Lurie, 2007).

In the 70th years a wide availability and diffusion of fungicides made less rentable the use of heat in post-harvest treatments which always were linked to supplementary energy costs (Schirra, 2005). However, it was just in the years of widest diffusion of synthetic fungicides that the advantages to combine postharvest treatments with warming of active solutions were observed. This improved efficiency was associated to a better distribution, penetration and cuticle diffusion of fungicides in the commodity tissues (Cabras et al., 1999; Schirra et al., 2002a). Usually the cuticle showed a relatively low permeability to most agricultural chemicals (Baur & Schönherr, 1995; Schreiber & Schönherr, 2009) but we have to consider the presence of cracks or other breaks in the cuticle surface that increases with fruits maturation (Faust & Shear, 1972; Freeman et al., 1979; Barthlott, 1990; Bianchi, 1995). These cracks are important for the fungicide uptake in plant tissues but, when the application is combined with thermotherapy, there is also a partial fusion and distribution of cuticular waxes that cover these cracks and reduce the ways of fungal infection (Gleen et al., 1985; El-Otmani et al., 1989; Roy et al., 1994; 1999; Bally, 1999). Many studies have demonstrated this evidence in different commodities, including citrus (Schirra & D'hallewin, 1997; Schirra et al., 1998a; Porat et al., 2000), cactus pear (D'hallewin et al., 1999; Schirra et al., 1999), sweet pepper (Fallik et al., 1999), and melon (Fallik et al., 2000).

Evaluations and experiments of thermotherapy combination have been reported on the application of a widely used fungicide to control moulds in citrus fruit (Eckert & Ogawa,

1985; Eckert & Eaks, 1988). Treatments with Sodium ortho-phenylphenate (SOPP) were performed after warming at 40 °C to control citrus moulds on 'Pineapple' and 'Valencia' oranges (Hayward & Grieson, 1960). The same fungicide was used at a concentration of 2%, with 1% hexamine and 0.2% sodium hydroxide in water at 22.8 and 37.8 °C. In this case the cold solution was more effective to control mould decay than the heated solution because of the phytotoxic effect of the SOPP (McCornack & Hayward, 1968).

Heated and unheated fungicides were tested on apples to control blue mould development (Spalding et al., 1969).

With the aim of controlling *Monilinia fructicola,* a solution of 2,6-diclhoro-4-nitroaniline was positively used at 51.5 °C for one and a half minute dip on peaches, plums, and nectarines (Wells & Harvey, 1970). These results were reconsidered and the research on thermotherapy was newly promoted in the 80th years (Carter, 1981a) due in part to the new cultural approach to the environment problems, the higher care for the human health and the difficulties to accept the use of fungicides and other chemical controlling pests largely used in the food industry (Schirra, 2005).

Actually we have to consider that few postharvest fungicides (Tomlin, 1997; Adaskaveg & Förster, 2010) are admitted to postharvest treatments of horticultural commodities and that their effectiveness is largely compromised by the natural selection of resistant strains of the main postharvest pathogens, as a consequence of their repeated use (Schmidt et al., 2006). For this reason, the research of alternatives to chemical pesticides for the control of postharvest disorders is considered as a prioritary objective for many researcher involved in the test of many methods: thermotherapy, use of radiations, gas and vapour, use of microbiological resources (like antagonist or toxic strains), use of chemical compounds of not toxic salts and compounds generically recognized as safe (GRAS).

2.1 Applications of thermotherapy

The thermotherapy is possible thank to the transfer of thermal energy to the horticultural commodity by means of a fluid mean which can be liquid, water vapour or air saturated of humidity. Hot dip are made in water with temperatures ranging from 43 to 53 °C and time of dipping comprised between 1-3 minutes and 2 hours (Ben-Yehoshua et al., 2000; Fallik, 2004). A different kind of dip treatment is the brushing of horticultural commodities under a hot water flow. Water temperature in this case are higher than in the case of hot dip, ranging between 48 and 60-63 °C during shorter periods of 15-30 seconds until 1 minute.

Treatments with air or hot vapour showed some technical difficulties and need more time but often their effectiveness is more prolonged with respect to hot dip (Mulas & Schirra, 2007).

As a further possibility among others, applications of radiofrequency waves may be useful for thermotherapy of horticultural commodities as actually in the wood industry or for the treatment of cereals (Johnson et al., 2003).

2.2 Control of physiological disorders

Fruits and vegetables of tropical and subtropical origin like citrus are particularly sensitive to chilling injury. Also this physiological disorder may be reduced or avoided by means of the thermotherapy, like the exposition of the commodities to hot air at 37 °C during 3 days under saturated humidity (Lafuente & Zacarias, 2006), or by mean of applications of hot dip

(Wild & Hood, 1989; McDonald et al., 1991; Wang, 1993; Gonzales-Aguilar et al., 1997; Mulas, 2006). More indications are available in order to control chilling injury of horticultural commodities of the temperate regions, like apples, by means of the thermotherapy (Lurie, 1998; Mulas & Schirra, 2009).

The effectiveness of thermotherapy to control oxidative stress in chilled fruits seems also linked directly to the enhanced activity of enzymes controlling antioxidant responses (Sala, 1998; Sala & Lafuente, 1999; 2000).

2.3 Control of pests

Many markets require horticultural commodities absolutely free of pesticide residues but at the same time also free from eggs, larvae and adults of insects. Also in this direction, alternative methods to the use of chemical fumigants, like methyl-bromide (forbidden in many countries), have been developed. Optimal times and heating temperatures for disinfestations depend on species and cultivars. For citrus, temperature of 44 °C for 90 minutes or 46 °C for 50 minutes are indicated as effective, but it is true that not all the citrus cultivars withstand these treatments. Thus, symptoms of senescence have been observed after this kind of treatments in bloody and 'Valencia late' oranges (Mulas et al., 2001; Schirra et al., 2004; 2005a).

Good results have been obtained against some coleoptera infesting dates by heating them at 55 °C for 2 hours and 30 minutes (Rafaeli et al., 2006).

2.4 Direct control of postharvest pathogens

Hot dip is effective controlling main postharvest pathogens of horticultural commodities both as a consequence of the fruit surface washing leading to reduced inoculums, as well as because of the thermal inactivation of the microorganisms. Hot dips and high temperature conditioning were effective to improve shelf quality of late crop cactus pear fruit (Schirra et al., 1996a).

Furthermore, we have to consider the possibility that thermotherapy may stimulate the biosynthesis or elicitation of endogenous compounds having fungicide effects (Lurie, 1998; Schirra et al., 2000a).

Evidence of these effects on fruit surface have been showed in a research carried out on 'Montenergrina' mandarins (Montero et al., 2010), where hot dip treatments at 60 °C with brushing significantly reduced the number of tangerines affected by decay.

In a recent study, hot water dip at 48 °C for 12 or 6 minutes showed good control of postharvest brown rot on 'Roig' peach and 'Venus' nectarine, without visual symptoms of heat damage and significant loss of fruit quality (Jemric et al., 2011). The elicitation of compounds involved in the plant defence as a consequence of hot air treatments at 38 °C during 24-72 hours was further demonstrated on tomato cherry fruit by means of higher lignin deposition, and higher activation of phenylalanine ammonia-lyase and β-1,3-glucanase (Zhao et al., 2009; 2010).

3. Combination of thermotherapy with fungicides

In spite of the beneficial effects of thermotherapy to the control of microbial postharvest disorders of fresh horticultural commodities, this treatment is not able to have the same effectiveness of fungicides, particularly when the marketing is made after a period more or less long of storage (Schirra, 2005; Mulas, 2008). In some countries, thermotherapy is applied to control mould decay in commodities from organic cultivation or when the postharvest fungicide treatment is completely forbidden (Fallik, 2004). In other cases, the chemical defence is always of fundamental importance against mould decay and there is not a convenience to use other methods (Dezman et al., 1986; Papadopoulou-Mourkidou, 1991).

3.1 Old fungicides

After the first experiences in the 60th years, many investigations have been promoted in the 70th and 80th years, but a new start of researches on old fungicides was observed in the 90th.

Mango decay was controlled by combination of hot dips with Benomyl (Spalding & Reeder, 1972; 1978). Brown rot and rhyzopus of peaches, nectarines and stone fruits were controlled by different combination of Benomyl or other fungicides and thermotherapy (Smith, 1971; Wells & Gerdts, 1971; Wells, 1972; Jones & Burton, 1973; Smith & Anderson, 1975; Wang & Anderson, 1985).

The control of rots of guavas was tested by application of heated Benomyl and Guazatine dips (Wills et al., 1982). Rotting and browning of lytchee fruit were controlled by hot benomyl and plastic film (Scott et al., 1982; Wong et al., 1991).

New experiments were carried out to demonstrate the synergistic effect of heat and sodium ortho-phenylpenate to inactivate *Penicillium* spores and suppress decay in citrus fruits (Barkai-Golan & Appelbaum, 1991). The efficacy of hot water and Carbendazim treatments was tested against the brown rot of apple (Sharma & Kaul, 1990).

3.2 Thiabendazole and Imazalil

The most used fungicides in postharvest of horticultural commodities are Imazalil and Thiabendazole. Both are synthetic compounds with different mode of action that can be applied in waxes or water (Brown et al., 1983; Brown, 1984). Imazalil is very effective against the green mould (*Penicillium digitatum*), including benzimidazole-resistant strains, while Thiabendazole is effective to a wide range of pathogens and often may have beneficial effects also against chilling injury (Schiffman-Nadel et al., 1972; 1975; Brown & Dezman, 1990; McDonald et al., 1991; Schirra et al., 2000b).

Many studies demonstrate that the efficacy of Imazalil and/or Prochloraz increases when applied in hot water in a number of fruits such as mango (Spalding & Reeder, 1986; Johnson et al., 1990; Prusky et al., 1990; Dodd et al., 1991; Coates et al., 1993; McGuire & Campbell, 1993; Waskar, 2005); melons (Carter, 1981b; Mayberry & Kartz, 1992), taro corms (Quevedo et al., 1991), red tamarillos (Yearsley et a., 1988), and in citrus fruit (Ansari & Feridoon, 2008).

That is because of the best cuticle and tissue penetration by mean of active compounds (Schirra et al., 2002a; 2008a). Wide demonstration of these effects has been reported in the case of the Imazalil and Thiabendazole use against *Penicillium* mould of citrus (Mulas & Schirra, 2007; Dore et al., 2009). The two fungicides resulted effective in some cases with very low concentration, up to 50 mg ·L[-1], if the application was made in hot water (50 °C) (Schirra & Mulas, 1993; 1994; 1995a; b; c; Schirra et al., 1995; 1997a).

Treatments of citrus with Imazalil at 490 mg ·L^{-1}, in water at 37.8 °C resulted more effective in mould control than in wax mixture at 4200 mg ·L^{-1} at ambient temperature (Smilanick et al., 1997).

The residue control of Imazalil in lemons after applications at 50 °C of concentrations from 250 to 1500 mg ·L^{-1} demonstrated that fruit submitted to hot dip contained about 4.5-fold higher fungicide concentration (Schirra et al., 1996b). Good results were obtained with hot dip in Imazalil mixtures of lemons (Schirra et al., 1997b) and 'Marsh' grapefruits (Schirra et al., 1998b).

Thiabendazole residues are dependent on pH of solutions (Wardowski et al., 1974), not influenced by duration of the treatment (Cabras et al., 1999), and correlated to the amount of fungicide applied (Schirra et al., 1998c)

Synergic effect of thermotherapy with fungicide application was not always so clear, and in the past there is evidence of some negative combination of the two treatments, like in the case of 'Tarocco' oranges when the preharvest treatment with Thiabendazole was effective in the postharvest control of *Penicillium* moulds and against chilling injury, but the combination with a curing treatment at 37 °C during 48 hours resulted favourable to the decay by *Phytophthora citrophtora* (Schirra et al., 2002a).

Positive effects of Thiabendazole in combination with hot dip have been confirmed by treatments on 'Eureka' lemons at 0, 25, 50 e 100 µg ·mL^{-1} in water containing 200 µg ·mL^{-1} of sodium hypochlorite and 0.2 ml ·L^{-1} of Triton X-100, at temperature of 16, 27, 38, or 49 °C after artificial inoculation with a Thiabendazole resistant strain of *Penicillium digitatum* (M6R) (Smilanick et al., 2006a). In fact, mould control by only dip temperature was of 50% at 49 °C, while the addition of only 100 µg ·mL^{-1} of Thiabendazole was effective for an almost total control of the pathogen.

A further confirmation of the synergic effect between postharvest low dose treatment with Thiabendazole and warming of the dip solution at 52 °C was observed with the control of pathogens and, partially, of the chilling injury in cactus fruit of the 'Gialla' cultivar (Schirra et al., 2002b).

In tamarillos, Imazalil was effective against *Collectotricum gleosporioides* or *C. acutatum* both in water and in wax if fungicide application followed a hot dip treatment in water at 50 °C for 10 minutes (Yearsley et al., 1987).

Good results were also recorded for the control of *Penicillium* mould and of chilling injury in citrus fruits by the fungicide Imazalil when this compound was used after molecular complexation with β-cyclodestrine at 100 mg ·L^{-1} in water dip at 50 °C (Schirra et al., 2002c). Different works made with other fungicides demonstrate the effectiveness of β-cyclodestrine since it is a carrier of active molecules that prevents their degradation (Szejtli, 1988; Kenawy et al., 1992; Lezcano et al., 2002).

3.3 New fungicides

Most of the positive effects of synergy between thermotherapy and low dose applications of fungicides has been confirmed with the experimental use of a new generation of fungicides so called "natural mimetic"(Gullino et al., 2000; Ragsdale, 2000; Leroux, 2003). These compounds are also defined like generally recognized as safe (GRAS) with respect to the previously used fungicides. They showed higher effectiveness at low doses, a more favourable toxicological and eco-toxicological profile and different mechanisms of action with respect to the old generation of fungicides (Heye et al., 1994; Errampalli & Crnko, 2004). Therefore, these new fungicides may be very useful as an alternative to traditional fungicides which are ineffective against resistant strains of pathogens (Schirra, 2005).

Among others it is important to consider the experiences realized with the strobilurine-like compounds Azoxystrobine and Trifloxistrobine (Margot et al., 1998; Reuveni, 2000; 2001; Barlett et al., 2002; Wood & Hollomon, 2003; Schirra et al., 2006), which have their site of action in the fungal mitochondrion and are quickly destroyed in soil and groundwater (Sudisha et al., 2010). Preharvest application of Azoxystrobine were effective to control *Alternaria alternata* in mandarin cultivars 'Minneola' and 'Nova' (Oren et al., 1999) and citrus scab and melanose in seedlings of rough lemon and grapefruit (Bushong and Timmer, 2000).

Azoxystrobine in postharvest was applied to 'Star Ruby' grapefruits and oranges by 3 minutes dip at 50 °C and low concentration (50 mg·L^{-1}) (Schirra et al., 2002d; 2010). The long ability of Azoxystrobine residues to remain constant during cold storage and in darkness was confirmed by other studies on apples (Ticha et al., 2008) and peppers (Garau et al., 2002), while rapid decline of residues was observed with preharvest treatments (Angioni et al., 2004). However, Azoxystrobine showed good natural decay control but was less effective against artificially inoculated *P. digitatum*, while Trifloxistrobine was highly effective against blue and green mould after inoculation and in association with hot dip at 50 °C for 3 minutes (Schirra et al., 2006).

Further experiences were made with Fludioxonil (synthetic analogous of pyrrolnitrine), which after fruit inoculation with *Penicillium* provided the same results as Imazalil at 100 mg·L^{-1} in water dip at 50 °C and 400 mg·L^{-1} at 20 °C (Schirra et al., 2005b), and with Pyrimethanil (anilinepyrimidine) that resulted effective at 400 mg·L^{-1} at 20 °C or 100 mg·L^{-1} at 50 °C against *P. digitatum* and *P. italicum* (D'Aquino et al., 2006). Pyrimethanil inhibits elongation of mycelium and the secretion of cell wall degradation enzymes (Daniels & Lucas, 1995; Milling & Richardson, 1995; Rosslenbroich & Stuebler, 2000; Sholberg et al., 2005; Kanetis et al., 2008a). Pyrimethanil increased notably their effectiveness when coupled with thermotherapy (D'Aquino et al., 2006; Smilanick et al., 2006b).

Cyprodinil was positively tested in combination with hot water dip against apple moulds (Errampalli & Brubacher, 2006) and inoculated *Penicillium digitatum* moulds on 'Valencia' oranges (Schirra et al., 2009a).

The possibility to control many postharvest pathogens has been investigated on a wide range of horticultural commodities (Smilanick et al., 2006b; Zhang, 2007; Kanetis et al., 2008a; Montesinos-Herrero & Palou, 2010) and recently Azoxystrobine, Fludioxonil and Pyrimethanil have been authorized for postharvest treatments in U.S.A. (Kanetis et al., 2007; Förster et al., 2007).

The mode of action of Fludioxonil is that of a mutagen-activated protein chinase pathway that stimulates glycerol synthesis (Kanetis et al., 2008b).

Recently, fludioxonil was also positively tested in association with thermotherapy on mango (Swart et al., 2009), pomegranates (Palou et al., 2007; D'Aquino et al., 2009), apple (Errampalli, 2004; Errampalli et al., 2005), citrus (Schirra et al., 2005), nectarines, apricots and peaches (D'Aquino et al., 2007), and 'Precoce di Fiorano', 'Coscia' and 'Spadona estiva' summer pears after inoculation with *Penicillium expansum* and *Botrytis cinerea* (Schirra et al., 2008b; 2009b).

Fig. 2. Stored mandarins treated with Imazalil at 100 mg ·L⁻¹ and 50 °C and 400 mg ·L⁻¹ at 20 °C against *P. italicum* (Schirra et al., 2005b; courtesy of Dr. Mario Schirra).

Fig. 3. Stored mandarins treatment with pyrimethanil at 100 mg ·L⁻¹ and 50 °C against *P. italicum* (D'Aquino et al., 2006; courtesy of Dr. Mario Schirra).

Fig. 4. Wounding of 'Coscia' pears before inoculation with *Penicillium expansum* (upper), apparatus for water dip at 50 °C (middle), and mould development after shelf life (above) (Schirra et a., 2008b; pictures of the author).

Fig. 5. Stored 'Coscia' pears after inoculation with *Penicillium expansum* (upper), water dip at 50 °C (middle), and treatment with fludioxonil at 100 mg ·L⁻¹ and 50 °C (above) (Schirra et a., 2008b; pictures of the author).

4. GRAS compounds

Researchers have focused their interest on GRAS compounds in order to find alternatives to traditional fungicides or to enhance their effectiveness by means of synergistic effects when combined with thermotherapy. Among others, it is interesting to point out calcium chloride preharvest applications in combination with 2,4-D, hot dip and fungicides on 'Satsuma' mandarins (Yildix et al., 2005); the ethanol used for the control of *B. cinerea* after inoculation on table grape (Karabulut et al., 2004; Gabler et al., 2005); the acetic acid as an alternative to ethanol or water vapour to control *B. cinerea* on kiwi fruit (Lagopodi et al., 2009); the sodium carbonate and bicarbonate, which were effective against *Penicillium* mould in citrus fruits dipped for 150 seconds in water at 45 °C containing a 3% of the salt (Palou et al., 2001). This is also the case of the bicarbonate, which resulted useful on citrus fruit in order to increase the effectiveness of Imazalil to inhibit germination of *P. digitatum* (Smilanick et al., 2005), and of Thiabendazole against the resistant strain of *P. digitatum* M6R in combination with dip at 49 °C during 60 seconds and addition of sodium hypochlorite at 200 µg·mL^{-1} (Smilanick et al., 2006). Good results have been also obtained by application of Imazalil or Thiabendazole in combination with potassium sorbate (Smilanick et al., 2008).

Potassium phosphite (2 mg·mL^{-1}) in combination with hot dip at 50 °C for 3 minutes induced a three-fold reduction in blue mould incidence   and was as effective as Thiabendazole after six months of storage at 2 °C of 'Elstar' apples (Amiri & Bompeix, 2011).

The positive effect of carbonate and bicarbonate addition to solutions of Thiabendazole used to control *P. digitatum* was demonstrated also on clementine fruits, 'Nova' mandarins, 'Valencia late' oranges with a higher penetration of the fungicide in the fruit tissues (Schirra et al., 2008c). 'Montenegrina' tangerines were exposed to postharvest thermotherapy and sodium carbonate and bicarbonate treatments in combination with carnauba wax application, which resulted in fruit protection against mould decay (Montero et al., 2010).

On 'Satsuma' mandarins, the combination of thermotherapy and fungicides was effective against resistant strains of pathogens, and the combination of thermotherapy with antagonist microorganisms was effective against the *Rhizopus* decay on strawberry (Zhang et al., 2005). On the other hand, the synergic effect of thermotherapy with the use of antagonist strains of yeasts and sodium bicarbonate was demonstrated against *Colletotrichum acutatum* and *P. expansum* in stored apples (Mulas & Schirra, 2007), as well as of thermotherapy in combination with *Cryptococcus laurentii* against *P. italicum* and *Rhizopus stolonifer* on peach, and with *Rhodotorula glutinis* against *P. italicum* on pear storage (Zhang et al., 2007; 2008).

Prevention of spoilage caused by fungi in cherry tomato was provided by heat treatment at 38 °C (24-72 hours) followed by *Pichia guilliermondii* application (Zhao et al., 2009; 2010). Other results showed that the combined application of hot air at 38 °C for 36 hours and the same yeast antagonist *Pichia guilliermondii* was effective in the control of postharvest anthracnose rot of loquat fruit (Liu et al., 2010).

The residue determination of pesticides, or of other chemicals used in combination with thermotherapy, is an essential condition to guarantee treatment effectiveness avoiding to overcome levels of tolerance. This is the only way to maintain the safety of horticultural commodity, which is associated with the absence of chemical residues (Schirra, 2005).

5. Main effects on quality of commodities

In spite of the numerous applications of thermotherapy to many horticultural commodities, there are generalized recurrences of not complete control of mould decay if the treatments are limited to the physical means (Lurie, 2006). At the present state of knowledge, it seems more appropriate the use of thermotherapy as a synergic tool of the available fungicides and GRAS compounds (Schirra, 1995; Mulas & Schirra, 2007).

Among quality traits is important to point out that thermotherapy may build-up volatile compounds as shown in 'Tarocco' (Schirra et al., 2002b) and other bloody oranges (Mulas et al., 2001; Schirra et al., 2004; 2005a). Increases in endogenous ethanol and acetaldehyde production that change fruit taste have been reported in this fruits, but studies for a complete definition of gas exchange during and after thermotherapy are in a phase of development (Mulas et al., 2004; 2006; 2008a; 2008b).

5.1 Contra-indications

Many experiments of thermotherapy were designed to optimize the protocol to avoid decay or physiological disorders, but damages derived from this kind of treatments have been observed (Lurie, 2006). Negative effects have been described on apple fruit after brushing with water at 60 or 65 °C for 15 seconds (Fallik et al., 2001). Heat-damages manifested as increased electrolyte leakage from biological membranes and surface browning may occur in cactus pear fruit (*Opuntia ficus-indica* Miller L.) after brushing with water at 65-70 °C for 10-30 seconds (Dimitris et al., 2005). Negative effects of thermotherapy were also observed on strawberry (Wszelaki & Mitcham, 2003). Short hot-water rinsing and brushing treatment for 20 seconds at 55, 60, and 65 °C significantly reduced the epiphytic microbial population on fruit surface of strawberry cultivar 'Feng xiang', as well as decay development and weight loss (Jing et al., 2010). However, about 60% of the fruit treated at 65 °C showed heat damages.

The effects of treatment temperatures within the range from 20 to 75 °C were studied on 'Navelate' oranges after dip during 150 seconds (Palou et al., 2001). Any negative effect was recorded until 45 °C, while 17 and 28% of fruit treated at 53 or 55 °C showed slight or medium symptoms of heat damages on 100% of fruit surface.

The contemporary evolution of chilling injury and heat damage has been described in 'Satsuma' mandarins, as well as the involvement of antioxidant enzymes, vacuolar ATPase, and pyrophosphatase (Ghasemnezhad et al., 2007).

Heat damages may produce symptoms in the internal tissues of different commodities even in total absence of any external alteration. This is the case of the internal browning, which was observed in avocados, citrus, peaches, nectarines, and lytchees (Zhou et al., 2002; Follet & Sanxter, 2003; Lurie, 2006). Other symptoms, like low pulp colour evolution, anomalous softening, lack of starch hydrolysis, and development of internal cavities have been observed on mangoes and papaya (Jacobi et al., 2001; Lurie, 2006). Lytchee fruits of the cultivar 'McLean's Red' after water dip at 50 °C for 2 minutes or water dip at 55 °C for 1 minute showed superficial scald, while the treatment at 60 °C for 1 second was less harmful (Sivakumar & Korsten, 2006). Dragon fruits (*Hylocereus undatus*) were tolerant to hot air disinfestations treatments until a temperature of 46.5-48.5 °C for 20 minutes as measured in the fruit centre. Because the lack of significant differences for bracteas and peduncle turgidity, fruit general appearance and presence of mould decay, peel colour and pulp firmness, total soluble solids concentration, acidity, taste and pulp brightness, fruit quality was kept immediately after the treatment, after 4 weeks storage at 5 °C in propylene bags, and after the shelf life period at 20 °C (Hoa et al., 2006).

Some field variables, like seasonal temperatures and rainfall may influence the thermotherapy effectiveness, particularly in those commodity very sensitive to chilling injury or to heat scald like citrus fruit. Some studies demonstrate thermotherapy effectiveness in the control of chilling injury, but also that the same treatment may be harmful depending on the harvest date (Schirra et al., 1997; Lafuente et al., 2005; Lafuente & Zacarias, 2006).

5.2 Effects on maturation and senescence

Postharvest thermotherapy treatments slow down maturation of climacteric fruits (Fallik, 2004; Lurie, 2006). Ethylene biosynthesis inhibition by heat treatments slows pulp softening and favours low colour and aromatic compound development in apples and kiwis treated at 38 °C, while the treatment at 39 °C for 90 minutes slow down colour development in tomatoes (Ali et al., 2004).

Investigations on 'Caldesi 2000' nectarines and 'Royal Glory' peaches showed that water dip at 46 °C reduced pulp softening of fruits sealed in thin polyethylene bags and stored at 0 °C for one or two weeks (Malakou & Nanos, 2005). This effect is the result of the combination of hot water treatment, modified atmosphere application and package, particularly in white pulp nectarines, which maintain functional cell membranes because cell wall hydrolytic enzyme inactivation, mainly polygalacturonase. Hot air treatment may also change the organoleptic characteristics of peaches 'Dixiland' by decreasing total acidity and increasing red pigments in pulp and peel (Budde et al., 2006).

Usually the effects of postharvest heat treatments are reversible, if the application is not too long, and then the physiological damage is avoided. This is the case of the tomatoes treated with water at 42 °C that showed regular biosynthesis of aromatic compounds and of lycopene (Mulas & Schirra, 2007)

High temperature may induce temporary inhibition of polygalacturonase in mango and tomato, as well as low activity of other enzymes involved in softening. Changes in other characters linked to maturation may be reduced in non-climacteric fruits such as strawberries with slow colour development and pulp softening (Lurie, 2006), which may be associated with low acidity (Vicente et al., 2002).

Water dip at 45 °C improves strawberry resistance to pathogens but determined external damages and reduction of the solubility of cell wall polysaccharides (Lara et al., 2006).

Heated vapour treatment at 52.5 °C or 55 °C during 18-27 minutes of table grape of the 'Sultanina' cultivar did not affect weight loss, berry firmness, colour, and total soluble solid and acid content modification. Nevertheless, treatment at higher temperature (58 °C) or for a more prolonged period (55 °C, 30 minutes) reduced fruit quality since it increased weight loss and berry browning (Lydakis & Aked, 2003).

Among effects of heat treatments on tissue senescence of horticultural commodities, it has been shown that the natural yellowish of broccoli is delayed both after exposing it to water dip at 45 °C for 10 minutes or after air conditioning at 50 °C for 2 hours (Funamoto et al., 2003).

Other symptoms of senescence, like geotropism deviation of spears of asparagus, and sprouting of onion, garlic and potato have been controlled by dip treatments in water at 50-55 °C for 2-4 minutes (Cantwell et al., 2003; Lurie, 2006). Pineapple fruit treatment at 38-60 °C for 60 minutes was effective to control internal browning during cold storage (Weerahewa & Adikaram, 2005). Sapote fruit [*Pouteria sapota* (Jacq.) H.E. Moore & Stearn], exposed to disinfestations with dip treatment at 60 °C for 60 minutes showed less pulp browning respect to untreated control (Diaz Perez et al., 2001).

Postharvest treatments of early ripening pears of 'Camusina' and 'Precoce di Fiorano' cultivars with hot dip at 50 or 60 °C for 3 minutes or 1-Methylcyclopropene at 20 °C were useful to delay senescence and internal browning, while soy lecithin and calcium chloride resulted less effective (Mulas et al., 2008a). However, dip treatment at 60 °C determined heat damages in the peel of the two cultivars.

Water dip treatment for 3 minutes a 50 °C and with hot air at 37 °C for 48 hours were effective reducing chilling injury on bloody oranges of the cultivars 'Tarocco', 'Moro', 'Doppio sanguigno', and 'Sanguinello'. Any treatment caused visible damages during quarantine storage of 16 days at 1 °C, subsequent storage of 3 weeks at 8 °C and a further week of simulated shelf life at 20 °C. However, while dip treatment was not influent on fruit firmness, taste, aroma, juice content and composition (total soluble solids, titratable acidity, ascorbic acid and ethanol content), hot air treatment negatively influenced fruit firmness, taste and chemical composition (Schirra et al., 2004).

Postharvest treatments of disinfestations against pests with humid air at 44 or 46 °C, measured inside the fruit during 100 and 50 minutes respectively, do not produce negative effects on 'Olinda' and 'Campbel' oranges (clones of 'Valencia late' cultivar) both for the external appearance and the internal composition (Schirra et al., 2005a). However, these protocols cannot be recommended to the bloody oranges because of the negative influence on fruit quality (off flavours development, fruit softening, high weight loss), and of the reduced resistance to moulds (Mulas et al., 2001).

5.3 Nutritional value

Many studies demonstrated that thermotherapy may influence the biosynthesis of antioxidant or nutraceutical compounds (Schreiner & Huyskens-Keil, 2006) and researches in this direction are increasing. In papaya (*Carica papaya*) fruit, for example, thermotherapy reduces chilling injury, slow down superoxide-dismutase and catalase activities and stops the increase of peroxidase activity (Huajaikaew et al., 2005). Hot air treatment at 34 °C and 50% R.H. during 24 hours of the tomato 'Rhapsody' did not affect to the antioxidant properties of the fruit that developed a normal colour during storage at 10 °C (Soto-Zamora et al., 2005). Otherwise, the fruit exposition to 38 °C during 24 hours in air o in an atmosphere containing a 5% of oxygen determined some negative effects, like loss of antioxidant properties and lack of colour evolution.

Dipping of mango fruit at 50 °C for 60 hours to kill pests may enhance carotene biosynthesis, reduce fruit shelf life, while the thermotherapy associated with cold storage slows carotene development (Talcott et al., 2005). More studies on the cultivar 'Kensington Pride' showed high effectiveness in maintaining fruit quality of thermotherapy in air at 40 °C for 8 hours, or with water dip at 52 °C for 10 minutes (Dang et al., 2008).

Broccoli (*Brassica oleracea* L.) treated at 48 °C for 3 hours with hot air showed slow senescence at 20 °C, better quality and a significant higher contents of chlorophyll, sugars, proteins and antioxidants (Costa et al., 2005). A slow degradation of chlorophyll and an increase in antioxidant properties has been also reported in spinach after water dip treatment at 40 °C for 3 minutes (Gomez et al., 2008).

Dip treatment at 45 °C for 4 minutes of pomegranate fruits (*Punica granatum* L. cultivar 'Mollar de Elche') also produced higher antioxidant activity, total phenols, ascorbic acid, anthocyans, sugars, and organic acids (Mirdehghan et al., 2006).

Studies on different horticultural commodities showed that hot dip at 35 °C for 12 hours (tomato), at 55 °C for 5 minutes (melon), and at 42 °C for 24 hours (mango) inhibited polyphenol-oxidase and peroxidase activities and reduced slow biosynthesis of antocyans, but maintain good nutraceutical properties in the fruits (Cisneros-Zevallos, 2003; Brovelli, 2006; Mulas & Schirra, 2007).

Good results have been obtained with water dip at 50 °C for 2 minutes in order to maintain nutritional and functional properties of kumquat fruit (Schirra et al., 2008d), and the quality of blueberry was also satisfactory after thermotherapy application (Fan et al., 2008).

6. Conclusions and future perspectives

In spite of the beneficial effects, thermotherapy is not sufficient alone to provide protection against postharvest disorders during long term storage (Tab. 1-4). This is because some of the induced mechanisms are only transient. However, it is very clear their synergic effect with the old fungicides and the need of maintaining also this possible supplement of effectiveness for the applications of the new fungicides. In fact there is yet some signal of pathogen resistance for the recently admitted postharvest fungicides.

Further investigations are necessary to optimize protocols for different horticultural commodities, cultivars and zone of production. A good direction to develop more studies is the possibility to combine the thermotherapy with other physical and chemical treatments. In the Tables 1-4 a synthesis of the main effects and indication of the termotherapy is proposed to stimulate new ideas and investigations.

The heat effect on temporary inhibition or enhancement of the enzyme activity, as well as on slowing maturation and senescence of commodities, are some of the evidences more critical to get a general insight of the complex physiological consequences of thermotherapy. The influence of this practice on the biosynthesis of ethylene and of phytochemicals of nutraceutical value, and on the development of aromatic and off volatiles are also some directions for future researches (Cisneros-Zevallos, 2003; Mulas et al., 2006; 2008b; Lafuente et al., 2011).

Commodity	Treatment	Temperature (time)	Effect	References
Apple	Hot humid air	38 °C (4 days)	Mould control. Synergic to antagonist strains and sodium bicarbonate.	Mulas & Schirra, 2007
Apple	Brushing	60-65 °C (15 seconds)	Heat damages	Fallik et al., 2001
Apple	Hot humid air	38 °C (30-120 hours)	Low development of colour and aroma.	Lurie, 2006
Blueberry	Water dip	60 °C (15-30 seconds)	Mould control.	Fan et al., 2008
Broccoli	Water dip. Hot humid air.	45 °C (10 minutes). 50 °C (2 hours)	Delayed yellowish.	Funamoto et al., 2003
Broccoli	Hot humid air.	48 °C (3 hours)	Delayed senescence	Costa et al., 2005
Cactus fruit	Water dip.	52 °C (3 minutes.)	Chilling injury control. Synergic to Thiabendazole.	Schirra et al., 2002b
Cactus fruit	Brushing	65-70 °C (10-30 seconds)	Hot damages.	Dimitris et al., 2005
Citrus (orange, mandarins, lemons, grapefruits, kumquats)	Water dip	50-53 °C (2-3 minutes)	Chilling injury control. Partial control of moulds. Synergic to Thiabendazole and Imazalil.	Schirra & Mulas, 1993; 1994; 1995a; b; c; Schirra et al., 1995; 2002c; 2008c
Citrus	Water dip	50 °C (3 minutes)	Synergic to Azoxystribin, Fludioxonil and Pyirimethanil	Schirra et al., 2005b; 2006; 2010; D'Aquino et al., 2006;
Citrus	Water dip	45-49 °C (150-60 seconds)	Synergic to sodium carbonate and bicarbonate, Thiabendazole and sodium hypochlorite in the mould control.	Palou et al., 2001; Smilanick et al., 2006; Schirra et al., 2008b
Citrus	Water dip Hot humid air	52 °C (3 minutes) 37 °C (48 hours)	Ethanol, acetaldehyde accumulation. Loss of quality.	Mulas et al., 2004; 2006; Schirra et al., 2004
Citrus ('Valencia late' and "bloody oranges")	Hot humid air	44 °C (90 minutes) 46 °C (50 minutes)	Quarantine. Senescence on "bloody oranges".	Mulas et al., 2001; Schirra et al., 2004; 2005a
Citrus ('Tarocco')	Hot humid air.	37 °C (48 hours)	Synergic to Thiabendazole	Schirra et al., 2002a
Citrus (Navelate)	Water dip	53-75 °C (150 seconds)	Hot damages	Palou et al., 2001
Citrus ('Valencia late')	Water dip. Hot humid air.	50 °C (3 minutes) 38 °C (24 hours)	Increased respiration, ethylene production, ethanol and acethaldheyde accumulation	Mulas et al., 2008b
Citrus	Water dip	50 °C (30 seconds)	Synergic to potassium sorbate, Thiabendazole, Imazalil, Fludioxonil and Pyirimethanil	Smilanick et al., 2008

Table 1. Thermotherapy treatments and their different effects.

Commodity	Treatment	Temperature (time)	Effect	References
Dates	Hot humid air.	55 °C (2 hours and 30 minutes)	Quarantine.	Rafaeli et al., 2006
Garlic	Water dip	50-55 °C (4-2 minutes)	Avoid spruce	Lurie, 2006
Kiwi	Hot vapour.	47-53 °C (3-6 minutes)	Mould control. Synergic to ethanol	Lagopodi et al., 2009
Kiwi	Hot humid air.	38 °C (30-120 hours)	Low development of colour and aroma.	Lurie, 2006
Lemons	Water dip	49 °C (60 seconds)	Synergic to Thiabendazole	Smilanick et al., 2006
Lemons	Water dip	50 °C (2 minutes)	Synergic to Imazalil	Dore et al., 2009
Loquat	Hot humid air	38 °C (36 hours)	Synergic to antagonist strains.	Liu et al., 2010
Lytchee	Water dip	49 °C (20 minutes)	Heat damages.	Follet & Sanxter, 2003
Lytchee ('McLean's Red')	Water dip	50-60 °C (2 minutes -1 second)	Heat damages.	Sivakumar & Korsten, 2006
Mandarins ('Fortune') and other citrus	Hot humid air.	37 °C (3 days)	Chilling injury control.	Lafuente & Zacarias, 2006
Mandarins ('Dancy')	Hot humid air.	45-48 °C (1-4 hours)	Heat damages.	Lurie, 2006
Mandarins ('Satsumas')	Water dip	45-55 °C (2-5 minutes)	Heat damages. ATPase, pyrophosphatase and antioxidant enzymes.	Ghasemnezhad et al., 2007
Mandarins ('Fortune')	Hot humid air.	37 °C (1-2 days)	Chilling injury control without loss of flavonoids and vitamin C.	Lafuente et al., 2011
Mandarins ('Montenegrina')	Water dip. Brushing.	60 °C (30 seconds)	Synergic to Imazalil	Montero et al., 2010
Mangoes	Hot humid air. Hot vapour. Water dip	51,5 °C (125 minutes) 46-48 °C (3-5 hours) 42-49 °C (7-120 minutes)	Heat damages.	Jacobi et al., 2001
Mangoes	Water dip	50 °C (60 minutes)	Quarantine. Quick maturation.	Talcott et al., 2005
Mangoes ('Kensington Pride')	Water dip. Hot humid air.	52 °C (10 minutes) 40 °C (8 hours)	Mould control.	Dang et al., 2008
Mangoes	Water dip	42 °C (24 hours)	PPO and POD inhibition with antocyans storage.	Mulas & Schirra, 2007
Mangoes	Water dip	50 °C (30 seconds)	Synergic to Fludioxonil and Prochloraz	Swart et al., 2009
Melons	Water dip	55 °C (5 minutes)	PPO and POD inhibition with antocyan storage.	Mulas & Schirra, 2007
Nectarine	Hot humid air.	41-46 °C (24-48 hours)	Heat damages	Lurie, 2006
Nectarine ('Caldesi 2000' and 'Royal Glory')	Water dip	46 °C (25 minutes)	Slow pulp firmness loss.	Malakou & Nanos, 2005
Onions	Water dip	50-55 °C(4-2 minutes)	Avoid sprouting.	Lurie, 2006

Table 2. Thermotherapy treatments and their different effects.

Commodity	Treatment	Temperature (time)	Effect	References
Papaya	Hot humid air.	32,5 °C (10 days)	Heat damages	Lurie, 2006
Papaya ('Sunrise')	Hot humid air.	42 °C (6 hours)	Chilling injury control. Low SOD, CAT and POD action.	Huajaikaew et al., 2005
Peach, plums and nectarines	Water dip.	51.5 °C (1 minutes and 30 seconds.)	Synergic to the fungicide 2,6-D-4-NA	Wells & Harvey, 1970
Peach	Hot humid air.	37 °C (48 hours).	Mould control. Synergic to antagonist strains.	Zhang et al., 2007
Peach	Water dip. Hot humid air.	37-43 °C (1-3 hours.) 37-43 °C (8-24 hours).	Heat damages	Zhou et al., 2002
Peach ('Dixiland')	Hot humid air.	39 °C (44 hours)	Loss of total acidity and pigment increase.	Budde et al., 2006
Peaches, nectarines and apricots	Water dip.	48 °C (2 minutes)	Synergic to Fludioxonil	D'Aquino et al., 2007
Peaches and nectarines	Water dip.	48 °C (6-12 minutes)	Mould control.	Jemric et al., 2011
Pears	Water dip	46 °C (10-20 minutes)	Mould control. Synergic to antagonist strains.	Zhang et al., 2008
Pears ('Precoce di Fiorano', 'Coscia', 'Spadona estiva')	Water dip	50 °C (3 minutes)	Mould control. Synergic to the Fludioxonil	Schirra et al., 2008a
Pears ('Camusina' and 'Precoce di Fiorano')	Wate dip	50 °C (3 minutes)	Partial control of internal bowning	Mulas et al., 2008a
Pineapple	Water dip	38-60 °C (60 minutes)	Chilling injury control	Weerahewa e Adikaram, 2005
Pitaya	Hot humid air.	46,5-48,5 °C (20 minutes)	Quarantine	Hoa et al., 2006
Pomegranate ('Mollar de Elche')	Water dip	45 °C (4 minutes)	Increase of antioxidant activity, sugars and acids	Mirdehghan et al., 2006
Pomegranate ('Wonderful')	Water dip	49 °C (30 seconds)	Synergic to Fludioxonil.	Palou et al., 2007
Pomegranate ('Primosole')	Water dip	50 °C (3 minutes)	Synergic to Fludioxonil.	D'Aquino et al., 2009
Potatoes	Water dip	50-55 °C (4-2 minutes)	Avoid sprouting	Lurie, 2006
Sapote	Water dip	60 °C (60 minutes)	Quarantine. Browning control.	Diaz Perez et al., 2001
Spears of asparagus	Water dip	50-55 °C(4-2 minutes)	Avoid geotropism deviation	Lurie, 2006
Spinach	Water dip	40 °C (3 minutes and 30 seconds)	Delayed senescence and increase of antioxidants.	Gomez et al., 2008

Table 3. Thermotherapy treatments and their different effects.

Strawberry	Water dip	55 °C (30 seconds)	Mould control. Synergic to antagonist strains.	Zhang et al., 2005
Strawberry	Water dip	63 °C (12 seconds)	Heat damages.	Wszelaki e Mitcham, 2003
Strawberry	Hot humid air.	45 °C (3 hours)	Slow pulp firmness loss and colour development. Low acidity.	Vicente et al., 2002;
Strawberry ('Pàjaro')	Hot humid air. Water dip.	40-50 °C (30-75 minutes.) 45 °C (15 minutes).	Heat colour. Low hydrolysis of polysaccharides	Lara et al., 2006
Strawberry ('Feng xiang')	Hit water rinsing and brushing	60 °C (20 seconds)	Mould control.	Jing et al., 2008
Table grape	Water dip	50-60 °C (30-60 seconds)	Mould control. Synergic to ethanol.	Karabulut et al., 2004; Gabler et al., 2005
Table grape	Heated vapour	52,5-58 °C (18-30 minutes)	Heat damages	Lydakis & Aked, 2003
Tomato	Hot humid air.	39 °C (90 minutes)	Delayed colour development.	Ali et al., 2004
Tomato ('Rhapsody')	Hot air with 50% RH	34 °C (24 hours). 38 °C (24 hours.)	Quarantine. Loss of antioxidants and anomalous colour.	Soto-Zamora et al., 2005
Tomato	Water dip	35 °C (12 hours)	PPO and POD inhibition with storage of antocyans.	Mulas & Schirra, 2007
Tomato	Hot humid air.	38 °C (24 hours)	Synergic to antagonist strains.	Zhao et al., 2009; 2010

Table 4. Thermotherapy treatments and their different effects.

7. Acknowledgements

The author want to thank Dr. Mario Schirra for the permission to print the pictures of the Figs. 2 and 3, as well as the Dr. Maria Teresa Lafuente for the critical contribution to the manuscript.

8. References

Adaskaveg, J. E. & Förster, H. (2010). New developments in postharvest fungicide registrations for edible horticultural crops and use strategies in the United States. In: Prusky, D. & Gullino, M.L. (eds.). Postharvest pathology. Springer, pp. 107-116.

Ali, M.S., Nakano, K. & Maezawa, S. (2004). Combination effect of heat treatment and modified atmosphere packaging on the colour development of cherry tomato. Postharvest Biology and Technology, 34, pp. 113-116.

Amiri, A. & Bompeix, G. (2011). Control of Penicillium expansum with potassium phosphite and heat treatment. Crop Protection, 30, pp. 222-227.

Angioni, A., Schirra, M., Garau, V.L., Melis, M., Tuberoso, C.I.G. & Cabras, P. (2004). Residues of azoxystrobin, fenhexamid and pyrimethanil in strawberry following field treatments and the effect of domestic washing. Food Additives and Contaminants, 21, pp. 1065-1070.

Ansari, N. A. & Feridoon, H. (2008). Postharvest application of hot water, fungicide and waxing on the shelf life of Valencia and the local orange cv. Siavarz. *Acta Horticulturae*, 768, pp. 271-277.

Bally, I.S.E. (1999). Changes in the cuticular surface during the development of mango (*Mangifera indica* L) cv. Kensington Pride. *Scientia Horticulturae*, 79, pp. 13-22.

Barkai-Golan, R. (2001). *Postharvest diseases of fruits and vegetables. Development and control.* Elsevier Science B. V. Amsterdam (The Netherlands).

Barkai-Golan, R. & Appelbaum, A. (1991). Synergistic effects of heat and sodium o-phenylphenate treatments to inactivate *Penicillium* spores and suppress decay in citrus fruits. *Tropical Science*, 31, pp. 229-233.

Barkai-Golan, R. & Phillips, D. S. (1991). Postharvest heat treatment of fresh fruits and vegetables for decay control. *Plant Disease*, 75, pp. 1085-1089.

Barlett, D.W., Clough, J.M., Godwin, J.R., Hall, A.A., Hamer, M. & Parr-Dobrzanski, B. (2002). The strobilurin fungicides. *Pest Management Science*, 58, pp. 649-662.

Barthlott, W. (1990). Scanning electron microscopy of the epidermal surface in plants. *In:* Claugher, D. (ed.). *Scanning Electron Microscopy in Taxonomy and Functional Morphology.* Clarendon Press. Oxford, (U.K.), pp. 69-94.

Baur. P. & Schönherr, J. (1995). Temperature dependence of the diffusion of organic compounds across plant cuticles. *Chemosphere*, 30, pp. 1331-1340.

Ben-Yehoshua, S., Peretz, J., Rodov, V., Nafussi, B., Yekutieli, O., Wiseblum, A. & Regev, R. (2000). Postharvest application of hot water treatment in citrus fruit: the road from the laboratory to the packing-house. *Acta Horticulturae*, 518, pp. 19-28.

Bianchi, G. (1995). Plant waxes. *In:* Hamilton, R.J. (ed.). *Waxes: chemistry, molecular biology and functions,* The Oily Press. Dundee, Scotland, (U.K.), pp. 177-222.

Brovelli, E.A. (2006). Pre- and postharvest factors affecting nutraceutical properties of horticultural products. *Stewart Postharvest Review*, 2, pp. 5.

Brown, G. E. (1984). Efficacy of citrus postharvest fungicides applied in water or resin solution wax. *Plant Disease*, 68, pp. 15-418.

Brown, G.E. & Dezman, D.J. (1990). Uptake of imazalil by citrus fruit after postharvest application and the effect of residue distribution on sporulation of *Penicillium digitatum. Plant Disease*, 74, pp. 927-930.

Brown, G.E., Nagy, S. & Maraulja, M. (1983). Residues from postharvest nonrecovery spray applications of imazalil to oranges and effects on green mold caused by *Penicillium digitatum. Plant Disease*, 67, pp. 954-957.

Budde, C.O., Polenta, G., Lucangeli, C.D., Murray, R.E. (2006). Air and immersion heat treatments affect ethylene production and organoleptic quality of 'Dixiland' peaches. *Postharvest Biology and Technology*, 41, pp. 32-37.

Bushong, P. M. & Timmer, L. V. (2000). Evaluation of postinfection control of citrus scab and melanose with benomyl, fenbuconazole and azoxystrobin. *Plant Disease*, 84, pp. 1246-1249.

Cabras, P., Schirra, M., Pirisi, F. M., Garau, V. L. & Angioni, A. (1999). Factors affecting imazalil and thiabendazole uptake and persistence in citrus fruits following dip treatments. *Journal of Agricultural and Food Chemistry*, 47, pp. 3352-3354.

Cantwell, M.I., Kang, J. & Hong, G. (2003). Heat treatment control sprouting and rooting of garlic cloves. *Postharvest Biology and Technology*, 30, pp. 57-65.

Carter, W.W. (1981a). Reevaluation of heated water dip as a postharvest treatment for controlling surface and decay fungi of muskmelon fruits. *HortScience*, 16, pp. 334-335.

Carter, W.W. (1981b). Postharvest treatments for control of stem-scar, rind, and decay fungi on cantaloup. *Plant Disease*, 65, pp. 815-816.

Cisneros-Zevallos, L. (2003). The use of controlled postharvest abiotic stresses as a tool for enhancing the nutraceutical content and adding value of fresh fruits and vegetables. *Journal of Food Science*, 68, pp. 1560-1565.

Coates, L.M., Johnson, G.I. & Cooke A. W. (1993). Postharvest disease control in mangoes using high humidity, hot air and fungicide treatments. *Annals of Applied Biology*, 123, pp. 441-448.

Costa, M.L., Civello, P.M., Chaves, A.R. & Martinez, G.A. (2005). Effect of hot air treatment on senescence and quality parameters of harvested broccoli (*Brassica oleracea* L. var. Italica). *Journal of Science and Food Agriculture*, 85, pp. 1154-1160.

Dang, K.T.H., Singh, Z. & Swinny, E.E. (2008). Impact of postharvest disease control methods and cold storage on volatiles, color development and fruit quality in ripe 'Kensington Pride' mangoes. *Journal of Agricultural and Food Chemistry*, 56, pp. 10667-10674.

Daniels, A. & Lucas, J. A. (1995). Mode of action of anilino-pyrimidine fungicide pyrimethanil. 1. *In vivo* activity against *Botrytis fabae* on broad bean (*Vicia faba*) leaves. *Pesticide Science*, 45, pp. 33-41.

D'Aquino, S., Schirra, M., Palma, A., Angioni, A., Cabras, P. & Migheli, Q. (2006). Residue levels and effectiveness of pyrimethanil vs imazalil when using heated postharvest dip treatment for control of *Penicillium* decay on citrus fruit. *Journal of Agricultural and Food Chemistry*, 54, pp. 4721-4726.

D'Aquino, S., Schirra, M., Palma, A., Tedde, M., Angioni, A., Garau, A. & Cabras, P. (2007). Residue levels and storage responses of nectarines, apricots and peaches after dip treatments with fludioxonil fungicide mixtures. *Journal of Agricultural and Food Chemistry*, 55, pp. 825-831.

D'Aquino, S., Schirra, M., Palma, A., Angioni, A., Cabras, P., Gentile, A. & Tribulato, E. (2009). Effectiveness of fludioxonil in control storage decay on pomegranate fruit. *Acta Horticulturae*, 818, pp. 313-318.

Dezman, D. J., Nagy, S. & Brown, G. E. (1986). Postharvest fungal decay control chemicals: treatments and residues in citrus fruits. *Residue Reviews*, 97, pp. 37-92.

D'hallewin, G., Schirra, M. & Manueddu. E. (1999). Effect of heat on epicuticular wax of cactus pear fruit. *Tropical Science*, 39, pp. 244-247.

Diaz Perez, J.C., Mejia, A., Bautista, S., Zaveleta, R., Villanueva, R. & Gomez, R.L. (2001). Response of sapote mamey [*Pouteria sapota* (Jacq.) H.E. Moore & Stearn] fruit to hot water treatments. *Postharvest Biology and Technology*, 22, pp. 156-167.

Dimitris, L., Pompodakis, N., Markellou, E. & Lionakis, S.M. (2005). Storage response of cactus pear fruit following hot water brushing. *Postharvest Biology and Technology*, 38, pp. 145-151.

Dodd, J.C., Bugante, R., Koomen, I., Jeffries, P. & Jeger, M.J. (1991). Pre- and postharvest control of mango anthracnose in the Philippines. *Plant Pathology*, 40, pp. 576-583.

Dore, A., Molinu, M.G., Venditti, T. & D'hallewin, G. (2009). Immersion of lemons into imazalil mixtures heated at 50 °C alters the cuticle and promoted permeation of imazalil into rind wounds. *Journal of Agriculture and Food Chemistry*, 57, pp. 623-631.

Eckert, J. W. & Ogawa, J. M. (1985). The chemical control of postharvest diseases: subtropical and tropical fruits. *Annual Review of Phytopathology*, 23, pp. 421-454.

Eckert, J. W. & Eaks, I.L. (1988). Postharvest disorders and diseases of citrus fruit. *In:* Reuther, W, Calavan, E.C. & Carman, G. E. (eds.). *The Citrus Industry*. Berkeley CA, University of California (U.S.A.), Vol 5, pp. 179-260.

El-Otmani, M., Arpaia, M.L., Coggins, C.W.Jr., Pehrson, Jr.J.E. & O'Connell, N.V. (1989). Developmental changes in 'Valencia' orange fruit epicuticular wax in relation to fruit position on the tree. *Scientia Horticulturae*, 41, pp. 69-81.

Errampalli, D. (2004). Effect of fludioxonil on germination and growth of *Penicillium expansum* and decay in apple cvs. Empire and Gala. *Crop Protection*, 23, pp. 811-817.

Errampalli, D. & Brubacher, N. R. (2006). Biological and integrated control of postharvest blue mold (*Penicillium expansum*) of apples by *Pseudomonas syringae* and Cyprodinil. *Biological Control*, 36, pp. 49–56.

Errampalli, D. & Crnko, N. (2004). Control of blue mold caused by *Penicillium expansum* on apples 'Empire' with fludioxonil and cyprodinil. *Canadian Journal of Plant Pathology*, 26, pp. 70–75.

Errampalli, D., Northover, J., Skog, L.; Brubacher, N.R. & Collucci, C. A. (2005). Control of blue mold (*Penicillium expansum*) by fludioxonil in apples (cv Empire) under controlled atmosphere and cold storage conditions. *Pesticide Managemement Science*, 61, pp. 591-596.

Fallik, E., Grinberg, S., Alkalai, S., Yekutieli, O., Wiseblum, A., Regev, R., Beres, H. & Bar-Lev, E. (1999). A unique rapid hot water treatment to improve storage quality of sweet pepper. *Postharvest Biology and Technology*, 15, 25-32.

Fallik, E., Aharoni, Y., Copel, A., Rodov, R., Tuvia-Alkalai, S., Horev, B., Yekutieli, O., Wiseblum, A. & Regev, R. (2000). A short hot water rinse reduces postharvest losses of 'Galia' melon. *Plant Pathology*, 49, pp. 333-338.

Fallik, E., Tuvia-Alkalai, S., Feng, X. & Lurie, S. (2001). Ripening characterization and decay development of stored apples after a short pre-storage hot water rinsing and brushing. *Innovative Food Science & Emerging Technologies*, 2, pp. 127-132.

Fallik, E. (2004). Prestorage hot water treatments (immersion, rinsing and brushing). *Postharvest Biology and Technology*, 32, pp. 125-134.

Fallik, E. & Lurie, S. (2007). Thermal control of fungi in the reduction of postharvest decay. *In:* Tang, J., Mitcham, E., Wang, S. & Lurie, S. (eds.). *Heat treatment for postharvest pest control: theory and practice*. CAB International, pp. 162-181.

Fan, L., Forney, C.F., Song, J., Doucette, C., Jordan, M.A., McRae, K.B. & Walker, B.A. (2008). Effect of hot water treatments on quality of highbush blueberries. *Journal of Food Science*, 73, pp. 292-297.

FAOSTAT (2011). www.fao.org. Date of consultation: June 15th, 2011.

Faust, M. & Shear, C.B. (1972). Fine structure of the fruit surface of three apple cultivars. *Journal of American Society for Horticultural Science*, 97, pp. 351-355.

Follet, P.A. & Sanxter, S.S. (2003). Lychee quality after hot water immersion and X-ray irradiation quarantine treatment. *HortScience*, 38, pp. 1159-1162.

Förster, H., Driever, G.F., Thompson, D.C. & Adaskaveg, J.E. (2007). Postharvest decay management for stone fruit crops in California using the "reduced-risk" fungicides fludioxonil and fenhexamid. *Plant Disease*, 91, pp. 209-215.

Freeman, B., Albrigo, L.G. & Biggs, R.H. (1979). Ultrastructure and chemistry of cuticular waxes of developing *Citrus* leaves and fruits. *Journal of American Society for Horticultural Science*, 104, 801-808.

Funamoto, Y., Yamauchi, N., Shigenaga, T. & Shigyo, M. (2003). Involvement of peroxidase in chlorophyll degradation in stored broccoli (*Brassica oleracea* L.) and inhibition of activity by heat treatments. *Postharvest Biology and Technology*, 28, pp. 39-46.

Gabler, F.M., Smilanick, J.L., Ghosoph, J.M. & Margosan, D.A. (2005). Impact of postharvest hot water or ethanol treatment of table grapes on gray mold incidence, quality, and ethanol content. *Plant Disease*, 89(3), pp. 309-316.

Garau, V. L., Angioni, A., Aguilera Del Real, A., Russo, M. T. & Cabras, P. (2002). Disappearance of azoxystrobin, cyprodinil, and fludioxonil on tomatoes in a greenhouse. *Journal of Agricultural and Food Chemistry*, 50, pp. 1929–1932.

Ghasemnezhad, M., Marsh, K., Shilton, R., Babalar, M. & Woolf, A. (2007). Effect of hot water treatments on chilling injury and heat damage in 'Satsuma' mandarins: antioxidant enzymes and vacuolar ATPase, and pyrophosphatase. *Postharvest Biology and Technology*, 48, pp. 364-371.

Gleen, G.M., Poovaiah, B.W. & Rasmussen, H.P. (1985). Pathways of calcium penetration through isolated cuticles of 'Golden Delicious' apple fruit. *Journal of American Society for Horticultural Science*, 110, pp. 166-171.

Gomez, F., Fernandez, L., Gergoff, G., Guiamet, J.J., Chaves, A. & Bartoli, C.G. (2008). Heat shock increases mitochondrial H_2O_2 production and extends postharvest life of spinach leaves. *Postharvest Biology and Technology*, 49, pp. 229-234.

Gonzales-Aguilar, G.A., Zacarias, L., Mulas, M. & Lafuente, M.T. (1997). Temperature and duration of water dips influence chilling injury, decay and polyamine content in 'Fortune' mandarins. *Postharvest Biology and Technology*, 12, pp. 61-69.

Gullino, M.L., Leroux, P. & Smith, C.M. (2000). Uses and challenges of novel compounds for plant disease control. *Crop Protection*, 19, pp. 1-11.

Hayward, F.W. & Grieson, W. (1960). Effects of treatment conditions on o-Phenilphenol residues in oranges. *Journal of Agricultural and Food Chemistry*, 8, pp. 308-310.

Heye, U.J., Speich, J., Siegle, H., Steinemann, A., Förster, B., Knauf Beiter, G., Herzog, J. & Hubele, A. (1994). CGA 219417–a novel broad-spectrum fungicide. *Crop Protection*, 13, pp. 541–549.

Hoa, T.T., Clark, C.J., Waddell, B.C. & Woolf, A.B. (2006). Postharvest quality of Dragon fruit (*Hylocereus hundatus*) following disinfesting hot air treatments. *Postharvest Biology and Technology*, 41, pp. 62-69.

Huajaikaew, L., Uthairatankij, A., Kanlayanarat, S. & Gemma, H. (2005). Effect of heat treatment on antioxidants and quality changes in papaya fruit stored at low temperatures. *Acta Horticulturae*, 682, 1063-1068

Jacobi, K.K., Macrae, E.A. & Hetherington, S.E. (2001). Postharvest heat disinfestation treatments of mango fruit. *Scientia Horticulturae*, 89, pp. 171-193.

Jemric, T., Ivic, D., Fruk, G., Matijas, H.S., Cvjetkovic, B., Bupic, M. & Pavkovic, B. (2011). Reduction of postharvest decay of peach and nectarine caused by *Monilinia laxa* using hot water dipping. *Food and Bioprocess Technology*, 4, pp. 149-154.

Jing, W., Tu, K., Shao, X.F., Su, Z.P., Zhao, Y., Wang, S. & Tang, J. (2010). Effect of postharvest short hot-water rinsing and brushing treatment on decay and quality of strawberry fruit. *Journal of Food Quality*, 33, pp. 262-272.

Johnson, G.I., Sangchote, S. & Cooke, A.W. (1990). Control of stem end rot (*Dothiorella dominicana*) and other postharvest diseases of mangoes (cv Kensington Pride) during short- and long-term storage. *Tropical Agriculture*, 67, pp. 183-187.

Johnson, J.A., Wang, S. & Tang, J. (2003). Thermal death kinetiks of fifth-instar *Plodia interpunctella* (*Lepidoptera: Pyralidae*). *Journal of Economic Entomology*, 96(2), pp. 519-524.

Jones, A.L. & Burton, C. (1973). Heat and fungicide treatments to control postharvest brown rot of stone fruits. *Plant Disease Reporter*, 57, pp. 62-66.

Kader, A.A. (2003). A perspective on postharvest horticulture. *HortScience*, 38(5), pp. 1004-1008.

Kanetis, L., Förster, H. & Adaskaveg, J.F. (2007). Comparative efficacy of the new postharvest fungicides azoxystrobin, fludioxonil, and pyrimethanil for managing citrus green mold. *Plant Disease*, 91, pp. 1502-1511.

Kanetis, L., Förster, H. & Adaskaveg, J.F.(2008a). Optimizing efficacy of new postharvest fungicides and evaluation of sanitizing agents for managing citrus green mold. *Plant Disease*, 92, pp. 261-269.

Kanetis, L., Förster, H., Jones, C.A., Borkovick, K.A. & Adaskaveg, J.E. (2008b). Characterization of biochemical and genetic mechanisms of fludioxonil and pyrimethanil resistance in field isolates of *Penicillium digitatum*. *Phytopathology*, 98, pp. 205-214.

Karabulut, O.A., Gabler, M.F., Mansour, M. & Smilanick, J.L. (2004). Postharvest ethanol and hot water treatments of table grapes to control gray mold. *Postharvest Biology and Technology*, 36, pp. 169-176.

Kenawy, E.R., Sherrington, D.C. & Akelah, A. (1992). Controlled release of agrochemical molecules chemically bonded to polymers. *European Polymers Journal*, 28, pp. 841-862.

Lafuente, M.T. & Zacarias, L. (2006). Postharvest physiological disorders in citrus fruit. *Stewart Postharvest Review*, 1: 2.

Lafuente, M.T., Zacarias, L., Sala, J.M., Sanchez-Ballesta, M.T., Gosalbes, M.J., Marcos, J.F., Gonzales-Candelas, L., Lluch, Y., Granell, A. (2005). Understanding the basis of chilling injury in Citrus fruit. *Acta Horticulturae*, 682, pp. 831-842.

Lafuente, M.T., Ballester, A.R., Calejero J. & González-Candelas, L. (2011) Effect of high temperature-conditioning treatments on quality, flavonoid composition and vitamin C of cold stored 'Fortune' mandarins. *Food Chemistry 12, pp.*: 1080-1086.

Lagopodi, A.L., Cetiz, K., Koukounaras, A. & Sfakiotakis, E.M. (2009). Acetic acid, ethanol and steam effects on the growth of *Botrytis cinerea* in vitro and combination of steam and modified atmosphere packaging to control decay in kiwifruit. *Journal of Phytopathology*, 157, pp. 79-84.

Lara, I., Garcia, P. & Vendrell, M. (2006). Post-harvest heat treatment modify cell wall composition of strawberry (*Fragaria x ananassa* Duch.) fruit. *Scientia Horticulturae*, 109, pp. 48-53.

Leroux, P. (2003). Modes d'action des produits phytosanitaires sur les organismes pathogèns des plantes. *Comptes Rendues de Biologies*, 326, pp. 9-21.

Lezcano, M., Al-Soufi, W., Novo, M., Rodrìguez-Nuñez, E. & Tato, J. V. (2002). Complexation of several benzimidazole-type fungicides with α- and β-cyclodextrins. *Journal of Agricultural and Food Chemistry, 50,* pp. 108-112.

Liu, F., Tu, K., Shao, X., Zhao, Y., Tu, S., Su, J., Hou, Y. & Zou, X. (2010). Effect of hot air treatment in combination with *Pichia guilliermondii* on postharvest anthracnose rot of loquat fruit. *Postharvest Biology and Technology, 58,* pp. 65-71.

Lurie, S. (1998). Postharvest heat treatments of horticultural crops. *Horticultural Review, 22,* pp. 91-121.

Lurie, S. (2006). The effect of high temperature treatment on quality of fruits and vegetables. *Acta Horticulturae, 712,* pp. 165-174.

Lydakis, D., & Aked, J. (2003). Vapour heat treatment of Sultanina table grapes. II: effects on postharvest quality. *Postharvest Biology and Technology, 27,* pp. 117-126.

Malakou, A. & Nanos, G.D. (2005). A combination of hot water treatment and modified atmosphere packaging maintains quality of advanced maturity 'Caldesi 2000' nectarines and 'Royal Glory' peaches. *Postharvest Biology and Technology, 38,* pp. 106-114.

Margot, P., Huggenberger, F., Amrein, J. & Weiss, B. (1998). CGA 279202: A new broad spectrum strobilurin fungicide. *In: Proceedings of the Brighton Crop Protection Conference, Pests and Diseases,* pp. 375-382.

Mayberry, K.S. & Hartz, T.K. (1992). Extension muskmelon storage life through the use of hot water treatment and polyethylene wraps. *HortScience, 27,* pp. 324-326.

McCornack, A.A. & Hayward, F.W. (1968). Factors affecting decay control of Dowcide A-hexamine treated citrus fruit. *Proceeding of Florida State Horticultural Society, 81,* pp. 290-293.

McDonald, R.E., Miller, W.R., McCollum, T.G. & Brown, G.E. (1991). Thiabendazole and imazalil applied at 53C reduce chilling injury and decay of grapefruit. *HortScience,* 26, pp. 397-399.

McGuire, R.G. & Campbell, C.A. (1993). Imazalil for postharvest control of anthracnose on mango fruit. *Acta Horticulturae, 341,* pp. 371-376.

Milling, R. J. & Richardson, C. J. (1995). Mode of action of the anilino-pyrimidine fungicide pyrimethanil. 2. Effects on enzyme secretion in *Botrytis cinerea. Pesticide Science, 45,* pp. 43- 48.

Mirdehghan, S.H., Rahemi, M., Serrano, M., Guillen, F., Martinez-Romero, D. & Valero, D. (2006). Prestorage heat treatment to maintain nutritive and functional properties during postharvest cold storage of pomegranate. *Journal of Agricultural and Food Chemistry, 54,* pp. 8495-8500.

Montero, C.R.S., Antes, R.B., Schwarz, L.L., Cunha dos Santos, L., Pires dos Santos, R. & Bender, R.J. (2010). Complementary physical and chemical treatments as an alternative to fungicide use to control postharvest decay incidence and fruit quality of Montenegrina tangerines. *Crop Protection, 29,* pp. 1076-1083.

Montesinos-Herrero, C. & Palou, L. (2010). Combination of physical and low-toxicity chemicals postharvest treatments for integrating disease management of citrus fruit: a review. *Stewart Postharvest Review, 1 (1),* pp. 1-11.

Mulas, M. (2006). Tratamientos postcosecha de los citricos con calor para el control del daño por frio y la mejor efficacia de los tratamientos fungicidas. *Todo Citrus, 8(35),* pp. 13-21.

Mulas, M. (2008). Uso de tratamientos térmicos en postcosecha de los frutos. In: L. Magali do Nascimento, J.D. De Negri & D. de Mattos Jr. *Tópicos em qualidade e pós-colheita de frutas*. Instituto Agronomico e Fundag, Cordeiropolis (Brasil), pp. 165-178.

Mulas, M. & Schirra, M. (2007). The effect of heat conditioning treatments on the postharvest quality of horticultural crops. *Stewart Postharvest Review*, 1: 2, pp. 1-6.

Mulas, M. & Schirra, M. (2009). Termoterapia e qualità post-raccolta dei prodotti ortofrutticoli. *Italus Hortus*, 16(4), pp. 53-65.

Mulas, M., Perinu, B., Francesconi, A.H.D., D'hallewin, G. & Schirra, M. (2001). Quality of blood oranges following heat treatments for disinfesting fruit fly. Proceedings of the International Conference *"Improving postharvest technologies of fruits, vegetables and ornamentals"*. Murcia (Spain), October 19-21, 2000, pp. 740-745.

Mulas, M., Fadda, A., Nieddu, M.A. & Schirra, M. (2004). Effetti della termoterapia sulla fisiologia di frutti di arancio di diversa conservabilità. *Italus Hortus*, 11(1), pp. 49-51.

Mulas, M., Mereu, V. & Schirra, M. (2006). Effetto dei trattamenti di termoterapia sulla produzione di etanolo e acetaldeide in frutti di agrumi. *Italus Hortus*, 13(5), pp. 38-42.

Mulas, M., D'Aquino, S., Palma, A., Ligios, G. & Schirra, M. (2008a). Effects of postharvest treatments with hot water, soy lecithin, calcium chloride, or 1-Methylcyclopropene (1- MCP) and cold storage on internal browning development in 'Camusina' and 'Precoce di Fiorano' pears. Proceedings of the International Congress *"Novel approaches for the control of postharvest diseases and disorders"*. Bologna (Italy), 3-5 may 2007, pp. 199-205.

Mulas, M., Mereu, V., Ligios, G. & Schirra, M. (2008b). Impact of heat treatments on respiration and ethylene production rates and on ethanol and acetaldehyde accumulation in the juice or their release by 'Valencia late' oranges during storage. Proceedings of the International Congress *"Novel approaches for the control of postharvest diseases and disorders"*. Bologna (Italy), 3-5 may 2007, pp. 206-215.

Narayanasamy, P. (2006). *Postharvest pathogens and disease management*. John Wiley & Sons, Inc. Hoboken, New Jersey (U.S.A.).

Oren, Y., Solel, Z., Kimki, M. & Sadovski, A. (1999). Controlling *Alternaria alternata* in the citrus varieties 'Minneola' and 'Nova'. *Phytoparasitica* (Abstr.), 27(2), pp. 152-153.

Palou, L., Smilanick, J.L., Usall, J. & Viñas, I. (2001). Control of postharvest blue and green molds of oranges by hot water, sodium carbonate, and sodium bicarbonate. *Plant Disease*, 85(4), pp. 371-376.

Palou, L., Crisosto, C.H. & Garner, D. (2007). Combination of postharvest antifungal chemical treatments and controlled atmosphere storage to control gray mold and improve storability of 'Wonderful' pomegranates. *Postharvest Biology and Technology*, 43, pp. 133–142.

Papadopoulou-Mourkidou, E. (1991). Postharvest-applied agrochemicals and their residues in fresh fruits and vegetables. *Journal of Association of Official Analytical Chemsit International*, 74, pp. 745-765.

Porat, R., Daus, A., Weiss, B., Cohen, L., Fallik, E. & Droby, S. (2000). Reduction of postharvest decay in organic citrus fruit by a short hot water brushing treatment. *Postharvest Biology and Technology*, 18, pp. 151-157.

Prusky, D., Fuchs, Y., Kobiler, I., Roth, I., Weksler, A., Shalom, Y., Fallik, E., Zauberman, G., Pesis, E., Akerman, M., Ykutiely, O., Weisblum, A., Regev, R. & Artes, L. (1990).

Effect of hot water brushing, prochloraz treatment and waxing on the incidence of black spot decay caused by *Alternaria alternata* in mango fruits. *Postharvest Biology and Technology*, 15, pp. 165-174.

Quevedo, M.A., Sanico, R.T. & Baliad, M.E. (1991). The control of postharvest diseases of taro corms. *Tropical Science*, 31, pp. 359-364.

Rafaeli, A., Kostukovshy, A. & Carmeli, D. (2006). Successful disinfestation of sap-beetle contaminations from organically grown dates using heat treatment. A case study. *Phytoparasitica*, 34(2), pp. 204-212.

Ragsdale, N.N. (2000). The impact of food quality protection act on the future of plant disease management. *Annual Review of Phytopathology*, 38, pp. 577-596.

Reuveni, M. (2000). Efficacy of trifloxystrobin (Flint), a new strobilurin fungicide, in controlling powdery mildews on apple, mango and nectarine, and rust on prune trees. *Crop Protection*, 19, pp. 335-341.

Reuveni, M. (2001). Activity of trifloxystrobin against powdery and downy mildew diseases of grapevines. *Canadian Journal of Plant Pathology*, 23, pp. 52-59.

Rosslenbroich, H. J. & Stuebler, D. (2000). *Botrytis cinerea* – history of chemical control and novel fungicides for its management. *Crop Protection*, 19, pp. 557-561.

Roy, S., Conway, W.S., Watada, A.E., Sams, C.E., Erbe, E.F. & Wergin, W.P. (1994). Heat treatment affects epicuticular wax structure and postharvest calcium uptake in 'Golden Delicious' apples. *HortScience*, 29, pp. 1056-1058.

Roy, S., Conway, W S., Watada, A.E., Sams, C.E., Erbe, E.F.m & Wergin, W.P. (1999). Changes in ultrastructure of the epicuticular wax and postharvest calcium uptake in apples. *HortScience*, 34, 121-124.

Sala, J. M. (1998). Involvement of oxidative stress in chilling injury 885 in cold-stored mandarin fruit. *Postharvest Biology and Technology*, 13, pp. 255-161.

Sala, J.M. & Lafuente, M.T. (1999). Catalase in the heat induced chilling injury of cold-stored hybrid Fortune mandarin fruits. *Journal of Agricultural and Food Chemistry*, 47, pp. 2410-2414.

Sala, J.M. & Lafuente, M.T. (2000). Catalase enzyme activity is related to tolerance of mandarin fruits to chilling. *Postharvest Biology and Technology*, 20, pp. 81-89.

Schiffman-Nadel, M., Chalutz, E., Waks, J. & Lattar, F.S. (1972). Reduction of pitting of grapefruit by thiabendazole during long-term cold storage. *HortScience*, 7, pp. 394-395.

Schiffman-Nadel, M., Chalutz, E., Waks, J. & Dagan, M. (1975). Reduction of chilling injury in grapefruit by thiabendazole and benomyl during long-term storage. *Journal of American Society for Horticultural Science*, 100, pp. 270-272.

Schirra, M. (2005). Postharvest pest management of horticultural crops by combined heat therapy and fungicide treatment. *Acta Horticulturae*, 682, pp. 2127-2132.

Schirra, M. & D'hallewin, G. (1997). Storage performance of Fortune mandarins following hot water dips. *Postharvest Biology and Technology*, 10, pp. 229-238.

Schirra, M. & Mulas, M. (1993). Keeping quality of 'Oroblanco' grapefruit-type as affected by hot dip treatments. *Advances in Horticultural Science*, 7, pp. 73-76.

Schirra, M. & Mulas, M. (1994). Storage of 'Monreal' clementines as affected by CaCl$_2$ and TBZ postharvest treatments. *Agricoltura Mediterranea*, 124, pp. 238-248.

Schirra, M. & Mulas, M. (1995a). Improving storability of 'Tarocco' oranges by postharvest hot-dip fungicide treatments. *Postharvest Biology and Technology*, 6, pp. 129-138.

Schirra, M. & Mulas, M. (1995b). 'Fortune' mandarin quality following prestorage water dips and intermittent warming during cold storage. *HortScience*, 30(3), pp. 560-561.

Schirra, M. & Mulas, M. (1995c). Influence of postharvest hot-water dip and imazalil-fungicide treatments on cold-stored 'Di Massa' lemons. *Advances in Horticultural Science*, 9 pp. 43-46.

Schirra, M., Mulas, M. & Baghino, L. (1995). Influence of postharvest hot-dip fungicide treatments on 'Redblush' grapefruit quality during long-term storage. *Food Science and Technology International*, 1 (1), pp. 35-40.

Schirra, M., Barbera, G., D'Aquino, S., La Mantia, T. & McDonald, R.E. (1996a). Hot dips and high temperature conditioning to improve shelf quality of late-crop cactus pear fruit. *Tropical Science*, 36, 159-165.

Schirra, M., Cabras, P., Angioni, A. & Melis, M. (1996b). Residue level of imazalil fungicide in lemons following prestorage dip treatment at 20 and 50 °C. *Journal of Agricultural and Food Chemistry*, 44, pp. 2865-2869.

Schirra, M., Agabbio, M., D'hallewin, G., Pala, M. & Ruggiu, R. (1997a). Response of 'Tarocco' oranges to picking date, postharvest hot water dips and chilling storage temperature. *Journal Agricultural and Food Chemistry*, 45, pp. 3216-3220.

Schirra, M., Cabras, P., Angioni, A., D'hallewin, G., Ruggiu, R. & Minelli, E. V. (1997b). Effect of heated solutions on decay control and residues of imazalil in lemons. *Journal of Agricultural and Food Chemistry*, 45, pp. 4127-4130.

Schirra, M., D'hallewin, G., Cabras, P., Angioni, A. & Garau, V.L. (1998a). Seasonal susceptibility of Tarocco oranges to chilling injury as affected by hot water and thiabendazole postharvest dip treatments. *Journal Agricultural and Food Chemistry*, 46, pp. 1177-1180.

Schirra, M., Cabras, P., Angioni, A., Ruggiu, R. & Minelli, E. V. (1998b). Synergistic actions of 50 °C water and imazalil dip treatments to preserve quality of late-season 'Marsh' grapefruits. *Advances in Horticultural Science*, 12, pp. 63-66.

Schirra, M., Angioni, A., Ruggiu, R., Minelli, E.V. & Cabras, P. (1998c). Thiabendazole uptake and persistence in lemons following postharvest dips at 50 °C. *Italian Journal of Food Science*, 10, 165-170.

Schirra, M., D'hallewin, G., Inglese, P. & La Mantia, T. (1999). Epicuticular changes and storage potential of cactus pear [*Opuntia ficus-indica* Miller (L.)] fruit following gibberellic acid preharvest sprays and postharvest heat treatment. *Postharvest Biology and Technology*, 17, pp. 79-88.

Schirra, M., D'hallewin, G., Ben-Yehoshua, S. & Fallik, E. (2000a). Host pathogen interaction modulated by heat treatment. *Postharvest Biology and Technology*, 21, pp. 71-85.

Schirra, M., D'hallewin, G., Cabras, P., Angioni, A., Ben-Yehoshua, S. & Lurie, S. (2000b). Chilling injury and residue uptake in cold stored 'Star Ruby' grapefruit following thiabendazole and imazalil dip treatments at 20 and 50°C. *Postharvest Biology and Technology*, 20, pp. 91-98.

Schirra, M., Cabras, P., Angioni, A., D'hallewin, G. & Pala, M. (2002a). Residue uptake and storage responses of 'Tarocco' blood oranges after preharvest thiabendazole sprays and postharvest heat treatment. *Journal of Agricultural and Food Chemistry*, 50, pp. 2293-2296.

Schirra, M., Brandolini, V., Cabras, P., Angioni, A. & Inglese, P. (2002b). Thiabendazole uptake and storage performance of cactus pear (*Opuntia ficus-indica* (L.) Mill. cv

Gialla) fruit following postharvest treatments with reduced doses of fungicide at 52 °C. *Journal of Agricultural and Food Chemistry*, 50, pp. 739-743.

Schirra, M., Delogu, G., Cabras, P., Angioni, A., D'hallewin, G., Veyrat, A., Marcos, J.F. & Candelas, L.G. (2002c). Complexation of Imazalil with β-cyclodestrin, residue uptake, persistence, and activity against *Penicillium* decay in Citrus fruit following postharvest dip treatment. *Journal of Agricultural and Food Chemistry*, 50, pp. 6790-6797.

Schirra, M., Cabras, P., Angioni, A. & Brandolini, V. (2002d). Residue levels and storage decay control in 'Star Ruby' grapefruit after dip treatments with azoxystrobin. *Journal of Agricultural and Food Chemistry*, 50, pp. 1461-1465.

Schirra, M., Mulas, M., Fadda, A. & Cauli, E. (2004). Cold quarantine responses of blood oranges to postharvest hot water and hot air treatments. *Postharvest Biology and Technology*, 31, pp. 191-200.

Schirra, M., Mulas, M., Fadda, A., Mignani, I. & Lurie, S. (2005a). Chemical and quality traits of 'Olinda' and 'Campbell' oranges after heat treatment at 44 or 46 °C for fruit fly disinfestation. *LWT - Lebensmittel-Wissenshaft und-Technologie*, 38, pp. 519-527.

Schirra, M., D'Aquino, S., Palma, A., Marceddu, S., Angioni, A., Cabras, P., Scherm, B. & Migheli, Q. (2005b). Residue level, persistence, and storage performance of Citrus fruit treated with fludioxonil. *Journal of Agricultural and Food Chemistry*, 53, pp. 6718-6724.

Schirra, M., D'Aquino, S., Palma, A., Angioni, A., Cabras, P. & Migheli, Q. (2006). Residues of the quinone outside inhibitor fungicide trifloxystrobin after postharvest dip treatments to control *Penicillium* spp. on citrus fruit. *Journal of Food Protection*, 69, pp. 1646-1652.

Schirra, M., D'Aquino, S., Palma, A., Angioni, A. & Cabras, P. (2008a). Factors affecting the sinergy of thiabendazole, sodium bicarbonate, and heat to control postharvest green mold of citrus fruit. *Journal of Agricultural and Food Chemistry*, 56, pp. 10793-10798.

Schirra, M.; D'Aquino, S.; Mulas, M.; Melis, R.A.M.; Giobbe, S.; Migheli, Q.; Garau, A.; Angioni, A. & Cabras, P. (2008b). Efficacy of heat treatments with water and fludioxonil for postharvest control of blue and gray molds on inoculated pears and fludioxonil residues in fruit. *Journal of Food Protection*, 71(5): 967-972.

Schirra, M., D'Aquino, S., Palma, A., Angioni, A. & Cabras, P. (2008c). Factors affecting the synergy of thiabendazole, sodium bicarbonate, and heat to control postharvest green mould of citrus fruit. *Journal of Agricultural and Food Chemistry*, 56,. 10793-10798.

Schirra, M., Palma, A., D'Aquino, S., Angioni, A., Minello, E.V., Melis, M. & Cabras, P. (2008d). Influence of postharvest hot water treatment on nutritional and functional properties of kumquat (*Fortunella japonica* Lour. Swingle cv. Ovale) fruit. *Journal of Agricultural and Food Chemistry*, 56, pp. 455-460.

Schirra, M., D'Aquino, S., Cabras, P. & Angioni, A. (2009a). Influence of post-harvest application rates of cyprodinil, treatment time and temperature on residue levels and efficacy in controlling green mould on 'Valencia' oranges. *Food Additives and Contaminants*, 26 (7), pp. 1033-1037.

Schirra, M., D'Aquino, S., Migheli, Q., Pirisi, F.M. & Angioni, A. (2009b). Influence of postharvest treatments with fludioxonil and soy lecithin co-application in controlling blue and gray mold and fludioxonil residues in Coscia pears. *Food Additives and Contaminants. Part A*, 26, pp. 68-72.

Schirra, M., Palma, A., Barberis, A., Angioni, A., Garau, V. L., Cabras, P. & D'Aquino, S. (2010). Postinfection activity, residue levels, and persistence of azoxystrobin, fludioxonil, and pyrimethanil applied alone or in combination with heat and imazalil for green mold control on inoculated oranges. *Journal of Agricultural and Food Chemistry*, 58, pp. 3661-3666.

Schmidt, L.S., Gosoph, J.M., Matgosan, D.A. & Smilanick, J.L. (2006). Mutation at β-tubulin codon 200 indicated thiabendazole resistance in Penicillum digitatum collected from California citrus packinghouses. *Plant Disease*, 90, pp. 765-770.

Schreiber, L. & Schönherr, J. (2009). Water and solute permeability of plant cuticles. Measurement and data analysis. Springer-Verlag, Berlin.

Schreiner, M. & Huyskens,-Keil, S. (2006). Phytochemicals in fruit and vegetables: health promotion and postharvest elicitors. *Critical Reviews in Plant Science*, 25, pp. 267-278.

Scott, K.J., Brown, B.L., Chaplin, G.R., Wilcox, M.E. & Bain, J. M. (1982). The control of rotting and browning of litchi fruit by hot benomyl and plastic film. *Scientia Horticulturae*, 16, pp. 253-262.

Sharma, R.L. & Kaul, J.L. (1990). Efficacy of hot water and carbendazim treatments in controlling brown rot of apple. *Indian Journal of Micology and Plant Pathology*, 20, pp. 47-48.

Shellie, K.C. & Robert, I.M. (2000). Postharvest disinfestation heat treatments: response to fruit and fruit fly larvae to different heating media. *Postharvest Biology and Technology*, 21, pp. 51-60.

Sholberg, P.L., Bedford, K. & Stokes, S. (2005). Sensitivity of *Penicillium* spp. and *Botrytis cinerea* to pyrimethanil and its control of blue and gray mold of stored apples. *Crop Protection*, 24, pp. 127-134.

Sivakumar, D. & Korsten, L. (2006). Evaluation of the integrated application of two types of modified atmosphere packaging and hot water treatments on quality retention in the litchi cultivar 'McLean's Red'. *Journal of Horticultural Science & Biotechnology*, 81, pp. 639-644.

Smilanick, J.L., Michael, I.F., Mansour, M.F.;,Mackey, B.E.; Margosan, D.A., Florers, D. & Weist, C. F. (1997). Improved control of green mold of citrus with imazalil with warm water compared with its use in wax. *Plant Disease*, 81, pp. 1299-1304.

Smilanick, J.L., Mansour, M.F., Margosan, D.A., Gabler, F.M. & Goodwine W.R. (2005). Influence of pH and NaHCO3 on effectiveness of imazalil to inhibit germination of *Penicillium digitatum* and to control postharvest green mold on citrus fruit. *Plant Disease*, 89, p.p. 640–648.

Smilanick, J.L., Mansour, M.F. & Sorensen, D. (2006a). Pre- and postharvest treatments to control green mold of citrus fruit during ethylene degreening. *Plant Disease*, 90(1), pp. 89-96.

Smilanick, J.L., Mansour, M.F., Gabler, F.M. & Goodwine, W.R. (2006b). The effectiveness of pyrimethanil to inhibit germination of *Penicillium digitatum* and to control citrus green mold after harvest. *Postharvest Biology and Technology*, 42, pp. 75-85.

Smilanick, J.L., Mansour, M.F., Mlikota Gabler, F. & Sorenson, D. (2008). Control of citrus postharvest green mold and sour rot by potassium sorbate combined with heat and fungicides. *Postharvest Biology and Technology*, 47, pp. 226–238.

Smith, W.L.Jr. (1971). Control of brown rot and rhyzopus of inoculated peaches with hot water and hot chemical suspensions. *Plant Disease Reporter*, 55, pp. 228-230.

Smith, W.L. & Anderson, R.E. (1975). Decay control of peaches and nectarines during and after controlled atmosphere and air storage. *Journal of American Society for Horticultural Science*, 100, pp. 84-86.

Soto-Zamora, G., Yahia, E., Brecht, J.K. & Gardea, A. (2005). Effect of postharvest hot air treatment on the quality and antioxidant level in tomato fruit. *Lebensmittel-Wissenshaft und Technologie*, 38, pp. 657-663.

Spalding, D.H., Vaught, H.C., Day, R.H. & Brown, G. A. (1969). Control of blue mold rot development in apples treated with heated and unheated fungicides. *Plant Disease Reporter*, 53, pp. 738-742.

Spalding, D.H. & Reeder, W.F. (1972). Postharvest disorders of mangos as affected by fungicides and heat treatments. *Plant Disease. Reporter*, 56, pp. 751-753.

Spalding, D.H. & Reeder, W.F. (1978). Controlling market diseases of mangos with heated benomyl. *Proceedings of Florida State Horticultural Society*, 91, pp. 186-187.

Spalding, D.H. & Reeder, W.F. (1986). Decay and acceptability of mangos treated with combinations of hot water, imazalil, and γ-radiation. *Plant Disease*, 70, pp. 1149-1151.

Sudisha, J., Niranjana, S.R., Sukanya, S.L., Girijamba, R., Lakshmi Devi, N. & Shekar Shetty, H. (2010). Relative efficacy of strobilurin formulations in the control of downy mildew of sunflower. *Journal of Pest Science*, 83, pp. 461-470.

Swart, S., H., Serfontein, J.J., Swart, C. & Labuschagne, C. (2009). Chemical control of postharvest disease of mango: the effect of fludioxonil and prochloraz on soft brown rot, stem end rot and antrachnose. *Acta Horticulturae*, 820, pp. 503-509.

Szejtli, J. (1988). Cyclodextrin Technology. *In*: Davies, J.E.D. (ed.). *Topics in Inclusion Science*. Kluwer Academic Publishers, Dordrecht (The Netherlands), Vol. 8, pp. 1-450.

Talcott, S.T., Moore, J.P., Lounds-Singleton, A.J. & Percival, S.S. (2005). Ripening associated phytochemicals changes in mangos (*Mangifera indica*) following thermal quarantine and low-temperature storage. *Journal of Food Science*, 70(5), pp. 337-341.

Ticha, J., Hajslova, J., Jech, M., Honzicek, J., Lacina, O., Kohoutkova, J, Kocourek, V., Lansky, M., Kloutvorova, J. & Falta, V. (2008). Changes of pesticide residues in apples during cold storage. *Food Control*, 19, pp. 247-256.

Tomlin, C. (1997). *The pesticide manual: a world compendium*. British Crop Protection Council, Farnham (U.K.).

Vicente, A.R., Martinez, G.A., Civello, P.M. & Chaves, A.R. (2002). Quality of heat-treated strawberry fruit during refrigerated storage. *Postharvest Biology and Technology*, 25, pp. 59-71.

Wang, C.Y. (1993). Approaches to reduce chilling injury of fruits and vegetables. *Horticultural Review*, 15. pp. 63-95.

Wang, C.Y. & Anderson, R.E. (1985) Progress on controlled atmosphere storage and intermittent warming of peaches and nectarines. In: *Symposium series*, Oregon State University, School of Agriculture. Vol. 1, pp. 221–228.

Wardowski, W.F., Hayward, F.W. & Dennis, J.D. (1974). A flood-recovery TBZ fungicide treatment system for citrus fruits. *Proceedings of Florida State Horticultural Society*, 87, pp. 241-243.

Waskar, D.P. (2005). Hot water treatment for disease control and extension of shelf life of 'Kesar' mango (*Mangifera indica* L.) fruits. *Acta Horticulturae*, 682, pp. 1319-1323.

Weerahewa, D. & Adikaram, N.K.B. (2005). Heat-induced tolerance to internal browning of pineapple (*Ananas comosus* cv. 'Mauritius') under cold storage. *Journal of Horticultural Science & Biotechnology*, 80(4), pp. 503-509.

Wells, J.M., & Harvey, J.M. (1970). Combination heat and 2,6-dichloro-4-nitroaniline treatments for control of *Rhizopus* and brown rot of peaches, plums, and nectarines. *Phytophathology*, 60, pp. 116-120.

Wells, J.M. & Gerdts, M. H. (1971). Pre- and postharvest benomyl treatments for control of brown rot of nectarines in California. *Plant Disease Reporter*, 55, pp. 69-72.

Wells, J.M. (1972). Heated wax-emulsions with benomyl and 2,6,-dichloro-4-nitroaniline for control of postharvest decay of peaches and nectarines. *Phytopathology*, 62, pp. 129-133.

Wild, B.L. & Hood, C.W. (1989). Hot dip treatment reduce chilling injury in long-term storage of 'Valencia' oranges. *HortScience*, 24, 109-110.

Wills, R.B.H., Brown, B.L. & Scott, K.J. (1982). Control of ripe fruit rots of guavas by heated benomyl and guazatine dips. *Australian Journal of Experimental Agriculture and Animal Husbandry*, 22, pp. 437-440.

Wong, L.S., Jacob, K.K. & Giles, J.E. (1991). The influence of hot benomyl dips on the appearance of cold stored lychee (*Litchi chinensis* Sonn.). *Scientia Horticulturae*, 46, pp. 245-251.

Wood, P.M. & Hollomon, D.W. (2003). A critical evaluation of the role of alternative oxidase in the performance of strobilurin and related fungicides acting at the Qo site of complex III. *Pesticide Management Science*, 59, pp. 499-511.

Wszelaki, A.L. & Mitcham, E.J. (2003). Effect of combination of hot water dips, biological control and controlled atmospheres for control of grey mold on harvested strawberry. *Postharvest Biology and Technology*, 27, pp. 255-264.

Yearsley, C.W., McGrath, H.J.W., Taucher, J.A. & Dale, J.R. (1987). Red tamarillos (*Cyphomandra betacea*): post-harvest control of fungal decay with hot water and imazalil dips. *New Zealand Journal Experimental Agriculture*, 15, pp. 223-228.

Yearsley, C.W., Huang, B.I., McGrat, H.J.W., Fry, J., Stec, M.G.H. & Dale, J. R. (1988). Red tamarillos (*Cyphomandra betacea*): comparison of two postharvest dipping strategies for the control of fungal storage disorders. *New Zealand Journal of Experimental Agriculture*, 16, pp. 359-366.

Yildiz, F., Kinay, P., Yildiz, M., Sen, F. & Karacali, I. (2005). Effects of preharvest applications of $CaCl_2$, 2,4-D and benomyl and postharvest hot water, yeast and fungicide treatments on development of decay on satsuma mandarins. *Journal of Phytopathology*, 153, pp. 94-98.

Zhang, J. (2007). The potential of a new fungicide fludioxonil for stem-end rot and green mold control on Florida citrus fruit. *Postharvest Biology and Technology*, 46, pp. 262-270.

Zhang, H., Zheng, X., Wang, L., Li, S. & Liu, R. (2005). Effect of yeast antagonist in combination with hot water dips on postharvest *Rhizopus* rot of strawberries. *Journal of Food Engineering*, 78, pp. 281-287.

Zhang, H., Wang, L., Zheng, X. & Dong, Y. (2007). Effect of yeast antagonist in combination with heat treatment on postharvest blue mold decay and *Rhizopus* decay of peaches. *International Journal of Food Microbiology*, 115, pp. 53-58.

Zhang, H., Wang, L., Huang, X., Dong, Y. & Zheng, X. (2008). Integrated control of postharvest blue mold decay of pears with hot water treatment and *Rhodotorula glutinis*. *Postharvest Biology and Technology*, 49, pp. 308-313.

Zhao, Y., Tu, K., Su, J., Tu, S., Hou, Y., Liu, F. & Zou, X. (2009). Heat treatment in combination with antagonistic yeast reduces diseases and elicits the active defense responses in harvested cherry tomato fruit. *Journal of Agricultural and Food Chemistry*, 57, pp. 7565-7570.

Zhao, Y., Tu, K., Tu, S., Liu, M., Su, J. & Hou, Y. (2010). A combination of heat treatment and *Pichia guilliermondii* prevents cherry tomato spoilage by fungi. *International Journal of Food Microbiology*, 137, pp. 106-110.

Zhou, T., Xu, S.Y., Sun, D.W. & Wang, Z. (2002). Effects of heat treatment on postharvest quality of peaches. *Journal of Food Engineering*, 34, pp. 17-22.

9

Role of MAP Kinase Signaling in Secondary Metabolism and Adaptation to Abiotic/Fungicide Stress in *Fusarium*

Emese D. Nagygyörgy, László Hornok and Attila L. Ádám
Agricultural Biotechnology Center and Department of Plant Protection,
Szent István University, Mycology Group of the Hungarian Academy of Sciences, Gödöllő,
Hungary

1. Introduction

Phosphorylation by protein kinases including mitogen activated protein kinases (MAPKs) is a major signal transduction mechanism used by eukaryotic cells to regulate different functions, virtually almost all activities that define their phenotypic behavior. Considering their diverse cellular roles, it was not surprising to find that significant portions of eukaryotic genes are devoted to code for protein kinases. For example, the genome of *Saccharomyces cerevisiae*, the budding yeast contains 130 distinct protein kinase encoding genes, representing approximately 2% of the entire yeast genome (Hunter and Plowman, 1997). The human genome contains 518 protein kinase genes comprising 1.7% of the genome (Manning *et al.*, 2002).

The MAPK signal transduction pathways constitute a cascade of phosphorylation events that transmit extracellular signals from membrane-bound receptors to the nucleus. MAPKs are highly selective for phosphorylation of serine/threonine residues lying immediately N-terminal to a proline residue within a peptide substrate (Hanks and Hunter, 1995; Brábek and Hanks, 2004). MAP kinase cascades control almost all aspects of fungal growth, development, sexual and asexual reproduction, metabolism, proliferation and stress tolerance.

Two of the three MAP kinase pathways of filamentous fungi, the HOG1 (*h*igh *o*smolarity *g*lycerol according to yeast nomenclature) MAPK and the CWI (*c*ell *w*all *i*ntegrity) MAPK (homologous to Slt2/Mpk1 in yeast) pathways take part in abiotic stress tolerance including tolerance to salt, osmotic, oxidative, heat, cold, arsenite and citric acid stressors. Recently the HOG MAPK pathway has also been indicated in tunicamycin induced endoplasmic reticulum stress in *S. cerevisiae* (Torres-Quiroz *et al.*, 2010). The third MAPK route, the so-called *p*athogenicity *M*AP *k*inase (PMK) pathway is homologous to the mating/filamentation Fus3/Kss1 MAPK pathway of the yeast. PMK is required for the infection process including penetration into the host cells and invasive growth. This pathway is also involved in the yeast-to-hyphal transition in dimorphic species. As far as studies on *Fusarium* species are concerned, the Fmk1 and the Gpmk1 MAP kinases of *Fusarium oxysporum* (Di Pietro *et al.*, 2001) and *Fusarium graminearum* (Jenczmionka *et al.*, 2003), respectively were also found to be essential for pathogenicity. These PMK-type MAP kinases regulates the expression of several genes encoding cell-wall degrading hydrolytic enzymes (Jenczmionka and Schäfer , 2005).

Although PMK-type MAP kinases were generally regarded to have not much role in stress adaptation, recent studies demonstrated that the pathogenicity MAPK pathway controls the oxidative stress response in *Cochliobolus heterostrophus* (Izumitsu *et al.*, 2009). These observations highlight a more complex nature of stress signaling in filamentous fungi as it was anticipated previously (Aguirre *et al.*, 2006).

We present here a functional analysis of Δ*Fvhog1* and Δ*Fvmk2* CWI MAPK mutants of *F. verticillioides* by comparing their sensitivity to different oxidative stressors. *Fusarium verticillioides* (teleomorph: *Gibberella moniliformis*) is a world-wide occurring pathogen of maize that synthesizes a range of secondary metabolites, including fumonisins and carotenoids. To the best of our knowledge this is the first report on the comparison of oxidative stress tolerance of different MAPK mutants of the same filamentous fungus species. We also found that both Δ*Fvhog1* and Δ*Fvmk2* CWI MAPK gene-disruption mutants of *F. verticillioides* resulted in increased sensitivity to methylglyoxal, a toxic glycolytic by-product suggesting a double MAPK regulation of the cellular response to this compound. Secondary metabolite production is also regulated by different MAPK pathways in fungi: we provide here additional information on the recent findings available for fusaria. As the highly conserved fungicide signaling by fludioxonil is not dependent on the histidine kinase-HOG1 MAPK route in all filamentous fungi (Liu *et al.*, 2008) we compared the fludioxonil and hydrogen peroxide sensitivity of three *Fusarium* species. The extreme sensitivity *of F. graminearum* to fludioxonil and hydrogen peroxide was not associated with substantial changes in HOG MAPK mediated osmotic stress tolerance. We also found that Δ*hog1* mutants of two other *Fusarium* species showed fludioxonil tolerant and hydrogen-peroxide sensitive phenotypes, similarly to other filamentous species.

2. The role of HOG1 MAPK signaling in stress and fungicide tolerance of *Fusarium* species

Orthologues of the yeast HOG1 pathway genes have been identified either by functional or *in silico* analysis in several *Fusarium* species, including *F. graminearum, Fusarium proliferatum, F. oxysporum,* and *F. verticillioides* (Di Pietro *et al.*, 2001; Ochiai *et al.*, 2007; Ádám *et al.*, 2008a, b; Rispail *et al.*, 2009; Rispail and Di Pietro, 2010). In *F. proliferatum,* the HOG1 MAPK pathway plays a pivotal role in stress tolerance: this route takes part in salt, osmotic, heat, UV and oxidative (hydrogen peroxide) stress responses, but it is not required for invasive growth, sexual and asexual sporulation (Ádám *et al.*, 2008a, b). Osmotic stress caused a considerably higher rate of cell death in the Δ*Fphog1* MAPK gene disruption mutants as compared to the wild type strain. More importantly, when the fungi were subjected to osmotic (4% NaCl) stress, levels of reactive oxygen species (ROS), mitochondrial membrane permeability transition, nuclear disintegration and DNA fragmentation, four independent markers of programmed cell death (PCD) all showed significant increases in the Δ*Fphog1* mutants in comparison to the wild type strain suggesting that an important role of the functional Hog1 MAPK gene is attenuating apoptotic phenotypes under stress conditions (Ádám *et al.*, 2008a). Fig. 1. shows intense cell death symptoms and accumulation of ROS indicated by blue stained morphologically abnormal cells and green fluorescent cells (indicated by arrow), respectively in a Δ*Fphog1* gene-disruption mutant subjected to salt stress after adding 4%

(w/v) NaCl to the culture medium. Similarly to Δ*hog1* mutants of other fungi, like *Neurospora crassa, Aspergillus nidulans, C. heterostrophus* and *Colletotrichum lagenarium* the Δ*Fphog1* MAPK mutants of *F. proliferatum* became tolerant to phenylpyrrole and dicarboximide fungicides (Zhang *et al.*, 2002; Noguchi *et al.*, 2004; Yoshimi *et al.*, 2005; Ádám *et al.*, 2008b; Hagiwara *et al.*, 2009). Although the exact mode of action of these compounds is still unclear, the finding that heterologous expression of Hik1, the histidine kinase (HK) gene of *Magnaporthe oryzae* in the yeast, *S. cerevisiae* that contain only one HK gene, Sln1 confers susceptibility in this otherwise fludioxonil-resistant organism, suggests that class III HKs, located upstream of the HOG1 MAPK cascade are possible targets of this fungicide (Motoyama *et al.*, 2005). The class III HKs responsible for elevated osmo-tolerance and increased fludioxonil sensitivity in filamentous fungi are not the orthologues of Sln1 of the yeast (Catlett *et al.*, 2003). Inactivation of Fhk1, a class III HK in *F. oxysporum* resulted in osmo-sensitivity and resistance to phenylpyrrole and dicarboximide fungicides (Rispail and Di Pietro, 2010). The increased tolerance of the Δ*Fphog1* mutants of *F. proliferatum* to fludioxonil and vinclozoline (Ádám *et al.*, 2008b) suggests that functional HK-HOG1 MAPK pathway is required for sensitive response to these fungicides in *Fusarium* species. *In silico* analysis of HKs by reciprocal BLASTP searches in *Fusarium* genome sequences led to the identification of Fhk1 (FOXG_01684) orthologues both in *F. verticillioides* (hypothetical protein FVEG_08048) and *F. graminearum* (hypothetical protein FGSG_07118) (Nagygyörgy and Ádám, unpublished).

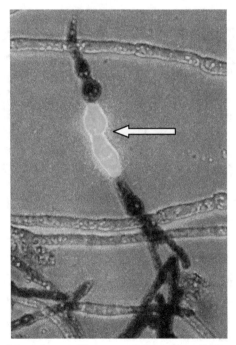

Fig. 1. Double staining of *Fusarium proliferatum* Δ*Fphog1-24* gene-disruption MAP kinase mutant with 2,7-dichlorodihydrofluorescein diacetate (DCHFDA) and Evans blue after NaCl (4% w/v) exposition. Intensive green fluorescence (indicated by arrow) and dark blue discoloration of the cells indicate accumulation of reactive oxygen species (ROS) and cell death, respectively.

HKs have five HAMP (*h*istidine kinase, *a*denylate cyclase, *m*ethyl binding protein and *p*hosphatase) repeats: mutations in these sequences are responsible for the increased osmo-sensitivity and fungicide resistance of *N. crassa*, *C. heterostrophus*, *Alternaria brassicicola* and *Botrytis cinerea* (Ochiai *et al.*, 2001; Yoshimi *et al.*, 2004; Motoyama *et al.*, 2005; Viaud *et al.*, 2006). Recent microarray analyses have further shown that the transcriptional response to fludioxonil depends on a Hog1 orthologue in *A. nidulans*. This response overlaps, in part with the transcriptional response to hyperosmotic stress but depends on factors other than the AtfA transcription factor responsible for conidial stress tolerance (Hagiwara *et al.*, 2009). Thus the identification of transcription factor(s), that are located downstream of Hog1 MAPK and influence gene expression response to fludioxonil requires further studies.

Although fungicide signaling by fludioxonil is highly conserved in filamentous fungi, response to this compound is still not entirely dependent on the HK-HOG1 MAPK route in all species. For example, the Δ*sak1* (Hog1 orthologue) knockout mutants of *B. cinerea* maintained their sensitive phenotype to fludioxionil (Liu *et al.*, 2008) indicating the complex nature of this signaling pathway. On-going research of our laboratory on fludioxonil sensitivity of three *Fusarium* species with available genome sequences (*F. graminearum* PH-1/NRRL 31084, *F. oxysporum* 4287 and *F. verticillioides* FGSC 7600; http://www.broadinstitute.org/annotation/genome/Fusariumgroup/MultiHome.html) led to somewhat surprising results (Fig. 2).

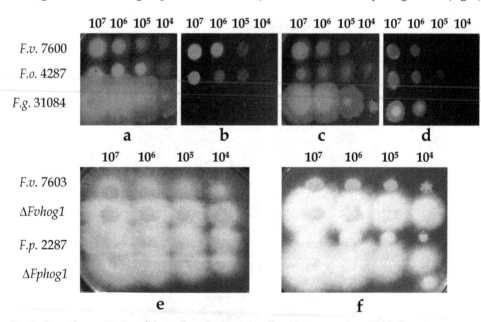

Fig. 2. Growth sensitivity of three *Fusarium* species: *Fusarium oxysporum* 4287, *Fusarium verticillioides* FGSC 7600 and *Fusarium graminearum* PH-1/NRRL 31084 against different stressors (a-d) and sensitivity of Δ*hog1* mutants of *F. verticillioides* FGSC 7603 and *F. proliferatum* FGSC 2287 (Δ*Fvhog1-14* and Δ*Fphog1-24*, respectively) against fludioxonil (e-f). Five-five µl of indicated concentrations of conidia (cells/ml) was spotted on (a, e) complex medium (CM) agar plates and CM agar plates supplemented with (b, f) 10 µg/ml fludioxonil, (c) 4% (w/v) NaCl and (d) 2 M sorbitol. Incubation time was 3 days for all plates at 24 °C.

F. graminearum showed increased sensitivity to this fungicide as compared to the other two species (Fig. 2a and b), but fludioxonil sensitivity was not accompanied with substantial changes in osmo-sensitivity of this fungus (Fig. 2c and d). On the other hand, Δ*hog1* mutants of *F. proliferatum* and *F. verticillioides* showed fludioxonil tolerance (Fig. 2e and f), paralleled with elevated osmo-sensitivity. Although these results do not exclude the involvement of the HK-HOG1 MAPK pathway in these phenotypes, further studies are needed to a better understanding of fungicide stress responses in *Fusarium* species.

3. The role of MAPK pathways in secondary metabolism of *Fusarium* species

In addition to its role in stress and fungicide tolerance, the HOG1 MAPK pathway plays also an important role in the regulation of secondary metabolism in different *Fusarium* species (Ochiai *et al.*, 2007; Kohut *et al.*, 2009). Disruption of either Fgos2 (a HOG-type MAPK orthologue) or Fgos4 (encoding a MAPK kinase) or Fgos5 (encoding a MAPK kinase kinase) blocked trichotecene production in *F. graminearum* and substantially reduced expression of the trichotecene gene cluster. On the other hand, amounts of aurofusarin were increased in all three types of mutants (Ochiai *et al.*, 2007). Deoxynivalenol production is controlled by Mgv1 CWI MAPK in *F. graminearum* (Hou *et al.*, 2002). Nitrogen depletion induced the production of fumonisin B1, a polyketide derivative mycotoxin and increased the expression of fuminisin biosynthesis genes in *F. proliferatum*. Under nitrogen starvation (absence of any N-source) conditions deletion of Fphog1, a HOG-type MAP kinase gene resulted in further increases in FUM1 and FUM8 gene expression, as well as fumonisin B1 production suggesting that this response is mediated via the HOG-type MAPK pathway in *F. proliferatum* (Kohut *et al.*, 2009). In a more recent study Fvmk1, a PMK-type MAPK gene was identified as a positive regulator of fumonisin B1 production in *F. verticillioides* (Zhang *et al.*, 2011). On the contrary, fumonisin B1 production was not regulated by cAMP signaling either in *F. proliferatum* (Kohut *et al.*, 2010) or *F. verticillioides* (Choi and Xu, 2010). This signaling route regulates, however negatively regulates the production of bikaverin, another polyketide metabolite in *Fusarium* species (Kohut *et al.*, 2010; Choi and Xu, 2010; García-Martínez *et al.*, 2011). Moreover, the production of another secondary metabolite such as carotenoids is upregulated in *Fusarium* species not only by cAMP signaling (García-Martínez *et al.*, 2011) but other regulatory elements related to sexual reproduction (Ádám *et al.*, 2011).

4. Complexity of oxidative stress signaling in fungi: Role of the HOG1 and the CWI MAPK pathways

Previous research with different species indicated that, besides the HOG MAPK pathway (Aguirre *et al.*, 2006; Du *et al.*, 2006; Ádám *et al.*, 2008a), the CWI MAPK pathway (Krasley *et al.*, 2006; Valiante *et al.*, 2007) also has a role in oxidative stress tolerance of fungi. To compare the particularities of the two pathways we generated both Δ*hog1* (Δ*Fvhog1*) and CWI MAPK (Δ*Fvmk2*) gene-disruption mutants in a single fungus species, *F. verticillioides* as we have described earlier (Ádám *et al.*, 2008a). These mutants were tested for oxidative stress tolerance in conidial dilution assay using hydrogen peroxide, menadione, diamide and methylglyoxal as stressors. Although all of these compounds elicit finally oxidative stress, the mechanisms they do this are different. Hydrogen peroxide induces lipid peroxidation, protein and DNA damage directly or indirectly and contributes to the formation of hydroxyl radicals (OH·) via the Fenton reaction (Thön *et al.*, 2007). Menadione is a redox cycling reagent that acts by generating superoxide radicals (O_2^-) using NADPH as a cofactor. Diamide causes depletion of the reduced glutathione pool and perturbation of the redox balance of cells; this compound also reacts with sulfhydryl groups of proteins in a reversible way (Pócsi *et al.*, 2005; Thön *et al.*, 2007). Methylglyoxal is a highly toxic natural glycolytic by-product interacting with proteins in a reversible way and, at higher concentrations (8-10 mM) it can deplete the glutathione pool in yeast cells (Aguilera *et al.*, 2005).

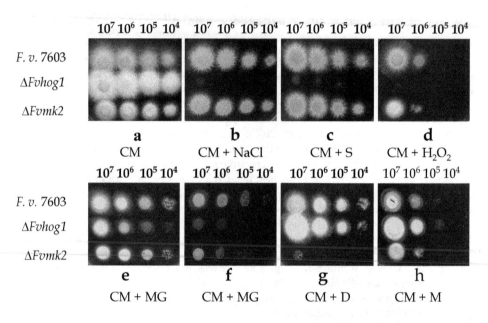

Fig. 3. Differential sensitivity of Δ*Fvhog1-14* and Δ*Fvmk2-16* CWI MAPK mutants of *Fusarium verticillioides* FGSC 7603 against oxidative stressors. Five-five μl of indicated concentrations of conidia (cells/ml) was spotted on (a) complex medium (CM) agar plate and CM agar plates supplemented with (b) 4% (w/v) NaCl, (c) 2 M sorbitol (S), (d) 2 mM hydrogen peroxide (H_2O_2), (e) 5 mM methylglyoxal (MG), (f) 10 mM methylglyoxal (MG), (g) 0,5 mM diamide (D) and (h) 0,03 mM menadione (M). Incubation time was 3 days for all plates at 24 °C.

The Δ*Fvhog1* mutant of *F. verticillioides* was highly sensitive to the osmotic stressors, sodium chloride and sorbitol (Fig 3a, b and c), similar to our former results with *F. proliferatum* Δ*hog1* mutants (Ádám *et al.*, 2008a). On the contrary, the Δ*Fvmk2* mutant showed no elevated osmo-sensitivity (Fig 3a, b and c). The Δ*Fvhog1* mutant was sensitive not only to osmotic stressors but, as well as to hydrogen peroxide and methylglyoxal. However, this mutation caused no change in menadione and diamide sensitivity (Fig. 3a, d, e, f g and h). This is a first report on the involvement of HOG MAPK pathway in methylglyoxal tolerance of a filamentous species. Formerly Aguilera *et al.* (2005) reported on methylglyoxal sensitivity of Δ*hog1* mutants of *S. cerevisiae*. Δ*Hog1* mutants of filamentous species showed different oxidative stress sensitivity. Mutants of *A. fumigatus* lacking the MAP kinase Δ*sakA*/Δ*hog1* were more sensitive to H_2O_2 and menadione compared to wild type strain (Du *et al.*, 2006).

In *F. oxysporum* the *Fhk1* HK mutant, deficient in an upstream element of the HOG1 MAPK pathway is sensitive to menadione induced oxidative stress but not to H_2O_2 (Rispail and Di Pietro, 2010).

Results obtained with the Δ*Fvmk2* CWI MAPK mutants were completely different: they showed no elevated osmo-sensitivity, but their sensitivity to methylglyoxal and diamide increased as compared to the wild type strain (Figs. 3a, b, c, d, e, f, g and h). According to these results, the cellular response to methylglyoxal is regulated by both the HOG1 and CWI MAPK pathway, but the other stressors are signaled separately either by the HOG1 or CWI MAPK pathway. As both methylglyoxal and diamide interact mainly with glutathione metabolism, the possible role of CWI MAPK is to maintain glutathione pools under stress conditions. As regarding the oxidative stress sensitivity of other species, CWI MAPKs played a fluctuating but positive regulatory role in stress tolerance. An exceptional case is *A. fumigatus*: the deletion of MpkA CWI MAPK gene resulted in increased H_2O_2 tolerance and sensitivity to menadione and diamide (Valiante *et al.*, 2007). In the case of *S. cerevisiae* Δ*slt2*/Δ*mpk1* mutants were sensitive to H_2O_2 (Krasley *et al.*, 2006). But in *C. albicans* and *S. pombe*, deletion mutants of Δ*mkc1* and Δ*pmk1*, respectively were sensitive to diamide but not to H_2O_2 and menadione (Navarro-Garcia *et al.*, 2005; Madrid *et al.*, 2006).

5. Sensitivity of different *Fusarium* species to hydrogen peroxide

We compared the hydrogen peroxide sensitivity of three *Fusarium* species, *F. graminearum* PH-1/NRRL 31084, *F. oxysporum* 4287 and *F. verticillioides* FGSC 7600 with available genome sequences (http://www.broadinstitute.org/annotation/genome/Fusariumgroup/ Multi Home.html). *F. gaminearum*, a causal agent of head blight of wheat and stalk/cob rot of maize was the most sensitive to this oxidative stressor both in a decimal conidium dilution assay (Valiante *et al.*, 2007) and in radial growth test (Ádám *et al.*, 2008a). Mycelial growth and conidial germination of this fungus was more strongly inhibited by 5-50 mM and 2 mM H_2O_2 concentrations as compared to *F. oxysporum* and *F. verticillioides* (Fig 4A, 4B), other two plant pathogenic species causing vascular wilt of a wide range of plants and maize cob rot, respectively. In a previous study (Nicolaou *et al.*, 2009), oxidative stress tolerance of 18 fungal species originating from different ecological niches and phylogenetic positions were compared and plant pathogenic species, like *Ustilago maydis*, *F. graminearum* and *M. grisea* were found to be relatively sensitive to oxidative stressors, including hydrogen peroxide. This result was somewhat surprising as both plant and animal pathogens that are exposed to massive oxidative and/or nitrosative stress by the host cells in many host-pathogen interactions (Brown *et al.*, 2009) would have been expected to acquire improved levels of oxidative stress tolerance during their evolution. When the hydrogen peroxide sensitivity tests were extended to other two *Fusarium* species, *F. fujikuroi* MP-C and *F. proliferatum* FGSC 2287, causing the bakane disease of rice and crown and root rot of a wide range of plants, respectively they also showed higher levels of hydrogen peroxide tolerance as compared to that of *F. graminearum*. The increased H_2O_2 sensitivity of *F. graminearum* can be putatively explained by the long saprophytic phase in the life cycle of this species. During the saprophytic phase, *F. graminearum* lives and propagates on dead tissues and in this niche, the fungus is much less exposed to oxidative stress influence than the other species, that spend much of their life cycle inside living plant tissues either as endophytes or vascular colonizers.

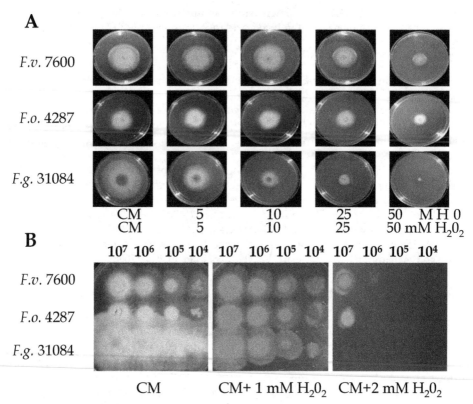

Fig. 4. Growth sensitivity of three *Fusarium* species: *Fusarium verticillioides* FGSC 7600, *Fusarium oxysporum* 4287 and *Fusarium graminearum* PH-1/NRRL 31084 against hydrogen peroxide (H₂O₂) in radial growth test (A) and decimal conidial dilution assay (B). In (B) five-five μl of indicated concentrations of conidia (cells/ml) was spotted on complex medium (CM) agar plate and CM agar plates supplemented with the indicated concentrations of hydrogen peroxide (H₂O₂). Incubation time was 5 days for plates in (A) and 3 days for plates in (B) at 24 °C.

In another approach we have studied the role of cAMP signaling in oxidative stress response of different *Fusarium* species. Previous studies in *N. crassa* demonstrated that cAMP signaling and HOG1 MAPK signaling play opposite role in respect to oxidative stress response: disturbances in the cAMP and the HOG1 pathways result in increased and decreased H₂O₂ tolerance, respectively. In *F. proliferatum* and *F. verticillioides,* disruption of Acy1, the adenylyl cyclase gene resulted in enhanced resistance to heat shock and oxidative stress (Kohut *et al.*, 2010; Choi and Xu, 2010). However, in contrast to these data, the *acyA⁻* mutants of *F. fujikuroi* MP-C were more sensitive to H₂O₂ than the wild type (García-Martínez *et al.*, 2011). This finding suggests the high versatility of the cAMP signaling route even in closely related fungi. These differences are particularly important, if we consider that heat shock and oxidative stress pathways have at least partially overlapping signaling routes and regulated by the same transcription factors in yeast (Ikner and Shiozaki, 2005).

6. Conclusion

Functional analysis of orthologous signal transduction genes in different filamentous fungal species highlighted the complex nature of stress signal transduction. This is especially true for oxidative stress signaling, where all fungal MAPK cascades, the HOG1, CWI and PMK MAPK pathways participate and interact in this regulatory network depending on the fungal species. One of the oxidative agents, mehylglyoxal, a toxic by-product of glycolysis is signaled either by the HOG1 MAPK or CWI MAPK pathway in *F. verticillioides*. All these MAPK cascades are also involved either in positive or negative regulation of secondary metabolite production including mycotoxins in different *Fusarium* species. The high versatility of oxidative stress and secondary metabolite signaling by the above-mentioned MAPK pathways and the cAMP-PKA pathway in different *Fusarium* species denotes that stress signaling is exposed to rapid evolution to tune stress responses in a niche-specific manner, independently of the phylogenetic position of a given species.

7. Acknowledgement

This research was supported by an OTKA grant (National Scientific Research Council of Hungary, K 76067). We are indebted for support from the Office for Subsidized Research Units of the Hungarian Academy of Sciences. A.L.Á. is also grateful to the Hungarian-Spanish Bilateral Inter-Governmental S & T Project (OMFB 00666/2009).

8. References

Ádám, A.L., García-Martínez, J., Szűcs, E.P., Avalos, J., & Hornok, L. (2011). The MAT1-2-1 mating type gene upregulates photo-inducible carotenoid biosynthesis in *Fusarium verticillioides*. *FEMS Microbiology Letters* 318: 76-83.

Ádám, A.L., Kohut, G. & Hornok, L. (2008a). Fphog1, a HOG-type MAP kinase, is involved in multistress response in *Fusarium proliferatum*. *Journal of Basic Microbiology* 48: 151-159.

Ádám, A.L., Kohut, G. & Hornok, L. (2008b). Cloning and characterization of a HOG-type MAP kinase encoding gene from *Fusarium proliferatum*. *Acta Phytopathologica et Entomologica Hungarica* 43: 1-13.

Aguilera, J., Rodríguez-Vargas, S. & Prieto, J. A. (2005). The HOG MAP kinase pathway is required for the induction of methylglyoxal-responsive genes and determines methylglyoxal resistance in *Saccharomyces cerevisiae*. *Molecular Microbiology* 56: 228-239.

Aguirre, J., Hansberg, W. & Navarro, R. (2006). Fungal responses to reactive oxygen species. *Medical Mycology* S44: (Suppl. 1), 101-107.

Brábek, J. & Hanks, S.K. (2004). Assaying protein kinase activity. In: *Signal Transduction Protocols, Methods in Molecular Biology*, R.C. Dickson, M.D. Mendenhall (Eds.), Vol. 284, 79-90, Humana Press, DOI: 10.1385/1-59259-816-1:079, Totowa, New York, USA

Brown, A.J.P., Haynes, K. & Quinn, J. (2009). Nitrosative and oxidative stress responses in fungal pathogenicity. *Current Opinion in Microbiology* 12: 384-391.

Catlett, N.L., Yoder, O.C & Turgeon, B.G. (2003). Whole-genome analysis of two-component signal transduction genes in fungal pathogens. *Eukaryotic Cell* 2: 1151-1161.

Choi, Y.E. & Xu, J.R. (2010). The cAMP signaling pathway in *Fusarium verticillioides* is important for conidiation, plant infection, and stress responses but not fumonisin production. *Molecular Plant-Microbe Interactions* 23: 522-533.

Di Pietro, A., Garcia-Maceira, F. I., Meglecz, E., & Roncero, M. I. (2001). A MAP kinase of the vascular wilt fungus *Fusarium oxysporum* is essential for root penetration and pathogenesis. *Molecular Microbiology* 39: 1140-1152.

Du, C., Sarfati, J., Latgé, J.P. & Calderone, R. (2006). The role of the sakA (Hog1) and tcsB (sln1) genes in the oxidant adaptation of *Aspergillus fumigatus*. *Medical Mycology* 44: 211-218.

García-Martínez, J., Ádám, A.L. & Avalos, J. (2011). Adenylyl cyclase plays a regulatory role in development, stress and secondary metabolism in *Fusarium fujikuroi*. *26th Fungal Genetics Conference at Asilomar*, Fungal Genetics Reports, S58, p. 205, ISBN 1941-4765, Asilomar, California, USA, March, 15-20, 2011

Hagiwara, D., Asano, Y., Marui, J., Yoshimi, A., Mizuno, T. & Abe, K. (2009). Transcriptional profiling of *Aspergillus nidulans* HogA MAPK signaling pathway in response to fludioxonil and osmotic stress. *Fungal Genetics and Biology* 46: 868-878.

Hanks, S.K. & Hunter, T. (1995). Protein kinases 6. The eukaryotic protein kinase superfamily: kinase (catalytic) domain structure and classification. *FASEB Journal* 8: 576-596.

Hou, Z., Xue, C., Peng, Y., Katan, T., Kistler, H. C. & Xu, J.R. (2002): A mitogen-activated protein kinase gene (MGV1) in *Fusarium graminearum* is required for female fertility, heterokaryon formation, and plant infection. *Molecular Plant-Microbe Interactions* 15: 1119-1127.

Hunter, T. & Plowman, G. (1997). The protein kinases of budding yeast: six score and more. *Trends in Biochemical Sciences* 22: 18-22.

Ikner, A. & Shiozaki, K. (2005). Yeast signaling pathways in the oxidative stress response. *Mutation Research/Fundamental and Molecular Mechanisms of Mutagenesis* 6: 13-27.

Izumitsu, K., Yoshimi, A., Kubo, D., Morita, A., Saitoh, Y. & Tanaka, C. (2009). The MAPKK kinase ChSte11 regulates sexual/asexual development, melanization, pathogenicity, and adaptation to oxidative stress in *Cochliobolus heterostrophus*. *Current Genetics* 55: 439-448.

Jenczmionka, N. J, Maier, F. J., Lösch, A.P. & Schäfer, W. (2003). Mating, conidiation and pathogenicity of *Fusarium graminearum*, the main causal agent of the head-blight disease of wheat, are regulated by the MAP kinase Gpmk1. *Current Genetics* 43: 87-95.

Jenczmionka, N.J. & Schäfer, W. (2005). The Gpmk1 kinase of *Fusarium graminearum* regulates the induction of specific secreted enzymes. *Current Genetics* 47: 29-36.

Kohut, G., Ádám, A.L., Fazekas, B. & Hornok, L. (2009). N-starvation stress induced FUM gene expression and fumonisin production is mediated via the HOG-type MAPK pathway in *Fusarium proliferatum*. *International Journal of Food Microbiology* 130: 65-69.

Kohut, G., Oláh, B., Ádám, A.L., García-Martínez, J. & Hornok, L. (2010). Adenylyl cyclase regulates heavy metal sensitivity, bikaverin production and plant tissue colonization in *Fusarium proliferatum*. *Journal of Basic Microbiology* 50: 59-71.

Krasley, E., Cooper, K.F., Mallory, M.J., Dunbrack, R. & Strich, R. (2006). Regulation of the oxidative stress response through Slt2p-dependent destruction of cyclin C in *Saccharomyces cerevisiae*. *Genetics* 172: 1477-1486.

Liu, W., Leroux, P. & Fillinger, S. (2008). The Hog1-like MAP kinase Sak1 of *Botrytis cinerea* is negatively regulated by the upstream histidine kinase Bos1 and is not involved in dicarboximide and phenylpyrrole resistance. *Fungal Genetics and Biology* 45: 1062-1074.

Madrid, M., Soto, T., Khong, H.K., Franco, A., Vicente, J., Pérez, P., Gacto, M., & Cansado, J. (2006). Stress-induced response, localization, and regulation of the Pmk1 cell integrity pathway in *Schizosaccharomyces pombe*. *Journal of Biological Chemistry* 281: 2033-2043.

Manning, G., Whyte, D.B., Martinez, R., Hunter, T. & Sudarsanam, S. (2002). The protein kinase complement of the human genome. *Science* 298: 1912-1934.

Motoyama, T.K., Kodama, T., Ohira, A., Ichiishi, M., Fujimura, I., Yamaguchi, I. & Kubo, T. (2005). A two-component histidine kinase of the rice blast fungus is involved in osmotic stress response and fungicide action. *Fungal Genetics and Biology* 42: 200-212.

Navarro-García, F., Eisman, B., Fiuza, S.M., Nombela, C. & Pla, J. (2005). The MAP kinase Mkc1p is activated under different stress conditions in *Candida albicans*. *Microbiology* 151: 2737-2749.

Nikolaou, E., Agrafioti, I., Stumpf, M., Quinn, J., Stansfield, I. & Brown, A.J.P. (2009). Phylogenetic diversity of stress signalling pathways in fungi. *BMC Evolutionary Biology* 9: 44-59.

Noguchi, R., Banno, S., Ichikawa, R., Fukumori, F., Ichiishi, A., Kimura M., Yamaguchi, I. & Fujimura, M. (2007). Identification of OS-2 MAP kinase-dependent genes induced in response to osmotic stress, antifungal agent fludioxonil and heat shock in *Neurospora crassa*. *Fungal Genetics and Biology* 44: 208-18.

Ochiai, N., Fujimura, M., Motoyama, T., Ichiishi, A., Usami, R., Horikoshi, K. & Yamaguchi, I. (2001). Characterization of mutations in the two-component histidine kinase gene that confer fludioxonil resistance and osmotic sensitivity in the os-1 mutants of *Neurospora crassa*. *Pest Management Science* 57: 437-442.

Ochiai, N., Tokai, T., Nishiuchi, T., Takahashi-Ando, N., Fujimura, M. & Kimura, M. (2007). Involvement of the osmosensor histidine kinase and osmotic stress-activated protein kinases in the regulation of secondary metabolism in *Fusarium graminearum*. *Biochemical Biophysical Resaearch Communications* 363: 639-644.

Pócsi, I., Miskei, M., Karányi, Z., Emri, T., Ayoubi, P., Pusztahelyi, T., Balla, G. & Prade, R.A. (2005). Comparison of gene expression signatures of diamide, H_2O_2 and menadione exposed *Aspergillus nidulans* cultures - linking genome-wide transcriptional changes to cellular physiology. *BMC Genomics* 6: 182-201.

Rispail, N. & Di Pietro, A. (2010). The two-component histidine kinase Fhk1 controls stress adaptation and virulence of *Fusarium oxysporum*. *Molecular Plant Pathology* 11: 395-407.

Rispail, N., Soanes, D.M. & Di Pietro, A. (2009). Comparative genomics of MAP kinase and calcium calcineurin signalling components in plant and human pathogenic fungi. *Fungal Genetics and Biology* 46: 287-298.

Thön, M., Al-Abdallah, Q., Hortschansky, P. & Brakhage, A.A. (2007). The thioredoxin system of the filamentous fungus *Aspergillus nidulans:* impact on development and oxidative stress response. *Journal of Biologycal Chemistry* 282: 27259–27269.

Torres-Quiroz, F., García-Marqués, S., Coria, R., Randez-Gil, F. & Prieto, J.A. (2010). The activity of yeast Hog1 MAPK is required during endoplasmic reticulum stress induced by tunicamycin exposure. *Journal of Biological Chemistry* 285: 20088-20096.

Yoshimi, A., Tsuda, M. & Tanaka, C. (2004). Cloning and characterization of the histidine kinase gene Dic1 from *Cochliobolus heterostrophus* that confers dicarboximide resistance and osmotic adaptation. *Molecular Genetics Genomics* 271: 228–236.

Yoshimi, A., Kojima, K., Takano, Y. & Tanaka, C. (2005). Group III histidine kinase is a positive regulator of Hog1-type mitogen-activated protein kinase in filamentous fungi. *Eukaryotic Cell 4:* 1820-1828.

Valiante, V., Heinekamp, T., Jain, R., Härtl, A. & Brakhage, A.A. (2007). The mitogen-activated protein kinase MpkA of *Aspergillus fumigatus* regulates cell wall signaling and oxidative stress response. *Fungal Genetics and Biology* 45: 618-627.

Viaud, M., Fillinger, S., Liu, W., Polepalli, J.S., Le Pêcheur, P., Kunduru, A.R., Leroux, P. & Legendre, L. (2006). A class III histidine kinase acts as a novel virulence factor in *Botrytis cinerea*. *Molecular Plant-Microbe Interactions* 19: 1042-1050.

Zhang, Y., Choi, Y.E., Zou, X. & Xu, J.R. (2011). The FvMK1 mitogen-activated protein kinase gene regulates conidiation, pathogenesis, and fumonisin production in *Fusarium verticillioides*. *Fungal Genetics and Biology* 48: 71-9.

Zhang, Y. , Lamm, R. , Pillonel, C., Lam, S. & Xu, J.R. (2002). Osmoregulation and fungicide resistance: the *Neurospora crassa* os-2 gene encodes a HOG1 mitogen-activated protein kinase homologue. *Applied and Environmental Microbiology* 68: 532-538.

10

Molecular Characterization of Carbendazim Resistance of Plant Pathogen (*Bipolaris oryzae*)

S. Gomathinayagam[1], N. Balasubramanian[2],
V. Shanmugaiah[3], M. Rekha[4], P. T. Manoharan[5] and D. Lalithakumari[6]
*[1]Faculty of Agriculture and Forestry, University of Guyana, Berbice Campus, Tain,
[2]CIRN and Department of Biology, University of Azores, Ponta Delgada, Azores,
[3]Department of Microbial Technology, School of Biological Sciences,
Madurai Kamaraj University, Madurai, Tamil Nadu,
[4]Department of Biotechnology, Kalasalingam University, Krishnankovil, Tirunelveli, Tamil Nadu,
[5]Department of Botany, Vivekananda College, Thiruvedakam, Madurai, Tamil Nadu,
[6]Research and Development, Bio-Soil Enhancers, Hattiesburg, MS,
[1]Guyana
[2]Portugal
[3,4,5]India
[6]USA*

1. Introduction

Agricultural practices are often portrayed as significantly contributing to environmental problems. When organic pesticides were discovered they were hailed for their positive contributions to increased food production. The challenges we are facing is that the changing environmental ethics coincide with rapid population growth. It is expected that within the next 50 years there will be approximately 50% more people on earth to feed, house and cloth. The world is not only increasing in population, but also in affluence. Invasive pests are a significant threat to the statuesque of our natural ecosystems, our health and agriculture. Many of the most serious plant diseases have already been spread throughout the world, but there is still good reason to be very diligent in monitoring and excluding those diseases that are still endemic to limited geographic areas. Emergence of new diseases and disease epidemics must be expected. One of the most important contributions of plant pathology research to science, is the recognition that pest evolve at a rapid rate and that single gene changes can turn an obscure microbe into the cause of an epidemic. The challenge for plant pathology is to be able to predict when such genetic changes will happen. The ability to genetically modified plants, animals and microbes in precise ways using molecular biology and biotechnology provides the present century best hope for meeting the food, fiber and nutritional needs of the growing population of the world without further compromising the quality of our environment. This technology is so important for the future that care has to be taken that no compromise is taken on long-term value by meeting short-term goals.

With the modern biotechnology on hand, research area calls for more attention in basic and applied research programmes in product, mode of action, mechanism of action and application of fungicides for safety and economic use. Development of fungicide resistance is a threatening topic to farmers who practice with potential systemic fungicides. When a fungicide fails because of the development of resistance by the target organism, in practice it is very important to know whether the effectiveness of other fungicides has been affected as this will lead to many more problems.

The developing research on biocontrol practice needs supportive control of some chemicals in the control of diseases that cannot be controlled by biocontrol. It is high time to go further with in-depth research on fungicides, to combat and overcome fungicide resistance in practice. Systemic chemicals are best alternatives for any failure practices of plant protection, provided they are carefully monitored for their mode of action, method of application and understood more on the mechanism of development of fungicide resistance.

The present research is case study on Carbendazim fungicide resistance in many important pathogens. *In vivo* and *in vitro* fungicide resistant mutants have been isolated and characterized for growth, sporulation and pathogenicity. The complete fitness and competition tests of the resistant strains we carefully carried out in the presence and absence of fungicide. Level of resistance, using petri plate method was compared to spot diagnosis techniques using allele specific oligonucleotides. The test fungicide chosen was Carbendazim, a Benzimidazole compound whose mode of action is inhibition of microtubules assembly. Benzimidazole compounds are increasingly important and most widely known, owing to the excellent systemic control of much important plant diseases. Development of fungicide resistance is now one of the major problems in plant disease control, but could be easily delayed or prevented through careful practices. This requires stable information's on genetics of resistance, level of resistance and mechanisms of resistance. The Present research give an insight on to the above factors to know more information's on fungicide resistance. New molecular techniques have been assessed for effective evaluation of the level of fungicide resistance at field level.

2. Materials and methods

2.1 General methods

Glassware was first soaked in chromic acid cleaning solutions (10% potassium dichromate solutions in 25% sulphuric acid) for few hours and washed thoroughly in tap water. After a second wash in detergent solution, they were again washed thoroughly in tap water, rinsed in distilled water and air dried. Glassware, media and buffers were sterilized at 120°C, 15 Ib pressure / inch^2 for 20 min.

2.2 Chemicals

Chemicals and solvents used were of analar grade. Ethyl Methane Sulfate (EMS), calf thymus DNA, cesium chloride, agarose, λ DNA, restriction enzymes and polyethylene glycol (PEG) were purchased from Sigma chemicals Co., St. Louis, USA. Novozym –234 was purchased from Novo Industry, Denmark.

2.3 Fungicides

BASF (India) Ltd., Chennai generously supplied Carbendazim. Other fungicide used in the investigation was purchased from various sources.

2.4 Preparation of stock solution of fungicide

Stock solutions of fungicide were prepared by dissolving the fungicide in 1 mL acetone or ethanol and made up to a known volume with distilled water. Required concentrations were prepared by diluting the stock solution. Fungicide solutions were prepared just before use and the final concentration of the solvent did not exceed 0.5% in the medium. Sterilization of stock solution was made by filtering through sterilized millipore filter system (0.22µ).

2.5 Isolation of fungi

2.5.1 Test pathogen

The test pathogen *Bipolaris oryzae* causing brown leaf spot were chosen as test pathogen. The phytopathogenic strains of *Bipolaris oryzae* were collected from infected paddy leaves in paddy filed of Siddi Vinayaga farm, Bandikavnoor, Chennai. Infected leaves were sterilized in mercuric chloride (0.01%) and placed onto PDA, to which streptomycin (50 µg / mL) was added to suppress bacterial growth and incubated at 20ºC for 3 days leaf bits were removed. Hundred agar blocks containing germlings of single conidia were picked up with a sterile needle under microscopic observation, transferred individually to PDA slants and incubated until they form sporulating colonies.

2.5.2 Culture conditions of the fungus

Fungus stock culture was maintained on PDA slants at 28ºC and transferred to new media at regular intervals. Petri plate cultures were maintained as follows. To 3 days old culture, 5 mL of sterile distilled water was added and scrapped with an inoculation needle. The conidial suspension was transferred to a 250 mL conical flask containing 100 mL molten PDA (40ºC), mixed thoroughly and poured into sterile Petri dishes. After 3 days of incubation, using a sterile cork borer, mycelial discs (8 mm dia) were cut at random of periphery region and were used for further experiments throughout the investigation.

2.5.3 Preparation of conidial suspension

Conidial suspension of phytopathogen of *Bipolaris oryzae* were prepared by washing the well sporulated slant cultures in sterile water containing one drop of Tween 20 and filtered through two layers of cheese cloth to remove hyphal fragments. The spore suspension was washed twice with sterile water and resuspended in sterile water. The concentration of conidia was determined using a Haemocytomer.

2.5.4 Pathogenicity test on rice plants (Chevalier et at., 1991)

Pathogenicity of rice pathogens was tested on rice plants under green house conditions. Forty five days old Ponni and IR 50 paddy plants were first sprayed with mycelial suspension of the sensitive strains of pathogen (*B. oryzae*) means of automizer. The plants were covered with individual's polythene bags to provide adequate humidity and kept at room temperature. The inoculated plants were observed after 7 days for characteristic symptoms.

2.6 Morphological and physiological characterization

2.6.1 Growth of test pathogen (sensitive strain) on solid media

Before the assessment of growth on fungicide amended medium, the maximum growth and sporulation rate inducing medium was screened. Growth rate of pathogens was assessed on five different solid media viz. PDA, PDYEA, CDA, MEA and OMA. Mycelial disc (8 mm dia) was placed at the center of the Petri plate in an inverted position containing media and kept for incubation. Quadruplicates were maintained for each treatment. At every 48 h, the diameter of the mycelial growth was measured. After incubation, 10 mycelial discs (8 mm dia) were cut at random from the periphery and transferred to 10 mL sterile water in 100 mL flasks. The flasks were kept on an orbital shaker for 30 min. The number of conidia /mL in different media was counted using Haemocytometer.

2.6.2 Growth rate of test pathogen (sensitive strains) in PD broth

The growth rates of test pathogen were studied by inoculating 4 mycelial discs (8 mm dia) into 250 mL conical flask containing 100 mL PD broth. One set of flasks were incubated on an orbital shaker (120 rpm) while another's sets of flasks were incubated under static conditions for 8 days. Every day, 4 flasks were removed from each set, the mycelial mat was filtered and dried separately in preweighted Whatman No. 1 filter paper at 80°C and the dry weights estimated.

2.6.3 Sensitivity of test pathogen (wild strain) to Carbendazim on PDA

Sensitivity of phytopathogen to Carbendazim was tested using poison food technique (Carpenter, 1942) by inoculating 8 mm mycelial discs at the centre of the Petri plates containing different concentration of the fungicide (i.e. 0.5, 1, 2.5, 5, 10, 25, 50, 100, 250, 500 and 1000μM) amended with PDA. Four plates were maintained for each treatment. After incubation for 10 days, the diameter of mycelial growth was measured in all treatments and control. The percentange inhibition over control was calculated by plotting probit values of the percent inhibition of the growth against log concentration of the fungicide (Nageswara Rao, 1965) and ED_{50} dose was estimated.

2.6.4 Sensitivity of test pathogen to Carbendazim in PD broth

In PD broth, the sensitivity of phytopathogens to Carbendazim was carried out by inoculating 4 mycelial discs (8mm dia) in 250 mL flasks containing 100 mL of PD broth amended with different concentrations of fungicides. The flasks were kept on an orbital shaker (120 rpm) for 4 days. Quadruplicates were maintained for each treatment. Then the mycelial mat was than collected and dried on a reweighed, Whatman No. 1 filter paper and the dry weight estimated. The ED_{50} dosage was calculated by plotting probit values of the percent inhibition of the growth against log concentration of the fungicide.

2.7 Development of *in vitro* resistant mutant of test pathogen against Carbendazim

2.7.1 Collection of conidia

The fungus was grown in PD broth at 27°C on a reciprocal shaker for 5 days. The conidia were separated from mycelia by filtering through double layers of cheese cloth and the filtrate was centrifuged at 1000Xg for 10 min. The conidia were suspended in a known volume of sterile water and the final concentration was adjusted to 1 X 10^6 conidia /mL by using Haemocytometer.

2.7.2 Development of Carbendazim resistant mutants of test pathogen by UV – irradiation (Tanaka et al., 1988)

The in vitro mutants resistant to Carbendazim were developed by following the procedure of Tanaka et al., (1988). Aliquots of conidial suspension (1 X 10^3 conidia / mL) containing Tween 20 (0.02% V/V) was poured into a Petri plate and exposed to UV irradiation (254 nm Phillips TUV 51 W G15T8) at a distance of 30 cm for 15, 30, 45 and 60 min. After which, the conidia were incubated in the dark for 2 h to prevent photo reactivation.

Conidia treated with mutagens as described above, were plated separately on PDA amended with Carbendazim 5 times ED$_{50}$ concentration of the sensitive strain. Medium devoid of fungicide served as control. The Petri plates were incubated at 20°C for 8 days. survival rates was calculated from the ratio of colonies which grew on the fungicide unamended medium and the number of conidia inoculated. Mutation rate was determined from the ratio of colonies which grew on fungicide amended media and number of conidia inoculated.

2.7.3 Induction and isolation of Carbendazim resistant mutants of phytopathogen by Ethyl Methyl Sulfate (EMS) treatment (Tanaka et al., 1988)

Conidia (1 X 10^6 /mL) of the sensitive strain of test pathogens were suspended in 100 mL conical flask containing 10 mL freshly prepared EMS solution at a concentration of 5, 10, 25, 50, 100, 250, 500, 1000,2000 and 5000µm in 0.1 M sterile phosphate buffer (pH 7.0). All the flasks were kept on a shaker for 6 h at 27°C. Treated conidia were washed twice with sterile distilled water by centrifugation at 1000 X g for 10 min to remove traces on EMS. Subsequently, the conidial concentration was adjusted to 1 X 10^6 mL and plated on PDA amended with fungicide (five times the concentration of ED$_{50}$ value of sensitive strain). The seeded plates were incubated for 5 d at 20°C. The colonies (mutants) that survived were counted and isolated for further studies and percent survival was also calculated.

2.7.4 Development of resistant strains of test pathogen by adaptation technique

Adaptability of phytopathogen to Carbendazim was carried out by the method of Rana and Sengupta (1977). Conidia (1 X 10^3 / mL) of the sensitive strain were placed on PDA medium amended with 5 times higher the ED$_{50}$ value of fungicide and the plates were incubated for 5 d at 23°C. From the developed colonies, mycelial discs (8 mm dia) were cut from periphery using a sterile cork borer and then transferred to stepwise increasing concentration of the fungicide. All the laboratory mutants resistant to test fungicide Carbendazim were numbered and maintained.

2.7.5 Collection of field resistant strains

Field resistant mutants of B. oryzae were collected from diseased parts of plants treated continuously with Carbendazim.

2.7.6 Stability test for fungicide resistance in the mutants of test pathogen

Carbendazim resistant mutants were subcultured on fungicide free PDA medium for 10 generations. After 10 generations, the mutants were transferred to the respective fungicide (5 X the ED_{50} concentration) amended PDA medium. Stability rate was determined from the ratio of colonies which grew on the fungicide amended medium and number of colonies inoculated. The Carbendazim resistant mutant strains of test pathogens obtained by various mutagenesis techniques were designated.

2.8 Level of fungicide resistance laboratory mutants and field mutants resistant to Carbendazim

2.8.1 Level of resistance based on the mycelial growth on PDA medium

Young mycelial discs (8 mm dia) were aseptically transferred to the centre of the Petri plates containing PDA medium amended with different concentrations (always above 5 times the concentration of ED_{50} value of sensitive strain) of Carbendazim (5µM to 1000µM) and incubated at 20ºC for 25days. Medium devoid of fungicide served as control. Four replicates were maintained for each treatment. Percent inhibition of growth was calculated and the ED_{50} values were derived by probit analysis.

2.8.2 Level of resistance based on the mycelial growth and conidial production in PD broth

One mL conidial suspension (1 X 10^3 conidia / mL) of the resistant mutants was inoculated in 100 mL PD broth amended with different concentrations of Carbendazim (5 to 1000µM) and incubated orbital shaker (120 rpm) at 20ºC for 12 days. PD broth devoid of fungicide served as control. Four replicates were maintained for each treatment. After the incubation period, mycelia were filtered and dry weight was determined. Numbers of conidia were counted in each flask per replicate per treatment. Percent inhibition of mycelial dry weight and conidial production was determined and ED_{50} values derived by probit analysis.

2.8.3 Level of resistance based on germination and primary hyphal elongation in PD broth

One ml conidial suspension (1 X 10^3 conidia / mL) of the resistant mutants was inoculated in 25 mL PD broth amended with higher concentration of Carbendazim (100 to 1000µM) and incubated on an orbital shaker (120 rpm) at 20ºC. Flasks inoculated with Cassava Root meal (CRM) were incubated for 30 h. PD broth devoid of fungicide served as control. Four replicates were maintained for each treatment. Percent inhibition of germination and primary hyphal elongation were calculated and ED_{50} values of the fungicide were derived by probit analysis.

2.8.4 Morphological characterization of the resistant mutants of test fungi

Colony morphology, pigmentation and conidial production of the resistant mutants in fungicide unamended and amended medium were compared with that of the sensitive strain. The conidia was measured using a calibrated microscope (Carl Zeiss, Germany)

2.9 Biochemical characterization of sensitive strain and resistant mutants

2.9.1 Measurement of oxygen uptake by sensitive strain and resistant mutants of test pathogen

Oxygen uptake was measured (Johnson, 1972) polar graphically in clark type oxygen probe fitted to a YSI (Yellow Springs Instruments Ltd., Ohio., USA) model oxygen monitor. The probe was standardized with distilled water. The probe was inserted into the chamber and 100% saturation of air was set. For the measurements of oxygen, 1 g fresh mycelium suspended in 5 mL Potato Dextrose Yeast Extract (PDYE) broth was transferred into the chamber. The decrease in saturation percent was measured at 1 min interval for 10 min (Arditti and Dunn, 1969). Percent saturation of air measured was converted to oxygen concentration using Rawso's monogram (Welch, 1948).

2.9.2 Determination of electrolytes

One gram fresh mycelium was washed with sterile double distilled water and incubated in 100 mL PD broth. One set of flasks was amended with ED50 concentration of the carbendazim and incubated on an orbital shaker (150rpm). Mycelium incubated in PD broth devoid of fungicide served as control. Four replicates were maintained for each treatment. Mycelium was harvested at every 4 h interval for 24 and washed with excess of distilled water. Washed mycelium was transferred to 25 mL of sterile double distilled water and incubated on an orbital shaker (150rpm) for 1 h. After incubation, myceliums were filtered and the conductance of the ambient water was measured using CM 82T Conductivity Bridge with a dip electrolytic cell. Dry weight of mycelium was determined.

2.9.3 Extraction of protein

Wet mycelium (500 mg) was ground with equal amount of acid washed sand and 0.1 M sodium phosphate buffer (pH 7.0) in a pre-chilled mortar and pestle for 20 min at 4°C. The ground material was centrifuged at 15, 000 x g for 15 min and the supernatant was made up to a known volume with the same buffer and dialyzed overnight against large volume of glass distilled water. The dialyzed extract was used for protein estimation.

2.9.4 Protein estimation

Protein was estimated by the method of Lowry et al., (1951) using Bovin Serum Albumin (BSA) as standard and the amount of protein was expressed as mg protein/g dry wt. of mycelium.

2.10 Electrophoretic protein pattern of sensitive strain an resistant mutant

Electrophoretic pattern of total protein of sensitive and resistant mutants was analysed by the method of Ornstein (1964).

2.10.1 Analysis of protein by SDS –PAGE (Laemmli, 1970)

The discontinuous buffer system of Laemmli (1970), which is a modification of Ornstein (1964), was used in the present study for the separation of proteins.

2.11 Molecular characterization of sensitive strain and resistant mutants of test pathogen

2.11.1 Extraction of degraded DNA and RNA

For quantification, DNA and RNA were extracted by the modified method of Scheneider (Munro and Fleck, 1966).

One gram fresh weight of the mycelium was treated with 5% trichloroacetic acid (TCA) for 60 min at 4°C. The supernatant containing the cold acid soluble compounds was discarded after centrifugation at 15,000 X g for 15 min at 4°C. The mycelium was their treated with 1 N perchloric acid (PCA) at 70°C in water bath for 20 min to hydrolyse the nucleic acid. The suspension was again centrifuged at 15, 000 X g for 30 min at 4°C and the supernatant was collected. The pellet was re extracted with 1 N PCA under the same conditions and the supernatants were pooled and used for DNA and RNA estimation.

2.11.2 Calorimetric determination of DNA

DNA was estimated by the improved method of Giles and Myers (1965) further modified by Lalithakumari et al (1975). The reaction mixture consisted of 2 mL of hydrolysate obtained after 1 N PCA treatment and 2 mL 1% freshly prepared diphenylamine in glacial acetic acid and was incubated at 30°C for 16 h in dark. The blue colour developed was measured at 595 nm and 700 nm in a Hitachi 150 – 20 spectrophotometer and the difference in optical densities were calculated. The solution without hydrolysate mixture was used for blank. The quantity of DNA was estimated using the standard curve prepared with calf thymus DNA.

2.11.3 Determination of rate of uptake of Carbendazim

One gram fresh mycelium grown in PDYE broth was harvested by filtration on a Buchner filter and washed thrice with distilled water. The mycelium was washed twice with incubation medium (25mM KH_2PO_4, 12.5 mM $K_2HPO_4.3H_2O$ buffer, pH 7.0, with 0.1 mM $CaCl_2. 2H_2O$ and 1% glucose) (De Waard and Nistelrooy, 1980) and resuspended in 50 mL of the same at 37°C for 30 min. After 30 min, ED_{50} of fungicide (25µm carbendazim) was amended in the mycelial suspension. Uninoculated medium served as control and four replicates were maintained for each treatment. At 10 min interval, 5 mL sample was filtered over Whatman No.1 filter paper. The mycelial residues were washed thrice with the incubation medium without fungicide and finally the incubation media were pooled and made up to a known volume. Similarly, 5 mL incubation medium was drawn from the control flask (incubation medium without fungicide) and made up to same volume as that of the test sample. Percent uptake of fungicide by each strain was determined from the difference in the amount of residual fungicide in the test and control medium. Sodium azide was added to the mycelial suspension 15 min prior to the addition of fungicide.

2.12 Polymerase Chain Reaction (PCR)

The polymerize chain reaction and then sequence analysis of the genomic DNA were used to rapidly characterize the sequence of β–tubulin DNA from the pathogens. Genomic DNA was prepared from each strain of test pathogens and subjected to PCR by using two generic β–tubulin primers. Constraints on primer design were that the amplified DNA had to contain codons 167 and 241, in which mutations were associated with resistance to benomly and that the primers had to anneal to a conserved region with minimal variation in the sequence. The 22-mer oligonucleotide A (5'- CAAACCATCTCTGGCGAACACG) and 22-mer oligonucleotide B (5' - TGGAGGACATCTTAAGACCACG) were used as primers. Primer A was identical in sequence to codons 22-28, and primer B was complementary in sequence to codons 359-365 of β–tubulin genes of *V.inaequalis* With these primers, a 1,191-bp fragment of the beta-tubulin gene was amplified. The primer was synthesized in Bangalore Gennie Pvt.Ltd, Bangalore.

The reactions were performed in a thermal cycler (35 cycles) with the repliprime DNA amplification system (Du Pont) according to the manufactures procedures. Each PCR reaction was performed in 25 μl (final volume) of reaction mixture. It consisted of 1 μl of DNA, 4 μl (1.02μl) 200μM dNTPs, 5.5μl (2.5 mM MgCl2), 5.5 μl (50 mM KCl), 5.5 μl (10 mM Tris HCl), Triton –100 X 0.05μl, 0.5 μM of each primer and 1 unit of Tag DNA polymerize (approximately 0.66μl) 1.79μl double distilled water. Negative controls were run in all the amplification reactions to detect contamination. In reactions involving primers A and B, 35 cycles were performed by pre –heating the sample for 5 min at 94ºC for each reaction as follows. 94ºC, 1 min; 55ºC, 1min; and 72º C, 2 min. Amplification products were analyzed for the expected 1,191-bp or 436-bp fragments by 1.0% agarose gel electrophoresis in 1X TBE buffer (0.1 M Tris-HCl, 0.1 M boric acid, 0.02 mM EDTA, pH 8.3. following electrophoresis, the DNA was visualized after staining with ethidium bromide.

2.12.1 Allele –Specific Oligonucleotide (ASO) analysis

The single spore isolate of test pathogen used this study was from a large collection of field strains and laboratory strains previously characterized in studies on the inheritance to benomly resistance negatively correlated cross-resistance to diethofencarb (Jones, et al., 1987). The PCR amplified β–tubluin DNA (25ng per sample) was denatured in 0.25 N NaOH for 10 min and then applied to a nylon membrane (Gen Screen-Plus, Du Pont, Boston, MA) in a dot blot manifold. The dot blots were incubated in prehybridization solution (1 M NaCl, 50 mM Tris –Hcl, pH 7.5, 10% dextran sulfate, 1% SDS, 0.2% Ficoll (MW 400,000), 0.2% polyvinylprrolidone (MW 40,000), 0.2% bovine serum albumin, 0.1% sodium pyrophosphate, and 0.25 mg/mL denatured salmon sperm DNA) for 2 h at 5ºC according to the manufacture's producer. An end-labeled ASO probe was then added to the prehybridization solution and incubated at 37ºC for at least 4 h. The blots were washed three times for 15 min each in 2 X SSC buffer (Saline – Sodium Citrate) (1 X SSC is 0.15 M NaCl, 0.015 M sodium citrate, pH 6.8) at room temperature. A high –stringency wash, three times for 2 min each in 2X SSC buffer, was then used to remove ASO probe with a single base pair mismatch from the blots. The optimum temperature for the high-stringency wash with each probe was determined empirically. The dot blots were exposed to X-ray film for 0.5-2 days at –70ºC.

Four ASO probes for detecting allelic mutations in the β-tubulin gene of *V. inaequalis* were synthesized in the macromolecular structure facility, Department of Biochemistry, Michigan State University, East Lansing, USA. The 18-mer oligo-nucleotides were designated as ASOS-LR, ASOMR, ASOHR and ASOVHR probes according to the specificity of each probe.

                                           198   200
ASOS-LR   5'C   TCT   GAC GAG ACA TTC TG3'
                                                              ▼
ASOMR   C   GAG ACA TAC TGC ▼ATT GA

ASOHR   C   TCT   GAC AAG ACA TTC TG
                                             ▼
ASOVHR   C   TCT   GAC GCG ACA TTC TG

3. Results

In recent years awareness of the importance of fungicide resistance in crop protection has been growing. Hence, in the present studies *B. oryzae* (Table 1) highly sensitive to benzimidazole of fungicide have been chosen to study the probability and level of resistance of Carbendazim in these pathogens and measured to overcome resistance by understanding the molecular mechanisms of resistance. All the test pathogen was tested for their virulence through pathogenicity test.

S. No	Organism	Place of collection	Disease caused
1	*Bipolaris oryzae*	Bandikkavanoor, Tamil Nadu, India. Agriculture field	Brown spot disease

Table 1.Test pathogen used in the present study

3.1 Growth kinetics of sensitive strains

3.1.1 Radial growth of sensitive strains of test pathogens on PDA

Growth kinetics of the test pathogen *B. oryzae*, (Table 2) shows that *B. oryzae* takes 12 day for complete growth

Day	Radial growth* (cm)
2	2.7
4	3.3
6	4.6
8	6.9
10	8.8
12	Plate fully covered

* Mean of 4 replicates

Table 2. Radial growth of sensitive strain of *B. oryzae* on PDA

3.1.2 Growth rate of sensitive strains of test pathogens in PD broth

Dry weight estimation in PD broth of test pathogen (Table 3) under static and shaken conditions always showed that under shaken condition the growth rate was fast with enhanced mycelial dry weight in all the test pathogen.

3.1.3 Effect of Carbendazim on the growth of test pathogens

Before understanding, the mechanism of resistance in different mutants, the base line sensitivity of the parent strains to test fungicide, Carbendazim was analyzed (Table 4). The ED_{50} value was calculated for all the test pathogen. The test pathogen the most sensitive to Carbendazim *B. oryzae* with ED_{50} values of 40 μM.

Day	Dry weight of the mycelium(mg)*	
	Static condition	Shaken condition
1	65.00	95.00
2	96.00	120.00
3	125.00	168.60
4	165.00	285.00
5	195.00	345.00
6	260.00	370.00
7	300.00	400.00
8	310.00	420.00
9	315.00	410.00
10	400.00	690.00

* Mean of four replications

Table 3. Growth rate of sensitive strain of *B. oryzae* in PD broth

Concentration (μM)	Mycelial dry weight (mg)	% inhibition
0	650	0
1	545	16.2
2.5	468	28.0
5	425	34.6
10	400	38.5
25	385	40.8
50	300	53.8
100	268	58.8
250	225	65.4
500	143	78.0
1000	50	92.3

ED_{50} value = 40 μM

Table 4. Effect of Carbendazim on the growth of *B. oryzae*

3.1.4 Induction of mutation of test pathogen

To understand the mechanisms of resistance EMS, UV and adaptation methods have been used to produce number of laboratory mutants (Table 5, Table 6, and Table 7). In the case of all the test pathogen nearly 200 mutants were produced and screened on PDA amended with five times the ED_{50} concentration of respective sensitive strains (according to test pathogen). Resistant colonies were picked up and stored in test tube slants amended with 5 times concentration of ED_{50} dose. The results in tables 7-9 clearly show that the percent survival rate depends on the test pathogen and the mutagenesis.

Concentration of EMS (μM)	No. of colonies survived	% of survival
50	450	90
100	400	80
250	375	75
500	175	35
1000	75	15
2500	10	2

Spore concentration: 0.5×10^3
Incubation   : 6 h

Table 5. Induction of mutation of *B. oryzae* by EMS

Exposure time	No. of colonies survived	% of survival
15	475	95
30	400	80
45	200	40
60	150	30
75	35	7

Spore concentration: 0.5×10^3
Incubation   : 6

Table 6. Induction of mutation of *B. oryzae* by UV-irradiation

Treatment	Concentration of Carbendazim (μM)	Mycelial growth in diameter(mm)
I	5	75
II	10	60
III	25	58
IV	50	56
V	100	48
VI	250	45
VII	500	38
VIII	1000	20
IX	2,000	15
X	5,000	10

Incubation time   : 6 d Medium   : PDA

Table 7. Development of resistant strains of *B. oryzae* by adaptation

3.1.5 Stability of test pathogen

The laboratory developed resistant mutants were subsequently tested for their stability for 10 generation in fungicide free medium (Table 8). All the tested fungicides which retained resistance after 10 generation only were taken for further studies. In *B. oryzae* the EMS mutants were more stable.

Mutants	No. of colonies examined	No. of stable colonies retained	Stability rate%
EMS	50	35	70.00
UV	45	20	44.44
AD	40	25	62.50

EMS　　: Carbendazim EMS resistant mutant
UV　　　: Carbendazim UV resistant mutant
AD　　　: Carbendazim Adapted resistant mutant

Table 8. Stability of fungicide resistance in *B. oryzae*

3.1.6 Pathogenicity test

All the stable mutants of the test pathogen were subsequently tested for pathogenicity. From all the tested mutants only those that developed symptoms of the diseases were chosen for further studies.

3.1.7 Categorization of level of fungicide resistance

The stable resistant mutants were screened to evaluate the level of resistance and then to categorize them as LR, MR, HR and VHR based on to their level of resistance. To group the selected mutants, they were grown on 5-10 times (LR), 10-15 times (MR), 15-20 (HR), 20-25 times and above as VHR based on ED50 value of the sensitive strain of test pathogen. The ED50 value of *B. oryzae* was 40 μm. accordingly various concentrations were prepared and 60 mutants in pathogen were screened (Table 9) and categorized.

	Concentration of Carbendazim (μM)									
	10	40	200	400	600	800	1000	1200	1400	1600
S	300	275	-	-	-	-	-	-	-	-
1	300	289	285	-	-	-	-	-	-	-
2	302	292	290	-	-	-	-	-	-	-
3	303	295	292	-	-	-	-	-	-	-
4	305	298	295	-	-	-	-	-	-	-
5	306	299	295	-	-	-	-	-	-	-
6	307	299	292	-	-	-	-	-	-	-
7	310	298	295	-	-	-	-	-	-	-
8	313	302	300	-	-	-	-	-	-	-
9	315	308	302	-	-	-	-	-	-	-
10	317	306	305	-	-	-	-	-	-	-
11	318	310	308	-	-	-	-	-	-	-
12	319	310	307	-	-	-	-	-	-	-
13	320	312	310	-	-	-	-	-	-	-
14	321	316	313	-	-	-	-	-	-	-
15	323	321	320	-	-	-	-	-	-	-
16	324	320	318	313	310	-	-	-	-	-

	Concentration of Carbendazim (μM)									
	10	40	200	400	600	800	1000	1200	1400	1600
17	325	318	315	312	308	-	-	-	-	-
18	326	319	315	309	302	-	-	-	-	-
19	327	315	313	310	306	-	-	-	-	-
20	328	319	316	309	302	-	-	-	-	-
21	330	321	320	318	315	-	-	-	-	-
22	331	325	320	318	313	-	-	-	-	-
23	334	326	320	315	313	-	-	-	-	-
24	336	332	330	328	327	-	-	-	-	-
25	338	332	328	325	318	-	-	-	-	-
26	340	338	332	330	328	-	-	-	-	-
27	343	332	328	320	318	-	-	-	-	-
28	346	340	333	328	325	-	-	-	-	-
29	347	341	313	329	320	-	-	-	-	-
30	348	338	332	330	328	-	-	-	-	-
31	349	340	338	335	330	-	-	-	-	-
32	350	342	340	338	329	-	-	-	-	-
33	352	350	348	345	340	-	-	-	-	-
34	351	350	347	340	338	-	-	-	-	-
35	353	350	348	340	337	335	330	-	-	-
36	355	350	347	345	340	338	335	-	-	-
37	357	352	348	345	342	337	332	-	-	-
38	360	353	345	342	340	336	333	-	-	-
39	362	353	350	348	345	340	339	-	-	-
40	363	355	350	347	343	340	340	-	-	-
41	365	355	350	348	345	340	338	-	-	-
42	368	360	355	352	348	345	340	-	-	-
43	370	361	360	355	350	350	348	-	-	-
44	371	362	358	357	355	350	349	-	-	-
45	373	365	359	355	350	348	345	-	-	-
46	376	370	369	365	360	357	355	-	-	-
47	377	371	370	366	363	360	358	-	-	-
48	380	375	372	370	368	363	360	-	-	-
49	382	378	375	370	368	365	360	359	355	.350
50	383	370	370	368	365	360	359	355	354	352
51	385	380	375	370	360	360	358	355	350	348
52	387	382	380	378	370	368	365	360	630	358
53	390	385	380	378	375	370	368	362	360	374
54	393	388	385	380	378	375	370	368	362	360
55	397	390	388	385	380	365	350	358	358	348
56	398	393	392	390	385	382	380	375	370	365
57	402	400	395	392	390	387	380	375	370	365
58	405	400	399	397	390	388	380	375	370	365
59	407	405	400	396	394	391	390	387	385	370
60	410	407	404	395	391	387	385	380	379	374

ED50 of sensitive strain 12μM

1 –10 time	15 mutants	LR
10 - 15 time	20 mutants	MR
15 –20 time	10 mutants	HR

Table 9. Categorization of level resistance based on mycelial growth of mutants of *B. oryzae*

3.1.8 Grouping of mutants of test pathogen

In *B. oryzae* 15 mutants were classified under group of LR, 19 under MR, 14 under HR and 12 as VHR were obtained (Table 10).

Name of strains	LR	MR	HR	VHR
Bipolaris oryzae	15	19	14	12

Table 10. Grouping of mutants of test pathogen

3.1.9 Sporulation of resistant strains of test pathogen

After grouping the resistant mutants as LR, MR, HR, and VHR, morphological characterization studies were carried out. Sporulation of the resistant mutants were evaluated for all the test pathogen and compared with the sensitive strain (Table 11). The sporulation was always more in sensitive strain than the resistant mutants. The rate of sporulation decreased according to level of resistance.

Strain	*B. oryzae* No. of conidia / mL
SEN	5.4×10^4
LR	4.9×10^4
MR	2.8×10^2
HR	2.2×10^2
VHR	1.2×10^2

Table 11. Sporulation of resistant strains of test pathogen

3.1.10 Rate of conidial germination and primary hyphal elongation of test pathogen

Following sporulation, the spore germination and primary hyphal length were measured and presented in Table 12 for the test pathogen. Interestingly the results show a distinct reduction in spore germination and primary hyphal elongation as the level of resistant increase. The time taken for spore germination and hyphal elongation was more in resistant mutants over the control. The test pathogen, *B. oryzae* showed similar results.

3.1.11 Oxygen uptake by the sensitive and resistant mutant of test pathogen

Physiological characterization was initiated with oxygen uptake (Table 13). The oxygen uptake in n mole/min/mg dry wt of mycelium showed a distinct reduction in all the mutants irrespective of the test pathogen.

Incubation	% of germination*				Primary hyphal length (μM)			
Strain	SEN	MR	HR	VHR	SEN	MR	HR	VHR
3	0	0	0	0	0	0	0	0
6	20	0	0	0	35	0	0	0
12	40	30	38	45	50	45	48	47
15	45	38	46	59	69	55	50	49
18	78	49	67	69	68	60	55	51
21	89	53	71	79	80	70	69	70
24	90	69	65	60	150	145	140	125

Shaken conditions
Incubation   : 10^3 conidia/ mL
Temperature   : 20 ºC
* Values are mean of 5 replicates.

Table 12. Rate of conidial germination and primary hyphal elongation of B. *oryzae* under shaken conditions (without fungicide)

Name of strains	B. *oryzae*
	Oxygen uptake n moles/min/mg/ dry wet. Mycelium
SEN	27.18
MR	22.12
HR	24.10
VHR	24.00

Table 13. Oxygen uptake by the sensitive and resistant mutant of test pathogen

3.1.12 Rate of efflux of electrolytes of sensitive and resistant mutants of test pathogen

Electrolytic leakage studies on various levels of mutants of test pathogens (Table 14) also showed decrease in efflux when compared to the sensitive strains. Though there was reduction in electrolytic leakage, it was increasing with incubation time.

3.1.13 Amino acid composition of sensitive and resistant mutants of test pathogens

Amino acid composition of sensitive and resistant mutants of test pathogen (Table 15) showed distinct variations in the quantity of amino acids. Trypsin, Cystine and Glycine were totally absent in sensitive and LR mutants of B. *oryzae*. In all the mutants of test pathogens, interestingly the Proline content was more and this was absent in sensitive strain.

Strains	Specific conductance+ μm hos / cm²/ h/g/dry wt of (mycelium) Incubation time (h)				
	2	4	8	16	24
Sensitive	2800	3680	3960	4850	4590
LR	2796 (0.14)	3560 (3.26)*	3920 (1.01)	4380 (9.69)	4580 (0.22)
MR	2786 (0.50)	3470 (5.70)	3890 (1.76)	4525 (6.70)	4420 (3.70)
HR	2775 (0.89)	3580 (2.71)	3770 (4.79)	4060 (16.23)	4360 (5.01)
VHR	2025 (27.67)	3660 (0.54)	3590 (9.34)	4050 (16.49)	4200 (8.49)
FR	2600 (7.14)	3000 (18.47)	3680 (7.07)	3960 (18.35)	4150 (9.59)

+ mean of 10 replication
* Figure in parenthesis indicate per cell (+) increase (or) (-) decrease over sensitive strain.

Table 14. Rate of efflux of electrolytes of sensitive and resistant mutants of B. oryzae

Amino acid	Composition (mg/g dry weight)					
	S	LR	MR	HR	VHR	FR
Arginine	0.48	0.49	0.55	0.65	0.60	0.55
Lysine	0.50	0.50	-	0.56	0.59	0.65
Aspartic acid	0.40	-	0.56	0.53	0.62	0.42
Glutamine	0.56	0.63	0.45	0.50	-	-
Asparagine	0.55	0.64	0.50	0.66	0.63	0.47
Serine	0.60	-	0.63	0.69	0.64	0.49
Valine	0.62	0.68	0.48	0.63	0.65	-
Phenylalanine	0.62	0.34	0.50	0.55	0.60	0.45
Proline	0.49	0.65	0.66	0.62	0.68	0.70
Histidine	0.48	Trace	-	0.60	0.59	0.75
Methionine	0.55	0.75	Trace	0.64	0.56	-
Leucine	Trace	Trace	0.35	0.65	0.57	0.78
Tryptophan	-	-	0.65	0.60	-	0.80
Cystine	-	-	0.66	0.61	0.60	0.82
Glycine	-	-	0.49	0.16	-	0.53
Ornithine	-	-	-	-	-	-
Threonine	-	-	-	-	-	-
Glutamic acid	0.49	0.60	0.62	0.63	0.65	0.54

*Mean of four replicates

Table 15. Amino acid composition of sensitive and resistant strains of B. oryzae

3.1.14 SDS gel electrophoresis of sensitive and resistant strains of test pathogen

Agarose gel electrophoresis of genomic DNA of all the levels of mutants of test pathogen did not show any variations among the levels of mutants and also between sensitive and resistant mutants.

3.1.15 GC% and Tm of resistant mutants of test pathogen

Results on the GC% (Table 16) and Tm showed an increase in GC% and as a result increase in melting temperature. Always the GC% was less in sensitive strains while all levels of mutants of test pathogens showed an increase in GC%.

S. No.	Name of the strains	GC% of undegraded DNA		Melting temperature (°C)
		UV absorbance	Thermal Denaturation	
1	Sensitive	60.12	60.18	80
2	MR (25)	65.14 (8.34)	64.18 (6.64)	88
3	HR (10)	63.80 (6.12)	64.19 (6.66)	90
4	VHR (5)	62.16 (3.40)	65.60 (9.00)	91

Figure in parenthesis indicate percent over sensitive strain
Mean of strain

Table 16. Base composition of DNA of sensitive and resistant strains of *B. oryzae*

S.No.	Strains	GC% of mt. DNA		Melting temperature (°C)
		UV absorbance	Thermal Denaturation	
1	Sensitive	50.00	60.60	89
2	MR (19)#	55.60 (11.20)	65.83 (8.63)	93
3	HR (14)	54.65 (9.30)	66.98 (10.52)	94
4	VHR (12)	56.80 (13.60)	68.80 (14.66)	95

Figure in parenthesis indicate percent over sensitive strain
Mean of strain

Table 17. Base composition of (GC%) of mitochondrial DNA of sensitive strain and resistant mutants of *B. oryzae*

Agrose gel electrophoresis of mitochondrial DNA of various levels of mutants of all the test pathogens when compared to control did not show variation at all but the mitochondrial DNA was subjected to Hind III restriction enzyme and the restriction enzyme digested profile of various levels of mutants showed distinct variation in the number of discrete bands. The profile pattern varied between the levels of mutants and test pathogens. The number of discrete bands developed varied among the levels of mutants.

In mitochondrial DNA GC% was assessed (Table 17) and always in all levels of mutants the GC% was more and consequently the melting temperature was always more than the control.

3.1.16 Cross resistance of resistant mutants of test pathogen

Fungicides	ED $_{50}$ value						Q – value				
	SEN	LR	MR	HR	VHR	FR	LR	MR	HR	VHR	FR
Benomyl	360	390	475	510	575	340	1.10	1.32	1.42	1.59	0.94*
Biloxozol	450	490	400	515	525	540	1.08	0.88*	1.14	1.16	1.20
Kitazin	65	125	150	210	200	220	1.92	2.30	3.23	3.07	3.38
Mancozeb	250	320	375	300	385	340	1.28	1.5	1.2	1.54	1.36
Fytolan	200	250	380	395	200	360	1.25	1.90	1.97	1.00	1.80
Ziram	150	200	140	285	300	295	1.33	0.93*	1.90	2.00	1.69
Dithane M-45	335	256	260	270	340	285	0.79*	0.78	0.81	1.01	0.85*
Edifenphos	102	240	250	308	330	310	2.4	2.45	3.01	3.23	3.04

* Negatively correlated

Table 18. Cross resistance in resistant mutants of *B. oryzae*

4. Discussion

Agriculture crops are under continuous attack of serious and noxious organisms. They are powerful challengers of nature and manmade technologies. To safeguard world food production, crop protection measures especially choice based intelligent and ecofriendly chemicals are indispensable as they are instant and with known mode of action.

However, control of pest and disease with chemicals also encounter several problems, when the pathogens resistance to the potential broad-spectrum chemical. The ability of a pathogen to develop stable resistance to any toxicant is the fundamental theory of survival of the fittest under unfavorable conditions. The evolution of organisms would have been impossible without this property.

In a fungal population that is originally sensitive to fungicide, forms may arise or already exist that are less sensitive to the fungicide. Such a decrease in sensitivity may be caused by genetic or non-genetic changes in the fungal cell. Decrease in sensitivity due to genetic changes in the pathogen is more serious and requires in-depth understanding to detect the resistance in the field and combat the resistance well in advance to avoid or delay build of resistance.

Progress in clarifying the biochemical mechanisms of resistance has been made with some systemic fungicides, viz. Carbendazim, Carboxamides and Organophosphorous compounds (Georgopoulos, 1977). However, it is usually better to act before the buildup of resistance starts. For this, one should collect information from experiments with test fungi in vitro about the chances of a resistance problem arising in practice. Laboratory mutants resistant to test fungicides can be developed through in vitro mutagenesis (mutagenic chemicals or Ultra violet irradiation). The wild or parent strains should be characterized throughout to compare the resistant mutants. Also, the wild strain should be thoroughly characterized for its sensitivity against the target fungicide. In the present studies paddy pathogen *B. oryzae* was taken as test pathogen.

Detailed chemical control work has recommended benzimidazoles as one of the best broad spectrum compound for the control of paddy pathogen, *B. oryzae*. Present study is an attempt to predict the disease control failure using modern molecular diagnostic method to detect under in vitro condition to combat the development of resistance.

Before proceeding to produce Carbendazim resistant strains the sensitivity of the test pathogens was carefully monitored and ED $_{50}$ value of carbendazim for all test pathogens *B. oryzae*. Laboratory mutants were developed through UV irradiation, EMS mutagenesis and adaptation. There are several reports on the development of resistant mutants through EMS, UV irradiation and adaptation (Sanchez et al., 1975, Davidse, 1981, Leach et al., 1982; Gangawane et al., 1988; Rana and Sengupta, 1977, Hillderbrand et al., 1988). Many of the laboratory mutants were stable with high level of resistance.

In the present studies, though resistance to Carbendazim in test pathogens was suspected to be site modification, it was made very clear through results that site modifications is quiet rare and is not frequently observed in fields under practice with target fungicides.

PCR amplification of β-tubulin and spot hybridization further confirmed that the resistance at various levels tested proved that there was no point mutation observed in all the test pathogens. Instead ample evidences are present to confirm that in majority of the test pathogens, the mechanism of resistance primarily and always observed in membrane modification which could easily be handled or controlled easily with negatively correlated chemicals.

Hence, the present studies had given an authentic molecular proof that all benzimodazole resistance observed need not be site modifications which is very difficult to manage but mostly membrane modification, clearly indicating the possibility that fungicide resistance at any level if predicted can easily be managed and it is recommended that selection of choice based chemical (alternate chemical) with careful monitoring will definitely give a stable and sustainable management of diseases successfully in the agricultural practices. This is an additional warranty for the management of uncontrollable diseases using biological control.

5. Summary

Many pathogens develop resistance under field conditions due to frequent application of fungicides. To evaluate the resistance risk, stable laboratory mutant of *B. oryzae* resistant to Carbendazim were developed under laboratory conditions. Important pathogen *B. oryzae* was selected for intensive studies on molecular mechanisms of the fungicide resistance to benzimidazole compound. For the precise understanding of resistant mutants the complete characterization of parent strains very carefully carried out. The growth kinetics of *B. oryzae* was estimated on PDA. Similarly the dry weights of the test pathogen were estimated under static and shaken conditions. The mycelial dry weight was more under shaken condition than the static condition. Test pathogens were screened for sensitivity against Carbendazim using different concentration. The ED_{50} value for *B. oryzae* was 40 µM.

Induction of laboratory mutants using EMS, UV irradiation and adaptation technique were carried out. More than 200 resistant colonies were taken to select resistant strains 200 colonies were screened on PDA amended with 5 times the ED_{50} value of Carbendazim. The respective sensitive strains and resistant colonies were picked up and stored in test tube slants amended with test fungicides. The percent survival was assessed for all the test pathogens. Stability of resistance in all the mutants were checked. All the stable mutants of test pathogens when tested for pathogenicity proved to be pathogenic by causing respective disease through artificial inoculation. All the stable resistant mutants were categorized as LR, MR, HR and VHR based on their level of resistance over ED_{50} values of sensitive strain. Selected mutants when grow on (5 – 10 times LR, 10 –15 times MR, 15-20 times HR and 20 – 25 times VHR). Sixty mutants in each pathogen were screened on the above concentration of Carbendazim and categorized as LR, MR, HR and VHR. Sporulation of the resistant mutants were evaluated for all test pathogens and compared. In sensitive strains more sporulation was observed than the sensitive strains. Agarose gel electrophoresis of genomic DNA of the all the levels of mutants of test pathogens did not show any variation. Cross resistance study showed that *B. oryzae* resistant mutants at all levels could be overcome by alternative use of Dithane-45 or Mancozeb. The PCR amplification of β-tubulin in DNA extract was very poor and in most of the resistant mutants DNA, β–tubulin could not be amplified.

6. Acknowledgment

Authors have to thanks to University of Guyana, Berbice Campus to financially support this book chapter and also special thanks to Prof. Daizal R.Samad, Director, Berbice Campus to fully support us to write a book chapter.

7. References

Arditti, J.and A. Dunn. 1969. Polarography and oxygen electrode. P.230-232. *In: Experimental plant physiology*, Appentix1. Hold Rinehart and Winstion, Inc;New York. Chevalier, M., Y. Yespinasse and S. Renautin. 1991. A microscopic study of the different classes of symptoms coded by the Vf gene in apple for resistance scab *(Venturia inaequalis)* , Pl.pathol.40:249-256.

Davidse, L.C. 1982. Benzimidazole compounds selectivity and protection.p.60-70. In : J. Dekker and S.G. Georgopoulos (eds.), *Fungicide resistance in crop protection*. Pudoc., Wageningen, The Netherlands.

Gangawane, L.V., S. Waghmare and B.M. Kareppa. 1988. Studies on benomyl resistance in Phytopthora derchsleri f. sp. cajani (Abstr.) *Indian phytopath*. 41: 299.

Georgopoulous, S.G., 1977. Development of fungal resistance to fungicides. P.439- 495.In: M.R Siegel and H.D. Sisler (eds). *Antifungal compounds* Vol.II. Marcel Dekker, New York.

Johnson M.J.1972. Continuous measurment of dissolved oxygen. In; Manometric and biochemical techniques (Eds) W.W.Umbrieit R.H.Burris and J.F.Stauffer.pp.126-132. Burgess Pubishing Co. Minnesota.

Laemmii.U.K.1970. Clevage of structural protein using the assembly of the head of bactriaphages. Nature (London).680-685.

Lalithakumari,.D, J.R.Decallonne and J. A. Meyer. 1975. M.Talpaert.1977. Interferance of bnomyl and methyl-2-benzimidazole carbamate (MBC) with DNA metabolism and nuclear divisions in Fusarium oxysporum, *Pestic.Biochem.Physiol*.7:273-282.

Leech, J., Lang, B.R., and Yoder B.C. 1982. Methods for selection of mutants and in votro culture of *Cochliobolus heterostrophus. J. Gen. Microbiol*. 128: 1719 – 1729.

Munro, H.N. and A.Fleck.1966. The determination of nucleic acids. In: Netherlands Biochemical analysis (Ed.) D. Glick. p.113-176. Inter Science Publishers, New York.

Rana;J.P. and P.K. Sengupta. 1977. Adaptationof two plants pathogenic fungi to some fungicide. *Z. Pflanzenkrankh. Pflanzensch*. 84:738-742

Sanchez, L.E., J.V. Leary and R.M. Endo. 1975.Chemical mutagenesis of Fusarium oxysporum f.sp. lycopersici: Non-selected changes in pathogenicity of auxotrophic mutants. *J. Gen. Microbiol*. 87:326-332.

Tanaka,C.,Y.Kubo and M.Tsuda.1988.Comparison of mutagens in *Cochliobo heterostrophus* mutagenesis. Ann.Phytopathol.Soc.Jpn.54: 503-509.

Welch, P.S. 1948. *Limnological methods*. P. 306. McGraw Hill, Inc.

11

Accuracy of Real-Time PCR to Study *Mycosphaerella graminicola* Epidemic in Wheat: From Spore Arrival to Fungicide Efficiency

Selim Sameh, Roisin-Fichter Céline,
Andry Jean-Baptiste and Bogdanow Boris
Platform Biotechnology and Plant Pathology, Institut Polytechnique LaSalle Beauvais,
GIS PhyNoPi: Groupement d'Intérêt Scientifique «Phytopathologie Nord-Picardie»,
France

1. Introduction

Mycosphaerella graminicola (Fuckel) J. Schroeter in Cohn (anamorph *Septoria tritici* Roberge in Desmaz.) is one of the most important pathogens on winter wheat in northern Europe (Leroux et al., 2007; Stammler et al., 2008b). This fungus is responsible for *Septoria* leaf blotch and causes 30–40% yield reduction under extreme conditions (Palmer & Skinner, 2002). The primary inoculum is commonly described by the arrival of *M. graminicola* ascospores (Shaw & Royle, 1989). At high humidity, spores germinate and germ tubes penetrate 12 hours post inoculation into the leaves, exclusively through the stomata. The internal colonization is still intercellular between mesophyll cells until 10–12 days post inoculation. Then, the host cells die in response to the necrotrophic mode of *S. tritici,* with the development of visual chlorosis and necrotic symptoms. The formation of brown/black pycnidia in the substomatal cavities of the necrotic spots results from asexual reproduction, which appears 14–21 days after inoculation (Kema et al., 1996). Pycnidiospores are the secondary inoculum and are responsible for the repetition of the infection in the upper foliar layers by the rain effect on the vertical spore transfer, called "splashing" (Shaw & Royle, 1993).

The application of fungicides is the most common method used for controlling this pathogen. Several families of fungicides exist and are used against *M. graminicola*, but their efficiencies differ and decline with use (Fraaije & Cools, 2008). The azole class of sterol demethylation inhibitors (DMIs), which includes triazoles, is able to control *S. tritici* targeting of the *CYP51* enzyme (14α-demethylase), and it has been used frequently over the past 25 years (Leroux et al., 2007; Stammler et al., 2008b). Over the past 20 years, however, significant changes in the sensitivity of *M. graminicola* strains toward DMI fungicides have been widely reported (Leroux et al., 2007), especially over the past five years in the case of triazoles.

Resistance to DMIs has been described by various mechanisms, one of which is point mutations in the *CYP51* gene, which reduces the affinity between DMIs and the 14α-demethylase enzyme (Leroux et al., 2007; Stammler et al., 2008b). To date, 17 DMI-resistant genotypes (R-types) characterized by amino acid alterations have been reported in the *M. graminicola CYP51* gene: L50S, V136A/C, A379G, S188N (Leroux et al., 2007); G510C, N513K, Y137F (Cools et al., 2005; Leroux et al., 2007); I381V (Cools et al., 2005; Leroux et al., 2007);

and Y459D/S/N/C, G460D, Y461H/S, ΔY459/G460 (Cools et al., 2005; Leroux et al., 2007). Each *CYP51* variant is selected differently by triazole fungicides, and there are two main categories: strains that contain the isoleucine amino acid at the position 381 of the *CYP51* protein sequence (I381) and strains with the point mutation (V381), where isoleucine is replaced by valine (Leroux et al., 2007; Stammler et al., 2008b).

As this disease has a long latent period, presymptomatic detection and quantification of *M. graminicola* and genotypes that are related to different DMI resistance levels are very important for the effective control of strains that are highly resistant to DMIs. This allows better timing, choice and dose of fungicide applications. However, it is not possible to achieve such detection by conventional methods, and hence fungicides cannot be applied until visible symptoms appear.

Molecular detection of disease in plants was developed using the polymerase chain reaction (PCR) technique, which offers a rapid and sensitive diagnostic method (Henson and French 1993). The association of specific chemistries and a fluorescent reporter molecule (TaqMan chemistry) with PCR has permitted the development of real-time PCR (Schena et al., 2004). This technique is an accurate and reliable tool for many phytopathological studies (Guo et al., 2006 & 2007; Fraaije et al., 1999; Schena et al., 2004). Real-time PCR analysis, also referred to as quantitative PCR (qPCR), can provide a deeper insight into host–pathogen interactions by detecting the primary inoculum, general infection level and genotypes, depending on the specific target gene. For *M. graminicola*, qPCR based on the specific and stable gene β-tubulin has been described as a diagnostic tool by (Bearchell et al., 2005) and on the *CYP51* gene for 14α-demethylase mutations by (Selim, 2009).

In the present study, we used qPCR to study the epidemiological context of *M. graminicola*, taking into account the effect of many factors, including external contamination by ascospores, cultivars resistance, leaf colonization stages and fungicide efficiency. The correlation between results of molecular qPCR analysis and visual symptoms observations were investigated.

2. Materials and methods

2.1 In vivo studies of *Septoria* leaf blotch using qPCR

2.1.1 Trapping of spores

The spore trap assay is described by (Fraaije et al., 2005). A spore trap was installed in the field, at a distance of 3 m from the wheat trial in order to avoid the capture of pycnidiospores that move vertically as a result of splashing raindrops. Spores were collected from a plastic film, and then they were ground using an MM 300 Mixer Mill (Qiagen, USA) in a 2-ml Eppendorf® Safe-lock micro test tube containing 250 mg white quartz (Sigma® S-9887) and one 5-mm stainless steel bead (Cat. No. 69989, Qiagen, France).

2.1.2 Evolution of *Septoria* leaf blotch and resistant wheat cultivars

The field sites were located at the Beauvais Agricultural Research Station of the Institut Polytechnique LaSalle-Beauvais, Beauvais, France. Based on the susceptibility to *M. graminicola*, one resistant cultivar (Maxwell) and four susceptible cultivars (Dinosor, Alixan, Trémie and Maxyl) were selected (Table 1). Rainfall, temperature and leaf wetness in the trial field were monitored daily (Figure 1a). Trials were performed during the 2008–2009 growing season using a completely randomized block design with three replicate plots.

Fertilization was according to plant requirements and protection against other diseases was carried out. The plot size was 2 × 12 m. Twenty leaves were randomly sampled from one foliar layer of the main stem. The foliar layer number (Fn) was determined by counting the position of the leaf from the flag leaf (F1). Disease evolution was determined by assessing visual symptoms and by qPCR analysis. The necrotic area related to Septoria blotch was recorded and then the leaves were stored at –80 °C until lyophilization.

Cultivar	MAXWELL	DINOSOR	ALIXAN	TREMIE	MAXYL
Producer	Saaten Union	Unisigma	Nickerson	Serasem	Momont
Year	2008	2005	2005	1992	2005
Susceptibility rating*	7	4	4	4	5

* Susceptibility rating (1–9): 1 (susceptible) to 9 (resistant).

Table 1. Wheat cultivars used in the resistance study against *Mycosphaerella graminicola*

2.1.3 *S. tritici* DMI-resistant genotypes

S. tritici DMI-resistant genotypes were determined for wheat leaf samples collected from the five wheat cultivars (listed above) grown under field conditions. Intentional mismatch primers (Table 2) and allele-specific qPCR were used as described by (Selim, 2009), to quantify the mutation proportions of *CYP51* in DNA samples.

Primers		Sequence (5′–3′)	Mutation	GenBank accession number	Amplicon size	T_m
SYBER Green						
MG	for	CCTCGCCGAACTTACGATCT		EF418622		58
MG	rev	CGCGGACTTCCTTTCCTTG			60	59
R6&7-	for	CCCTTCGTATTCACGCTCgAG	I381V	EF418630		60
R6	rev	GCTTACCAGGCCGTAGCCA		EF418628	276	60
R7-	rev	CCTTGCTTACCAGGCCGTt//GT	ΔY459/G460	EF418630	273	60
R7+	for	CAAAGAAACCCTTCGTATTCAtGG	A379G			60
R7+	rev	AATGGAGGCAGTCGGGAAA			167	59
TaqMan						
MG	for	GCCTTCCTACCCCACCATGT		AY547264	63	60
MG	rev	CCTGAATCGCGCATCGTTA				60
MG	P	FAM-TTACGCCAAGACATTC-MGB				69
W	for	AGTCGTCAAAGAAACCCTTCGTA		EF418622	106	58
W	rev	GGTGGTTGGAATGACGTATGC				59
W	P	Vic-CGCTCCAATCCACTC-MGB				68

Lower case nucleotide indicates the intentional mismatch, nucleotides in bold are located at the site of single nucleotide polymorphism (SNP), (for) and (rev) indicate forward and reverse orientation, (W) indicates wild-type specific primer, (T_m) indicates primer melt temperature, (//) and ΔY459/G460 indicate double deletion of tyrosine (Y) and glycine (G) at positions 459 and 460 of the amino acid sequence, respectively.

Table 2. Sequences of primers and probes used to determine moderate resistant (R6, R7– or R7+) or general (MG) genotypes of *M. graminicola*

2.2 *In vitro* studies of *Septoria* leaf blotch using qPCR

2.2.1 Plant material

Susceptible and resistant winter wheat cultivars Dinosor and Maxwell, respectively, were used. Grains were disinfected by incubation with 1% sodium hypochlorite for 5 minutes, with shaking, and then washed three times in autoclaved distilled water. Grain germination was realized in a 0.5% water–agar medium. After incubation in darkness; 24 hours at 20 °C, 48 hours at 4 °C and then 24 hours at 20 °C (Arraiano et al. 2001), grains were transferred into 500 ml pots containing an autoclaved soil mixture of horticultural compost, sand and silt–loam soil (1:1:2 v/v/v). Pots were incubated at 15 °C, for a 16-hour photoperiod, with 150 µmol of photon.m^{-2}.s^{-1}. Plantlets were watered twice a week with 50 ml distilled water per pot, and once a week with 50 ml water containing 25% Murashige and Skoog basal medium (Sigma® M 5519).

2.2.2 Inoculum preparation

M. graminicola isolates T0248, T0254 and T0256 were obtained from *S. tritici* collection strains held in the authors' laboratory. Sporidia (yeast-like cells) stored at –80 °C were activated by transfer to fresh potato dextrose agar medium (39 g.l^{-1}, Sigma, USA). After 10 days of incubation at 18 °C, with a 12-hour photoperiod, mycelia and spores were scraped off the surface and grown in a liquid yeast–sucrose medium (yeast extract 10 g.l^{-1}, sucrose 10 g.l^{-1}; Sigma, USA) for 7 days at 18 °C with permanent light (100 µmol of photon.m^{-2}.s^{-1}) and shaking (150 rpm). Spores were collected by centrifugation at 1500 rpm for 5 minutes at 15 °C, washed twice with sterile distilled water, and then suspended in 10 mM MgSO$_4$ (Sigma® M-9397) containing 0.1% Tween 20 surfactant. The concentration was adjusted to 10^5 spores.ml^{-1}.

2.2.3 Plantlet inoculation and leaf sampling

Twenty-one-day-old plantlets were inoculated either with 10 µl (one drop) of *M. graminicola* inoculum (10^5 spores.ml^{-1}) at the bottom part of the second leaf for microscopic observations, or by spraying a 3 ml covering over the whole plantlet to assess fungicide efficiency. The plants were enveloped in a transparent polyethylene bag for 3 days. The bag provided an atmosphere of saturated humidity around the inoculated leaves. Disease development was followed for 1 month after inoculation by sampling inoculated leaves at 0, 1, 2, 3, 5, 7, 12, 14, 16, 19, 23, 26, 28 and 30 days post inoculation (dpi). Six leaves per leaf layer were collected each time, three for the microscopic observations and three for qPCR analyses.

2.2.4 Fungicide treatments

JOAO (prothioconazole 250 g a.i. l^{-1}, Bayer CropSciences, Langenfeld, Germany) in its recommended dose (0.8 l.ha^{-1}) was tested for its efficiency in containing the development of *Septoria* disease under controlled conditions. Depending on the day of inoculation (d0), four modalities of fungicide application were tested: one preventive modality at day one before inoculation (d–1) and three curative modalities at 3, 7 and 10 days after inoculation (d+3, d+7 and d+10, respectively). Each plant was covered once with 3 ml of the fungicide solution, supplemented with one drop of Tween 20, using a TLC sprayer (Grace®, USA).

2.2.5 Microscopic observations

Microscopic observations were carried out according to (Shetty et al., 2003), with modifications. Leaves were cut and cleared by placing them overnight between two filter papers saturated with glacial acetic acid and absolute ethanol (1:3 v/v). They were then washed three times with distilled water and stored between two filter papers in a solution of lactoglycerol (lactic acid/glycerol/distilled water, 1:1:1, v/v/v) until observation. Coloration was carried out by incubating the leaf parts in 0.1% trypan blue in lactoglycerol for 1 hour at 50 °C.

Stained slides were microscopically assessed using a Leica DM 4500P research microscope (Leica Microsystems, Bensheim, Germany). Further magnification was achieved by analyzing the surface of cryofractured nonstained leaf fragments with an electron microscope (TM-1000 Tabletop Microscope; HITACHI High-Technology Corporation, Japan).

2.2.6 DNA extraction

All plant leaves analyzed by qPCR had less than 40% necrotic surface. Collected leaf samples were placed directly in liquid nitrogen and then lyophilized in a Virtis 12 XL lyophilizer for 48 hours. The dried samples were then ground using an MM 300 Mixer Mill (Qiagen, USA). DNA was extracted using the DNeasy 96 Plant kit (Qiagen, USA) according to the manufacturer's protocol. DNA was quantified by measurement of UV absorption at 260 nm (Cary 50 UV–Vis spectrophotometer; Varian, France).

2.2.7 Quantification of S. tritici using qPCR analysis

To quantify infection levels of S. tritici, primers and TaqMan minor groove binder probes were used to target a 63-bp fragment of the β-tubulin gene (GenBank accession no. AY547264), as described in (Bearchell et al., 2005). A TaqMan assay was carried out in 25 µl of a reaction mixture that contained the following: 12.5 µl Universal TaqMan PCR Master Mix (Applied Biosystems, USA), 0.3 µM of each primer, 0.2 µM probe, 200 ng DNA and water up to a volume of 25 µl. The conditions of qPCR determination were the following: 10 minutes at 95 °C, followed by 40 cycles of 15 seconds at 95 °C and 1 minute at 60 °C. All qPCR experiments were carried out using an ABI PRISM 7300 sequence detection system (Applied Biosystems, USA).

qPCR analysis of the M. graminicola β-tubulin gene was calibrated from 10^2 to 10^7 copies by serial dilution of the appropriate cloned target sequence.

3. Results

3.1 External contamination by ascospores

Data in figure 1a show that there was dry weather during the experimental season (2008–2009) and low temperatures of around 10 °C until the end of April. Results of spore capture showed a low level of external ascospore contamination (Figure 1b). Three periods of contamination were observed. The first period was during April, with two main peaks that represented 30 and 3000 ascospores per day. The second was the main period, which represented a continuous arrival of ascospores during May, with a stable number of about 100 spores per day. The third period was a classical period of ascospore production, which was at the end of the wheat growing season during July (about 30 spores per day).

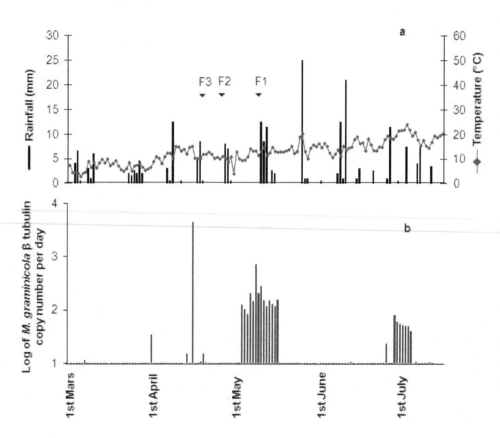

Fig. 1. (a) Weather conditions and (b) temporal dispersal of ascospores in *M. graminicola* DNA presented as the amount per day.

3.2 Evolution of fungal biomass

Evolution of *S. tritici* was characterized by simultaneously using visual symptoms and *M. graminicola* β-tubulin gene qPCR (Figure 2). Observation of necrotic areas on the top three leaf layers of the two cultivars Maxwell and Dinosor showed late disease development during June, with a "croissant"-like gradient from the top of the plant (flag leaf) to the bottom. The epidemic began in the susceptible cultivar Dinosor 15 days earlier than in the resistant cultivar Maxwell. By 16 June (Zadok's growth stage (GS) 85), Dinosor's leaf necrotic surface was 5, 22 and 25% for F1, F2 and F3, respectively, whereas Maxwell had not shown any symptoms. Two weeks later, all three leaf layers of Dinosor had more than 60% necrotic surface, whereas the values for Maxwell were 11.6, 23.5 and 67.5%, respectively.

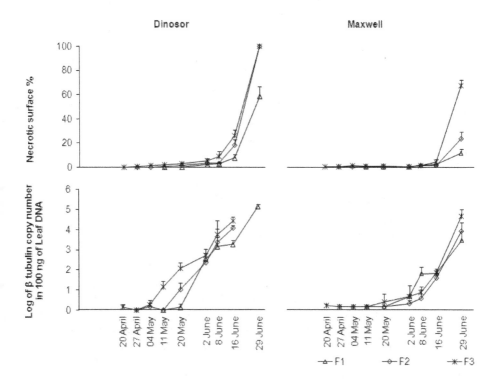

Fig. 2. Septoria blotch disease progression on Dinosor and Maxwell cultivars measured by necrotic surface observation and by qPCR.

The development of epidemics of *M. graminicola* was detected earlier with qPCR analysis (Figure 2) than when the same samples were observed visually. qPCR was not used for samples with > 40% of the surface area necrotic because of the negative effect observed on the accuracy of analysis. The *M. graminicola* β-tubulin gene on F3 was observed 2–4 weeks before symptoms appeared. There were five copies of the β-tubulin gene per 100 ng leaf DNA (CBT_{100ng}). The detection of *M. graminicola* DNA was two weeks later in the resistant cultivar than in the susceptible cultivar. The croissant gradient of the disease from the top of the plant to the bottom was also observed by qPCR analysis on all analysis dates except 2 and 8 June (GS 61 and 65 for Maxwell and GS 64 and 70 for Dinosor, respectively), when F1 qPCR values were higher than F2 and higher than or equal to F3. This increase was 14 days after the second period of the external contamination as determined by spore capture. This was characterized by the continuous arrival of ascospores (Figure 1b), combined with disease-favorable conditions, a high rate of precipitation (25 mm) and a temperature of 20–25 °C (Figure 1a).

3.3 Wheat resistance to Septoria leaf blotch

The resistance of wheat cultivars to *M. graminicola* was evaluated using visual symptoms assessment and qPCR analysis. Leaf samples were collected from the two leaf layers below the flag leaf, from the F3 leaf layer on 2 June at GS 61, 64, 65, 68 and 67, and from the F2 leaf layer on 16 June at GS 72, 71, 73, 73 and 73, of Maxwell, Dinosor, Alixan, Tremie and Maxyl, respectively. Results obtained for the *M. graminicola* DNA amount (Figure 3b), using either the F3 or F2 leaf layer, correlated well with the susceptibility rating given by the Arvalis Institut du Végétal (Arvalis Institut du Végétal 2008). Where two statistical groups were obtained, one presented the four susceptible cultivars (Dinosor, Alixan, Trémie and Maxyl) and the second presented the resistant cultivar (Maxwell). A close correlation was obtained between the results of the amount of DNA and the percentages of leaf necrotic area, especially when using the F2 leaf layer, where the discrimination between the two categories was clearer than with the F3 leaf layer (Figure 3a).

3.4 Distribution of R-types of *M. graminicola* to DMI

Quantification of the R-types of *M. graminicola* was not possible in the case of the Maxwell cultivar because of its low level of contamination. Therefore, the quantification of R-types (R6, R7–, R7+ and I381) was realized only for the susceptible cultivars Dinosor, Alixan, Trémie and Maxyl. Analyses of R-types were carried out on two dates and using two leaf layers: F3 leaf samples were used on 2 June and F2 leaf samples on 16 June. For F3, the averages of CBT_{100ng} were 1492, 2284, 1378 and 3606, and for F2 were they 3174, 5194, 3591 and 3642 for Dinosor, Alixan, Trémie and Maxyl, respectively. No significant differences were observed between R-types over the four cultivars used and over the two dates of analyses (Figure 3c). Regardless of the wheat cultivar, a high proportion of V381-strains ($\geq 94\%$) was found, with an average of 74.2, 0.1 and 19.7% for R6, R7– and R7+ genotypes, respectively, whereas, I381-strains represent only 6% of all *M. graminicola* populations.

3.5 Fungicide efficiency

The effect of fungicide application on the development of Septoria blotch disease was investigated in an *in planta* experiment over 21 days. Within this period, the nongreen leaf area reached 15% on nontreated-inoculated control plants. Between 3 to 15 dpi, disease dynamics on control plants was slow (CBT_{100ng} was 113–237 for Dinosor and 67–191 for Maxwell) but accelerated between 15 to 21 dpi to reach 1922 and 764 for Dinosor and Maxwell, respectively. The area under the disease progression curve (AUDPC) was calculated using qPCR analysis (CBT_{100ng}) for the period from 3 to 21 dpi. For control plants, AUDPC was 8633 and 4614 for Dinosor and Maxwell, respectively. The AUDPC for all fungicide treatments was lower than in the control. Figure 4 shows a significant negative effect of all fungicide treatments on the development of fungal DNA regardless of the plant cultivar. The most effective application occurred with the preventive treatment (d−1), where fungal DNA was strongly decreased on all dates of observations: protection levels of 79% and 85% were achieved for Dinosor and Maxwell, respectively. Microscopic observations showed that spores failed to germinate when fungicide was applied one day before inoculation (Figure 5). Curative treatment (d+3) was similar to preventive treatment; the level of protection for Dinosor and Maxwell was 73 and 71%, respectively. Less fungicidal impact was observed when the fungicide was applied seven or ten days after inoculation (d+7 and d+10); the level of protection was 45–60%.

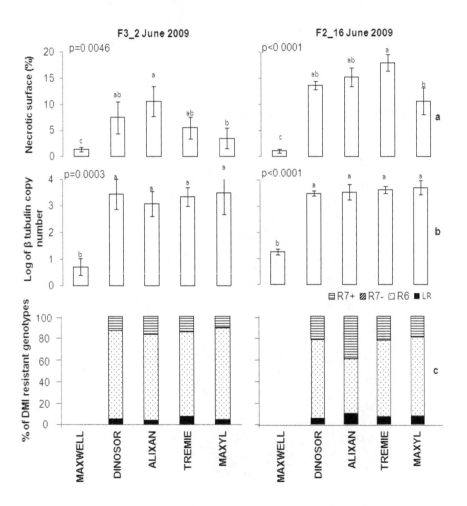

Fig. 3. (a and b) Evaluation of cultivar resistance using visual symptoms observations and qPCR. (c) Frequency of DMI-resistant genotypes of *M. graminicola*; low resistant (I381) and moderate resistant (R6, R7– and R7+). (Leaf samples were collected from leaf layers F3 and F2 on 2 and 16 June 2009, respectively.)

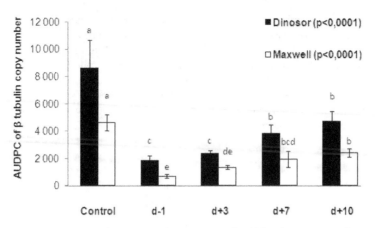

Fig. 4. Prothioconazole efficiency against Septoria leaf blotch on Maxwell and Dinosor: d–1, d+3, d+7 and d+10 are the fungicide treatments used one day before inoculation, and 3, 7 and 10 days after inoculation, respectively.

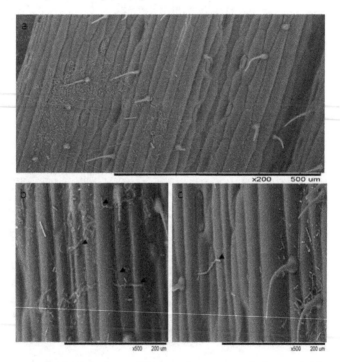

Fig. 5. Electron microscopic exposures of Dinosor leaves sampled at 15 dpi by a 10 µl-drop of *M. graminicola* conidiospores $10^5.ml^{-1}$: (a) nontreated control, where almost all spores germinated and stomata near from the point of inoculation were penetrated; and (b and c) wheat leaf treated with prothioconazole one day before inoculation (d–1), where spores failed to germinate and almost all spores degraded.

3.6 Validation of qPCR analyses

3.6.1 Relationship between qPCR analyses and *S. tritici* necrotic symptoms

The second leaf layer (F2) of the susceptible cultivar Maxyl was used to study the correlation between DNA amounts measured by qPCR for leaf samples with different levels of necrotic surfaces, from 0 to 100% (Figure 6). *M. graminicola* DNA was detected in samples without symptoms and the DNA amount increased proportionally with an increase in leaf necrotic surface. A high level of correlation (CF = 0.95) was observed up to 40% leaf necrotic surface. The amount of *M. graminicola* DNA decreased beyond 50% leaf necrotic surface and remained stable.

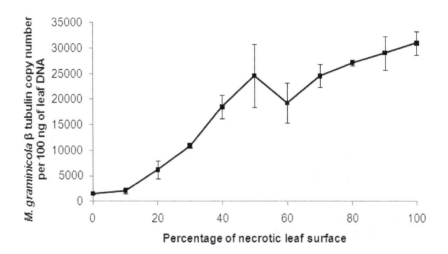

Fig. 6. Linear relationship between necrotic symptoms and qPCR analysis for cultivar Maxyl (F2 leaf layer)

3.6.2 Relationship between qPCR analysis and fungal colonization stages

To understand the relationship between the amounts of *M. graminicola* DNA measured by qPCR and leaf colonization stage, a *M. graminicola*–wheat pathosystem with one point of inoculation was developed. The base part of the second leaf of 21-day-old plantlets (GS 13) was inoculated using 10 µl (one drop) of *M. graminicola* inoculum (10^5 spores.ml^{-1}). Fungal colonization stages and the development of the amount of DNA are tabulated in Table 3.

One day after inoculation, 70% of spores had germinated and almost all germ tubes growing on the leaf surface were oriented toward the stomata (Figure 7a and b). Stomata located near the inoculation point were penetrated (Figure 7c and d). Appressorium-like structures (swellings of germ tube tips) were observed in contact with the ridges found at the guard cell lips (Figure 8a and b) and also sometimes over anticlinal walls, within the depressions between epidermal cells (Figure 8c, d and e). Electron microscopic observations of cryofractured leaf samples showed a direct penetration of leaf tissue in positions far from the stomata (Figure 8f). Intercellular colonization by mycelia growth was observed 5 and 7 dpi (Figure 7e). It was characterized by a progression from the inoculation point toward the top part of the leaf. This progression was fast at the beginning and became slower until 9 dpi. At 12 dpi, 55% of stomata were colored blue and cell death reached 20% of the leaf surface (Figure 7f). Disease development remained stable at this level until the appearance of symptoms at 26 dpi. This started from the leaf tip and progressed downward. Under the conditions used in this experiment, no pycnidium formation was observed by visual or microscopic observations.

Day post inoculation	Colonization state	CBT_g*
0	Inoculation	327
1	Germination	367
3	Penetration	443
5–7	Internal colonization	533–564
12	Cells degradation	683
26	Symptoms (2%)	1785

*Copy of *M. graminicola* β-tubulin gene per gram of fresh leaf.

Table 3. Relationship between *M. graminicola* colonization stages and results of qPCR analysis

The qPCR assay was performed for samples collected on the same dates as for samples used for the microscopic observations. *M. graminicola* DNA evolution was characterized by an increase in the number of β-tubulin gene copies per gram (CBT_g) of fresh leaf weight from 327 at the day of inoculation to 683 copies at 12 dpi. The amount of DNA then remained stable over 10 days. At 26 dpi, during the appearance of symptoms, an increase in the amount of DNA becomes important.

4. Discussion

Septoria leaf blotch in wheat has a long latent period (Verreet et al., 2000). Prompt information about epidemic diseases, external contamination (arrival of ascospores) and fungicide-resistant alleles can be used to modify disease management strategies, based on the optimal use of host resistance, chemical control and cultural practices (Fraaije et al., 2005). In the present study, we highlight the importance of the use of qPCR in the study of Septoria leaf blotch epidemics.

Ascospores of *M. graminicola* present the primary disease inoculation. These could be produced all year round under specific conditions, such as certain weather conditions (Cordo et al., 1999), and by mature pseudothecia (Kema et al., 1996). Results of spore traps have indicated that external contamination by ascospores affects the *Septoria* leaf blotch epidemic by increasing the level of contamination in the upper leaf layers (F1 and F2).

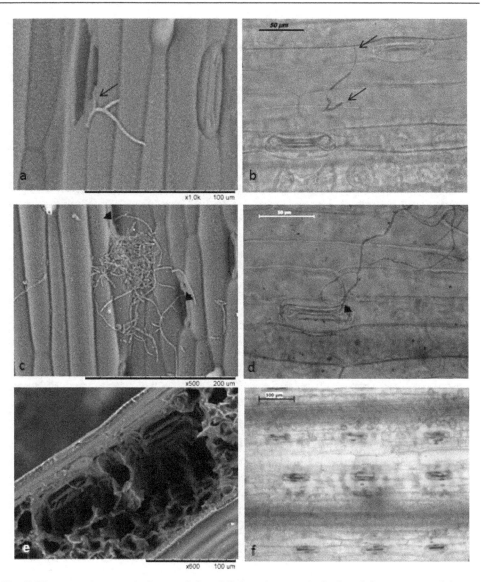

Fig. 7. Electron microscopic (a, c and e) and light microscopic (b, d and f) exposures of the
colonization stages of Dinosor leaves inoculated by a 10 μl-drop of *M. graminicola*
conidiospores ($10^5.ml^{-1}$): (a and b) spore germination at 1 dpi, 70% of spores germinated and
almost of germ tubes grew on the leaf surface oriented toward the stomata; (c and d) at 3
dpi, stomata located near the inoculation point were penetrated; (e) at 5 and 7 dpi, the
stomata were penetrated by passing through the stomatal guard cells and then intercellular
colonization by mycelia growth; (f) at 12 dpi, 55% of the stomata were colored blue and cell
death had reached 20% of the leaf surface. (Note the colonization of adjacent substomatal
chambers and the surrounding tissue of the substomatal chamber.)

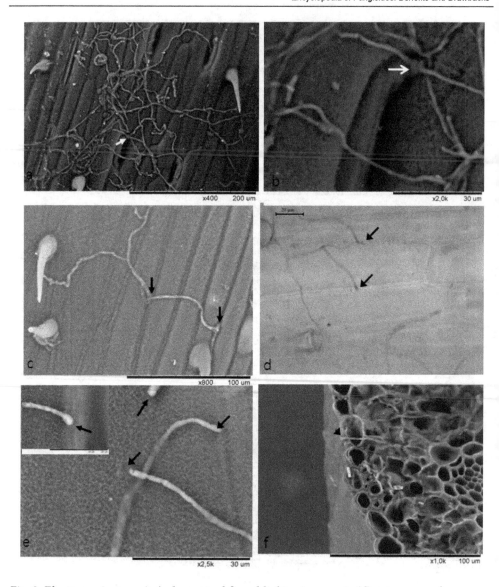

Fig. 8. Electron microscopic (a, b, c, e and f) and light microscopic (d) exposures of Dinosor leaves inoculated by a 10-μl drop of *M. graminicola* conidiospores (10⁵.ml⁻¹) sampled at 15 dpi: (a) stomata penetration by germ tubes; (b) two germ tubes with an appressorium-like structure penetrate a single stomata (arrows); (c, d and e) germ tube apical differentiations, which are similar to those seen over stomatal openings but here are associated with a leaf surface depression over an epidermal cell anticlinal wall (arrows); (f) cryofracture of *M. graminicola* colonized wheat leaf. Note that the site of penetration (arrowhead) is not per stomata, which proves the direct penetration by this fungus.

The position of these leaf layers allows them to capture more air spores than the leaf layers at the bottom of the plant. This external contamination could affect fungicide efficiency by presenting a potential reinfection process, which perhaps coming out of the potential period of fungicide. Therefore, the detection of this reinfection source provides an actual interpretation of the efficiency of chemical treatment. In ascospore traps, the limitation of qPCR is the absence of specific gene markers that can discriminate between conidiospores and ascospores. However, *M. graminicola* conidia have a limited spread of only a few meters (Boeger et al., 1993), whereas ascospores spread over longer distances (Fraaije et al., 2005). Therefore, the limitation of qPCR was now successfully eliminated by removing all plants within a 3 m diameter of the traps in order to avoid the capture of rain-splash-dispersed conidia, as described in (Fraaije et al., 2005).

The accuracy of the use of qPCR to evaluate the resistance level of wheat cultivars to *M. graminicola* was investigated. Generally, results of qPCR analysis have a high correlation with the results of visual disease assessment (0.95–0.99). Detection of *M. graminicola* using qPCR was achieved 2–4 weeks before symptoms appeared and was 2 weeks earlier in the susceptible cultivars than in the resistant one. However, the *M. graminicola* epidemiology was well characterized for the resistant cultivar by a low level of contamination and a longer period of incubation than the susceptible cultivars. Generally, results of qPCR analyses of F3 or F2 DNA samples correlated well with the susceptibility rating of cultivars that were previously identified by the (Arvalis Institut du Végétal, 2008). A close correlation was obtained between qPCR results and the percentages of leaf necrotic area, especially with the F2 leaf layer. However, under natural infection conditions, visual assessment is suitable for assessing the combined resistance to all pathogens involved, but it is not suitable for assessing resistance to individual pathogens in a disease complex (Loughman et al., 1996). Visual assessment lacks accuracy and specificity and may be confused with other diseases, stress-related symptoms or even normal plant development (Hollomon et al., 1999; Parker et al., 1995; Shimin, 2005). This problem was eliminated with qPCR, which has a high level of specificity of *M. graminicola* β-tubulin and other microorganisms on the leaf surface that interfere with *M. graminicola* DNA are avoided (Fraaije et al., 1999). Furthermore, the using of multiplex qPCR (Fraaije et al., 2001), permits the quantification of other wheat pathogens in the same DNA sample.

The effect of wheat genotypes on the distribution of DMI-resistant R-types of *M. graminicola* was also studied. As observed previously by (Selim, 2009), no significant effect of wheat cultivar was observed and the population structure was stable over all the analysis dates in the same season. The frequency of V381-genotypes was > 90%, which agrees with the frequency of > 50% reported in previous surveys carried out in 2005, 2006, 2007 and 2008 (Leroux et al., 2007; Stammler et al., 2008b), and > 70% in 2009 (Selim, 2009). However, the presymptomatic detection and quantification of *M. graminicola* R-types is very important for the effective control of strains that are highly resistant to DMIs, as it allows better timing, choice and dose of fungicide applications. For example, genotypes R6, R7– and R7+ are less sensitive to tebuconazole but are sensitive to prochloraz, whereas genotypes carrying substitution V136A (R5) are most resistant to prochloraz (Fraaije et al., 2007; Leroux et al., 2007; Stammler et al., 2008a).

In the absence of total wheat cultivar resistance to *Septoria* leaf blotch, DMIs remain the key fungicide agents against *M. graminicola* (Leroux et al., 2007; Mavrolidi & Shaw, 2005).

Their effects on *M. graminicola* are mostly attributed to the systemic action and inhibition of spore germination (Godwin et al., 1999). In the present study, qPCR analysis was used to evaluate the preventive and curative efficiency of DMI fungicides to control the more frequent genotypes of *M. graminicola* (R6, R7– and R7+). Results showed that prothioconazole treatments significantly decreased the amount of *M. graminicola* DNA by 80% in preventive treatment, and they were still 70% efficient when applied curatively at 3 dpi. Electron microscope observations showed that, after the preventive fungicide treatment (d–1), the spores failed to germinate. Although all fungicide treatments resulted in a significant DNA reduction, compared with a nontreated control, later applications of prothioconazole, at 7 to 10 dpi, resulted in a high loss of efficiency (> 50%). These results are in agreement with those reported by (Godwin et al., 1999) and (Guo et al., 2007), namely that prothioconazole has a significant inhibitive effect against spore germination and postgermination, and could have a good curative effect against *M. graminicola* when applied at up to 20% of the latent period. However, most fungicides work best when applied early in the infection cycle, prior to visual symptom expression (O'Reilly et al., 1988). Thus, a qPCR monitoring process could help determine when fungicide application would be needed, and hence it could be determined when spraying would be most economical (Guo et al., 2006).

For a good understanding of the *M. graminicola* epidemic, cytological investigations are very important; however, under field conditions they are difficult to carry out (Kema et al., 1996). Therefore, we studied the relationship between qPCR analysis and the *M. graminicola* infection processes under controlled conditions and by using a one-drop inoculation method. This method is efficient because it controls the number of spores at the beginning of infection, it reduces points of penetration that delay disease development and it eliminates the limitation of the qPCR method, which arises from the DNA quantification of dead and nongerminated spores (Allmann et al., 1995; Josephson et al., 1993). However, penetration of leaves occurred mostly through the stomata, which is in agreement with previous reports (Cohen & Eyal 1993; Duncan & Howard, 2000; Kema et al., 1996; Shetty et al., 2003). Where, appressorium-like swellings were produced over the stomata as well as periclinally and anticlinally, and all stomatal penetration took place from germ tubes with swellings. Direct penetrations have been demonstrated previously (Cohen & Eyal, 1993; Dancer et al., 1999; Rohel et al., 2001; Shetty et al., 2003) and was also in rare cases. It has been suggested that it was due to a secondary mechanism of invasion of the host (Cohen & Eyal, 1993); however, its trigger factors are not known. The time course of the wheat infection processes of *M. graminicola* has been described by (Kema et al., 1996) and (Duncan & Howard, 2000). A strong correlation between the results of qPCR and microscopic observations was found, and disease development determined using qPCR had a similar pattern to that previously revealed by ELISA (Kema et al., 1996) or by PCR/PicoGreen assay (Fraaije et al., 1999). Until the formation of necrotic lesions, the biomass increased only slightly (or even decreased), but then increased rapidly during necrosis and the formation of pycnidia.

Our data confirm that qPCR is an accurate and specific method for studying *Septoria* leaf blotch epidemics in wheat, and for evaluating cultivar resistance, fungicide efficiency and the appearance of fungicide-resistant genotypes. Further research is required to improve our understanding of *M. graminicola* epidemics, taking in account wheat genotype resistance and the effects of fungicide application.

5. Conclusion

Close correlations were obtained between qPCR analysis and leaf colonization stages as well as with visual observations when the leaf had less than 40% necrotic area.

Real-time PCR results showed that late ascospore arrival increase the amount of *M. graminicola* DNA in the upper leaf layers (F1 and F2) and affect the real evaluation of fungicide efficiency. Distribution of DMI-resistant populations of *M. graminicola* was not affected by wheat cultivars. Cultivar resistance determined by qPCR correlated well with the susceptibility rating given by disease symptoms evaluation. Direct penetration of leaf tissue was confirmed by electron microscopy and, coupled with qPCR results, prothioconazole showed a significant inhibitive effect against spore germination and postgermination. We concluded that qPCR is an accurate and specific quantitative method for detecting and quantifying *M. graminicola* leaf blotch, in wheat, spore arrival, fungicide efficiency and fungicide-resistant genotypes, and for assessing the resistance of cultivars.

6. Acknowledgments

This work was financially supported by Bayer CropScience in France (BCSF).

7. References

Allmann, M., Höfelein, E., Köppel, E., Lüthy, J., Meyer, R., Niederhauser, C., Wegmüller, B., & Candrian, U. (1995). Polymerase chain reaction (PCR) for detection of pathogenic microorganisms in bacteriological monitoring of dairy products. *Research in Microbiology*, 146, 86-97

Arraiano, L. S., Brading, P. A., & Brown, J. K. M. (2001). A detached seedling leaf technique to study resistance to *Mycosphaerella graminicola* (anamorph *Septoria tritici*) in wheat. *Plant Pathology*, 50, 339-346

Arvalis Institut du Végétal (2008). Blé tendre-maladies; comportement vis-à-vis des maladies. *Choisir*

Bearchell, S. J., Fraaije, B. A., Shaw, M. W., & Fitt, B. D. L. (2005). Wheat archive links long-term fungal pathogen population dynamics to air pollution. *Proceedings of the National Academy of Sciences of the United States of America*, 102, 5438-5442

Boeger, J. M., Chen, R. S., & McDonald, B. A. (1993). Gene flow between geographic population of *Mycosphaerella graminicola* (anamorph *Septoria tritici*) detected with restriction fragment length polymorphism markers. *Phytopathology*, 83, 1148-1154.

Cohen, L., & Eyal, Z. (1993). The histology of processes associated with the infection of resistant and susceptible wheat cultivars with *Septoria tritici*. *Plant Pathology*, 42, 737-743

Cools, H. J., Fraaije, B., & Lucas, J. A., 2005. Molecular examination of *Septoria tritici* isolates with reduced sensitivity to triazoles. In: 14th International Reinhardsbrunn Symposium, Modern Fungicides and Antifungal Compounds, at Friedrichroda, Germany, 103-114

Cordo, C. A., Simón, M. R., Perelló, A. E., & Alippi, H. E., 1999. Spore dispersal of leaf blotch pathogens of wheat (*Mycosphaerella graminicola* and *Septoria tritici*). In: CIMMYT, *Septoria* and *Stagonospora* diseases of cereals : a compilation of global research, at Mexico, 98-101

Dancer, J., Daniels, A., Cooley, N., & Foster, S. 1999. *Septoria tritici* and *Stagonospora nodorum* as model pathogens for fungicide discovery. In *Septoria on cereals: a study of pathosystem*, edited by J. A. Lucas, P. Bowyer and H. M. Anderson. Wallingford, UK: CABI Publishing

Duncan, K. E., & Howard, R. J. (2000). Cytological analysis of wheat infection by the leaf blotch pathogen *Mycosphaerella graminicola*. *Mycological Research, 104*, 1074-1082.

Fraaije, B. A., & Cools, H. J. (2008). Are azole fungicides losing ground against Septoria wheat disease ? Resistance mechanisms in *Mycosphaerella graminicola*. *Pest Management Science, 64*, 681-684

Fraaije, B. A., Lowell, D. J., Rohel, E. A., & Hollomon, D. W. (1999). Rapid detection and diagnosis of *Septoria tritici* epidemics in wheat using a polymerase chain reaction/PicoGreen assay. *Journal of Applied Microbiology, 86*, 701-708

Fraaije, B. A., Lowell, D. J., Coelho, J. M., Baldwin, S., & Hollomon, D. W. (2001). PCR-based assay to assess wheat varietal resistance to blotch (*Septoria tritici* and *Stagonospora nodorum*) and rust (*Puccinia striiformis* and *Puccinia recondita*) diseases. *European Journal of Plant Pathology, 107*, 905-917

Fraaije, B. A., Cools, H. J., Kim, S. H., Motteram, J., Clark, W. S., & Lucas, J. A. (2007). A novel substitution I381V in the sterol 14 a-demethylase (CYP51) of *Mycosphaerella graminicola* is differentially selected by azole fungicides. *Molecular Plant Pathology, 8*, 245-254

Fraaije, B. A., Cools, H. J., Fountaine, J., Lovell, D. J., Motteram, J., West, J. S., & Lucas, J. A. (2005). Role of ascospores in further spread of QoI-resistant cytochrome *b* alleles (G143A) in field populations of *Mycosphaerella graminicola*. *Phytopathology, 95*, 933-941

Godwin, J. R., Bartlett, D. W., & Heaney, S. P. 1999. Azoxystrobin: implications of biological mode of action, pharmacokinetics and resistance management for spray programmes against Septoria diseases of wheat. In *Septoria on cereals: a study of pathosystems*, edited by J. A. Lucas, P. Bowyer and H. M. Anderson. Wallingford, UK: CABI Publishing

Guo, J. R., Schnieder, F., & Verreet, J. A. (2006). Presymptomatic and quantitative detection of *Mycosphaerella graminicola* development in wheat using a real-time PCR assay. *FEMS Microbiology Letters, 262*, 223-229

Guo, J. R., Schnieder, F., & Verreet, J. A. (2007). A real-time PCR assay for quantitative and accurate assessment of fungicide effects on *Mycosphaerella graminicola* leaf blotch. *Journal of Phytopathology, 155*, 482-487

Henson, J. M., & French, R. C. (1993). The polymerase chain reaction and plant disease diagnosis. *Annual reviews of phytopathology, 31*, 81-109

Hollomon, D. W., Fraaije, B., Rohel, E., Butters, J., & Kendall, S. 1999. Detection and diagnosis of Septoria diseases: the problem in practice. In *Septoria on cereals: a study of pathosystems*, edited by J. A. Lucas, P. Bowyer and H. M. Anderson. Wallingford, UK: CABI Publishing.

Josephson, K. L., Gerba, C. P., & Pepper, I. L. (1993). Polymerase chain reaction detection of nonviable bacterial pathogens. *Applied and Environmental Microbiology, 59*, 3513-3515

Kema, G. H. J., Yu, D., Rkenberg, F. H. J., Shaw, M. W., & Baayen, R. P. (1996). Histology of the pathogenesis of *Mycosphaerella graminicola* in wheat. *Phytopathology, 86*, 777-786

Accuracy of Real-Time PCR to Study Mycosphaerella graminicola Epidemic in Wheat:
From Spore Arrival to Fungicide Efficiency

219

Leroux, O., Albertini, C., Gautier, A., Gredt, M., & Walker, A. S. (2007). Mutation in the CYP51 gene correlated with changes in sensitivity to sterol 14 a-demethylation inhibitors in field isolates of Mycosphaerella graminicola. Pest Management Science, 63, 688-698

Loughman, R., Wilson, R. E., & Thomas, G. J. (1996). Components of resistance to Mycosphaerella graminicola and Phaeosphaeria nodorum in spring wheat. Euphytica, 89, 377-385

Mavrolidi, V. I., & Shaw, M. W. (2005). Sensitivity distributions and cross-resistance patterns of Mycosphaerella graminicoma to fluquinconazole, prochloraz and azoxystrobin over a period of 9 years. Crop Protection, 24, 256-259

O'Reilly, P., Bannon, E., & Doyle, A. (1988). Timing of fungicide applications in relation to the development of Septoria nodorum in winter wheat. Proceeding of the British Crop Protection Conference-Pests and Disease, Farnham, UK: BCPC, 929-34

Palmer, C. L., & Skinner, W. (2002). Mycosphaerella graminicola: latent infection, crop devastation and genomics. Molecular Plant Pathology, 3, 63-70

Parker, S. R., Shaw, M. W., & Royle, D. J. (1995). The reliability of visual estimates of disease severity on cereal leaves. Plant Pathology, 44, 856-864

Rohel, E. A., Payne, A. C., Fraaije, B. A., & Hollomon, D. W. (2001). Exploring infection of wheat and carbohydrate metabolism in Mycosphaerella graminicola transformants with differentially regulated green fluorescent protein expression. Molecular Plant–Microbe Interactions, 14, 156-163

Schena, L., Nigro, F., Ippolito, A., & Gallitelli, D. (2004). Real-time quantitative PCR: a new technology to detect and study phytopathogenic and antagonistic fungi. European Journal of Plant Pathology, 110, 893-908

Selim, S. (2009). Allele-specific real-time PCR quantification and discrimination of sterol 14 a-demethylation-inhibitor-resistant genotypes of Mycosphaerella graminicola. Journal of Plant Pathology, 91, 391-400

Shaw, M. W., & Royle, D. J. (1989). Airborne inoculum as a major source of Septoria tritici (Mycophaerella graminicola) infections in winter wheat crops in the UK. Plant Pathology, 38, 35-43

Shaw, M. W., & Royle, D. J. (1993). Factors determining the severity of epidemics of Mycosphaerella graminicola (Septoria tritici) on winter wheat in the UK. Plant Pathology, 42, 882-899

Shetty, N. P., Kristensen, B. K., Newman, M.-A., MØller, K., Gregersen, P. L., & JØrgensen, H. J. L. (2003). Association of hydrogen peroxide with restriction of Septoria tritici in resistant wheat. Physiological and Molecular Plant Pathology, 62, 333-346

Shimin, T. (2005). Accurate assessment of wheat and triticale cultivar resistance to Septoria tritici and Stagonospora nodorum infection by biotin/avidin ELISA. Plant Disease, 89, 1229-1234

Stammler, G., Kern, L., Semar, M., Glaettli, A., & Schoefl, U., 2008a. Sensitivity of Mycosphaerella graminicola to DMI fungicides related to mutations in the target gene cyp51 (14 a-demethylase). In: 15th International Reinhardsbrunn Symposium, Modern Fungicides and Antifungal Compounds, at Friedrichroda, Germany, 137-142

Stammler, G., Carstensen, M., Koch, A., Semar, M., Strobel, D., & Schlehuber, S. (2008b). Frequency of different CYP51-haplotypes of *Mycosphaerella graminicola* and their impact on epoxiconazole-sensivity and -field efficacy. *Crop Protection, 27*, 1448-1456.

Verreet, J. A., Klink, H., & Hoffmann, G. M. (2000). Regional monitoring for disease prediction and optimization of plant protection measures: the IPM wheat model. *Plant Diseases, 84*, 816-826

12

Fungicides as Endocrine Disrupters in Non-Target Organisms

Marco F. L. Lemos[1,2], Ana C. Esteves[2] and João L. T. Pestana[2]
[1]ESTM and GIRM, Polytechnic Institute of Leiria
[2]Department of Biology and CESAM, University of Aveiro
Portugal

1. Introduction

In the past few decades concern has been growing about the possible consequences of environmental exposure to a group of chemicals (natural, synthetic, industrial chemicals or by-products) suspected to alter the functions of the endocrine system and consequently causing adverse health effects in an intact organism, its offspring, or (sub) population (European Commission, 2007), the Endocrine Disruptor Compounds (EDCs). Today, this concern is focused both on human health and on the impacts on wildlife and the environment, being already a priority in research and legislation within the European Union (European Commission, 1999, 2001, 2004, 2007), the US Environmental Protection Agency (Kavlock et al., 1996; U.S. EPA, 1998; Harding et al., 2006) and the World Health Organization (Damstra et al., 2002).

All vertebrate and invertebrate taxa use chemical signalling molecules (hormones). Changes of the endocrine function can occur when EDCs interfere with the synthesis, secretion, transport, action or elimination of natural hormones, which are responsible for homeostasis mechanisms, reproduction, growth and/or behaviour. These interferences can be caused by the direct binding of EDCs to hormone receptors – acting as hormone mimics (agonists) or as "anti-hormones" (antagonists) – or indirectly through modulation of endogenous hormone levels by interfering with biochemical processes associated with the production, availability, or metabolism of hormones or also by the modulation of their receptors (Rodriguez et al., 2007).

2. Endocrine active compounds and invertebrates

Although invertebrates dominate over 95% of the known animal species and represent more than 30 different phyla within the animal kingdom (Ruppert et al., 2003), potential effects of suspected EDCs on the various invertebrate endocrine systems have not been studied with comparable intensity as in vertebrates, especially in fish (e.g. Baker et al., 2009), reptiles (e.g. De Falco et al., 2007), amphibians (e.g. Kaneko et al., 2008), birds (e.g. Halldin et al., 2001), and mammals (e.g.Tabuchi et al., 2006).

Even though the issue of Endocrine Disruption (ED) in invertebrates received some scientific interest in the past, only a limited number of confirmed cases were reported (deFur et al., 1999). These are largely dominated by investigations on insect growth regulators (IGRs), which are designed to act as EDCs for insect pest control, and by studies on the antifouling biocide tributyltin (TBT) that has been shown to induce imposex and intersex in prosobranch snails (Matthiessen & Gibbs, 1998; Sousa, 2009b). Imposex has been associated with skewed sex ratios, reduced fecundity, population declines, and local extinctions of affected gastropod populations (Gibbs & Bryan, 1986). These are perhaps the most complete examples of ED studies in wildlife populations. Further examples for ED in invertebrates are scarce and limited to laboratory studies, where effects on endocrine regulated processes in marine and freshwater invertebrates (Porte et al., 2006), and in the soil compartment have been demonstrated for some compounds (Lemos et al., 2009, 2010a, 2010b, 2010c). Endocrine changes following exposure to certain compounds may therefore be missed or simply be immeasurable at present, even though there is increasing evidence indicating that invertebrates are susceptible to ED (Porte et al., 2006).

Consequently, there is no reason to suppose that far-reaching changes as demonstrated by TBT and its effects on prosobranch populations are in any sense unique within invertebrates (Matthiessen & Gibbs, 1998).

Additionally, since many chemicals have been considered as endocrine disrupters in vertebrates and chemical signalling systems and their basic mechanisms in the animal kingdom exhibit some degree of conservatism (McLachlan, 2001) we can presume that endocrine systems in invertebrates can be subject to modulation by identical or similar compounds as in vertebrates (Pinder and Pottinger, 1998). But, and as stated before, despite their abundance and variety, relatively little is known about their endocrine systems, making data obtained by studies on endocrine disruption rather difficult to interpret.

The "Endocrine Disruption in Invertebrates: Endocrinology, Testing, and Assessment" report (deFur et al., 1999) summarizes 56 studies where ED may have occurred in invertebrates, although non-endocrine mechanisms are also possible scenarios for the observed effects. Effects like reduced molting frequency, reduced fecundity, elevated ecdysteroid levels, delayed reproduction, reduced size of neonates, increased brood size, mortality, molting impairment, delayed maturation, impairment of reproduction, reduced egg production, delayed brood release, reduced elimination of testosterone metabolites, retardation of regenerative limb growth, suppression of ovarian growth, differential sex ratio and super-female induction, have been reviewed in this report. This includes several studies which comprise many compounds suspected of being hormonally active on aquatic crustaceans.

With the exception of TBT, effects in molluscs, that have been associated to a locally severe impact on community levels (e.g. Blaber, 1970), and IGRs in target terrestrial insects, there are only a few field examples of ED in invertebrates. Nevertheless, much more examples for ED affecting invertebrate populations and communities can be expected.

In fact, numerous studies provide strong evidence of effects on development, fecundity and reproductive output of invertebrates that can be attributed to substances acting as EDCs (Gibbs & Bryan, 1986; Matthiessen and Gibbs, 1998; Pinder and Pottinger, 1998; Oehlmann & Schulte-Oehlmann, 2003). So, carefully targeted monitoring programs are needed because effects in invertebrates are probably widespread but undetected (Fent, 2004).

3. Pesticides as Endocrine Disruptor Compounds

In June 2000 a list of 564 potential EDCs was published in two reports of BKH Consulting Engineers, Delft, and TNO Nutrition and Food Research, Zeist, both from The Netherlands (European Commission, 2000). This list of substances was compiled having in mind the compound persistence in the environment, its production volume, the scientific evidence of endocrine disruption and wildlife and human exposure. These criteria were used to categorise the candidate substances. From these, a group of 60 compounds considered to have endocrine disrupting activity (i.e. compounds for which endocrine activity has been shown in at least one *in vivo* study) and for which a high level of concern existed with regard to exposure, deserved a special attention. These 60 compounds were included in a high priority list of EDCs proposed by the EU Commission (European Commission DG ENV, 2002). This list includes industrial chemicals such as plasticizers (e.g. benzyl-n-butylphthalate, di-n-butylphthalate, bisphenol A) or flame retardants (e.g. PBBS) but also agrochemicals or crop protection products (e.g. lindane, vinclozolin, linuron, diuron, the common metabolite of linuron and diuron, 3,4-dichloroaniline, as well as triphenyltin compounds), and biocides with antifouling properties (tributyltin compounds).

4. Fungicides as endocrine-disrupters

The increased need for pest control has made pesticide use a major issue in environmental risk assessment. Within these compounds, despite some ecotoxicology studies reporting their low toxicity to non-target species (Jansch et al., 2005; Haeba et al., 2008), many fungicides have already been reported as affecting organisms and their molecular targets and effects are now well documented.

It is not surprising that several examples of endocrine disruption have been reported for terrestrial arthropods, as several pesticides have been specially tailored to affect insects endocrine systems (IGRs), eventually co-affecting several non-target invertebrates (deFur et al., 1999). In aquatic environments the examples are scarce and reduced to the effects of TBT in molluscs that have been associated with a locally severe impact at the marine community levels (Matthiessen & Gibbs, 1998).

For this review two compounds from the EU highest priority list were selected based on their environmental occurrence as well as the existing studies confirming ED effects: the fungicides tributyltin (androgen) and vinclozolin (anti-androgen). Focusing on these compounds and their effects on non-target invertebrates, an overview of terrestrial and aquatic compartments is here addressed.

5. Case studies

5.1 Tributyltin

Tributyltin (TBT) compounds are a subgroup of the trialkyl organotin family of compounds. Of all known organotins, some of the most toxic are tributyltin compounds like tributyltin oxide (TBTO; Fig.1A) and tributyltin chloride (TBTCl; Fig.1B) (Carfi 2008). TBT compounds are organic derivatives of tin (Sn^{4+}) characterized by the presence of covalent bonds between three carbon atoms and a tin atom (Antizar-Ladislao, 2008). While inorganic forms of tin are regarded as non-toxic, these more lipid-soluble organotins can be highly toxic (Gadd, 2000).

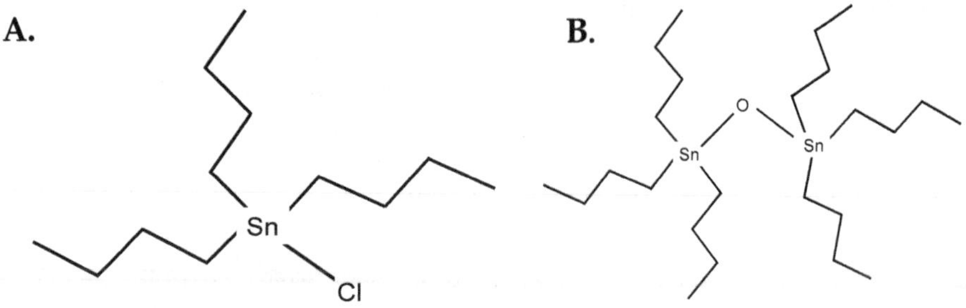

Fig. 1. Molecular structures of tributyltin chloride (A) and tributyltin oxide (B).

They are the main active ingredients in pesticides used to control a broad spectrum of organisms as they act as biocides for fungi, bacteria and insects (Mimura et al., 2008; Fent, 2006). Nevertheless, research undertaken since the early 1970s has shown that TBT is highly toxic to a larger number of non target aquatic organisms (Antizar-Ladislao, 2008). TBTs' properties were recognized in the 1950s and since then this compound has been extensively used for various industrial purposes such as slime control in paper, as a wood preservative, as a polyvinyl chloride (PVC) stabilizer, and as fungicide in agriculture (Mimura et al., 2008). In the 1970s, TBT paints widely replaced copper-based paints due to their superior performance in terms of efficacy and duration (Sonak, 2009). Since then, TBT has been used mostly as an antifouling agent in marine paint formulations to prevent the attachment of barnacles and slime on boat hulls and aquaculture nets (Kannan et al., 1998).

Due to its widespread use as an antifouling agent in boat paints, TBT is a common contaminant of marine and freshwater ecosystems. Its damaging consequences to marine ecosystems were recognized in early 1980s as the cause for the decline of some marine molluscs (Smith, 1981; Waldock & Thain, 1983; Bryan et al., 1986). As a result of field evidences of negative ecological impact of organotins, the European Union published a Directive (89/677/CEE) banning TBT application on ships smaller than 25 m. On the assumption that TBT concentrations in the open sea were too low to cause effects, there were not many restrictions on the use of organotins in larger ships. However, a similar impact in the open sea has been shown for TBT with incidence being correlated with shipping density (Santos et al., 2002). Thus, the International Maritime Organization banned the application of TBT-based paints in 2003 and called for a global agreement for total prohibition of the presence of organotins on ship hulls in 2008 (International Maritime Organization, 2001).

TBT from hulls and nets can be adsorbed onto suspended particles in the water, sediment and biota (Gadd, 2000). Subsequently it is readily incorporated into the tissues of filter-feeding zooplankton, invertebrates and eventually higher organisms such as fish and mammals where it accumulates (Antizar-Ladislao, 2008). Despite the present restrictions, TBT and its degradation products will not disappear immediately from the marine environment, and it can be expected that TBT will remain in waters and sediments for long periods of time because of the moderate to high persistency in anoxic sediments and its' widespread presence, as was confirmed by the work of Sousa and co-workers along the Portuguese coast (Sousa et al., 2009). Together with its lipophilicity, it tends to accumulate in oysters, mussels, crustaceans, molluscs, fish, and algae favouring the bioconcentration up the marine predators' food chain (Santos et al., 2009; Cruz et al., 2010).

Because of all the above mentioned, recently organotins have been considered as the most toxic substance ever introduced into the marine environment so far (Fent, 2006; Guo et al., 2010; Antizar-Ladislao, 2008; Sonak, 2009), and due to their use in a variety of industrial processes, their environmental fate and ecotoxicity as well as human exposure are topics of current concern.

In a study conducted by Guo et al (2010) on western clawed frog embryos (*Xenopus tropicalis*) the authors suggested that TBT might be the cause of several malformations. These include the loss of eye pigmentation, enlarged trunks and bent tails, in the presence of 50 ng/L of TBTCl after 24 hours of exposure. This is particularly relevant since the concentrations of TBT in open water, bays, estuaries, lakes and freshwater harbors commonly exceed this concentration with the highest TBT values found near marinas and seaports (Fent, 2006). At higher concentrations TBT may be lethal to several marine and freshwater species. Short and Thrower (1987) reported that the 96 hour LC50 for juvenile Chinook salmon (*Oncorhynchus tshawytscha*) is 1.5 µg/L.

TBT is known as an endocrine disruptor promoting adverse effects in organisms on diverse levels of biological organization (Guo et al., 2010). Matthiessen & Gibbs (1998) reported an interference with hormone metabolism, increasing the androgen levels of snails exposed to TBT. One of the best-documented and iconic adverse impacts of TBT in non-target organisms is imposex in molluscs. This pathology occurs when male sex characteristics and organs, such as penis, *vas deferens*, and seminiferous tubules are superimposed on normal female resulting in female sterilization, and even with spermatogenesis occurrence. Such pathology consequently has obvious impacts population dynamics (Gibbs et al. 1991). The first evidences linking TBT to imposex were reported in 1970 for the dog-whelk, *Nucella lapillus*, in the UK (Blaber, 1970). Since then, several studies have related TBT to the worldwide decline of marine mollusks (e.g. *Nassarius, Ilyanassa, Ocenebra* and *Urosalpinx*) in costal areas due to imposex (Gibbs et al., 1991). Administrating testosterone to the snail *Euchadra peliomphala* resulted in a stimulation of the development of male sex characters in female and castrated male gastropods (Takeda, 1980; Spooner et al., 1991). These authors also report increased testosterone titres in *N. lapillus* exposed to this fungicide. These findings have led to the hypothesis that the increased levels of testosterone in TBT-exposed organisms are responsible for the imposex development. The precise mechanism by which increased levels of testosterone are produced has not been fully described, but the weight of evidence suggests that TBT acts as a competitive inhibitor of cytochrome P450-mediated aromatase (Bettin et al., 1996). In laboratory tests, imposex was also reported from concentrations below 1 ng of TBT/L in the mud snail, *Ilyanassa obsoleta* (*Gooding* et al., 2003), and the dog-whelk, *N. lapillus* (Gibbs et al., 1988). There are still uncertainties regarding the sensitivity of the endocrine function in different genders and developmental stages in invertebrates exposed to EDCs (Rodriguez et al., 2007). Nevertheless, some authors agree that the hormonal impacts of TBT and EDCs in general may differ according to the specific life stage at which exposure occurs (e.g., embryolarval stages, gonadal development, etc.) (deFur et al., 1999; Lemos et al., 2009). Generally, the larvae/neonates of any tested species are more hormonally sensitive to tributyltin exposure than are the adults. For example, the adult female dog-whelk, *N. lapillus*, revealed imposex signs at concentrations as low as 5 ng of TBT/L while young and sexually immature females were more sensitive than adults with concentrations of 1 ng of TBT/L inducing the growth of penis and *vas deferens* tissue (Gibbs et al., 1987).

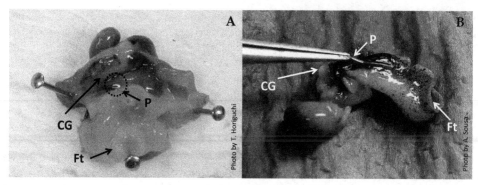

Fig. 2. Female gastropods exhibiting imposex: *Nucella lapillus* (A), and *Nassarius reticulatus* (B). CG - capsule gland; Ft - foot; P – penis. Photographs are courtesy of Prof. T. Horiguchi and Dr. A. Sousa.

Deviations from normal limb regeneration in the prawn, *Caridina rajadhari* (Reddy et al. 1991), and in the fiddler crab *Uca pugilator* (Weis et al., 1987), have been observed in laboratory experiments involving TBT exposure. Nevertheless, despite the findings and the known hormonal regulation of the molting process, it has been argued whether these effects are due to endocrine disruption rather than regular systemic toxicity. Unless specific parameters such as hormone levels are assessed no accurate conclusion can be drawn (Barata et al., 2004; Lemos et al 2009).

Levels in harbour and port waters prior to restrictions on TBT use in antifouling paints have shown levels higher than 500 ng/L in North American and European marinas. For example, one year before the UK ban (1986), TBT concentrations in Wroxham Broad and at the nearby River Bure boatyard were 898 ng/L and 1540 ng/L, respectively (Waite et al., 1989). They were significantly higher than in open surface waters, bays and estuaries where commonly values of up to 50 ng/L were observed (Fent, 2006). Albeit this regulation for the use of TBT and consequent general decrease in environmental TBT levels (Antizar-Ladislao, 2008), recent surveys still account for levels higher than those reported to elicit effects at a global scale: for instance, 32 ng TBT/L in South Korea (Sidharthan et al., 2002), 2–160 ng/L in Japanese coastal waters (Takeuchi et al., 2004), in the UK 10–78 ng/L were detected in marinas and harbors in 1998 (Thomas, 2001), and 200 ng/L in ferry ports in Corsica, Italy (Michel et al., 2001).

The impact of TBT in freshwater systems has been studied at a much lesser extent compared to estuarine and coastal environments. As still, endocrine effects have also been shown for freshwater molluscs. For example, after a 3 month exposure to a concentration of 50 ng of TBT/L, the giant ramshorn snail, *Marisa cornuarietis*, showed recognizable morphological characteristics of imposex development (Shulte-Oehlmann et al., 1995). Despite the scarce information (possibly reflecting a reduced concern about the impacts in this ecossystem), Schulte-Oehlmann (1997) reported concentrations in European lakes up to 930 ng of TBT /L in water and reaching 340 µg of TBT/g wet weight in sediments of River Elbe in Central Europe.

For the soil compartment, to our knowledge, there are no reported data on the effects of TBT to edaphic organisms.

5.2 Vinclozolin

Vinclozolin [Vz, 3-(3,5-dichlorophenyl)-5-methyl-5-vinyl-1,3-oxazolidine-2,4-dione; Fig.3] is a non-systemic dicarboxymide fungicide, manufactured by BASF and commercially sold under the names Ronilan®, Curalan®, and Ornilan®. It is efficient in controlling plants and fruit diseases caused by *Botrytis* spp., *Monilia* spp., and *Sclerotinia* spp. (Bursztyka et al., 2008) that affect crops such as lettuce, raspberries, beans and onions (Price et al., 2007). This fungicide is widely used in the United States of America and throughout Europe. In Britain, as well as in Germany, up to 50 tonnes of Vz are used each year and it was estimated that in 2002, in the USA, 2,330,738 US dollars were spent on this compound for crop protection (Gianessi & Reigner, 2005).

When sprayed as Ronilan®, at the maximum recommended application rate, the concentration of Vz in the soil is 1 mg active ingredient (a.i.)/kg (assuming that 70 % of the fungicide will reach the surface and is homogeneously distributed over the top 5 cm soil layer and the soil bulk density is 1.4 kg/dm³) (Lemos et al., 2009). Vinclozolin has a low to moderate persistence in soil, with reported half-lives from 28-43 days in the laboratory up to 34-94 days in the field and 6-12% of the original compound is present after 1 year (U.S. EPA, 1991; IUPAC, 2006). Despite this low-persistency of Vz, a reported increase of mortality with exposure time (Lemos et al., 2009) may be due to the increased concentration of its' three major metabolites (out from 15): 2-[[(3,5-dichlorophenyl)-carbamoyl]oxy]-2-methyl-3-butenoic acid (M1), 3',5'-dichloro-2-hydroxy-2-methylbut-3-enanilide (M2), and 3,5-dichloroaniline (DCA), which are reported to be more active than the parent compound (Kelce et al., 1997; Anway et al., 2005; Kavlock and Cummings, 2005) and have half-lives ranging from 179 to >1000 days (U.S. EPA, 2000). It has been reported that these metabolites may be produced both spontaneously in the presence of aqueous buffers and by biotransformation of Vz (Bursztyka et al., 2008). M1 and M2 are able to bind to the androgen receptor (Vinggaard et al., 1999) and competitively inhibit the binding of androgens to the human androgen receptor (Kelce et al., 1997; Kavlock and Cummings, 2005). Concerning the parent compound, it is known that it inhibits vertebrate testosterone-induced growth of androgen-dependent tissues (Kang et al., 2004).

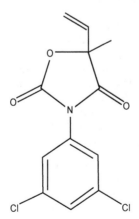

Fig. 3. Molecular structure of vinclozolin.

On plant leaves, Vz is detectable as the parent compound but does not wash off, since it is more soluble in oil than in water. This implies that Vz residues are commonly found on food (Szeto et al., 1989). Metabolites have also been found in human food (Gonzalez-Rodriguez et al., 2008).

Vinclozolin endocrine disruptor effects include induction of Leydig cell tumors, reduction of ejaculated sperm counts and prostate weight, and delayed puberty in exposed rats. One major concern is that Vz causes transgenerational effects. F1 to F4 generations of male rats exposed to Vz at the time of gonadal sex determination developed prostate disease, kidney disease, immune system abnormalities, spermatogenesis abnormalities, breast tumour development, and blood abnormalities as hypercholesterolemia, which have been associated with an alteration in the epigenetic programming of the male germ line (Anway et al., 2006; Anway & Skinner, 2008). Similar effects have been shown for pregnant rat exposed to Vz, where a transgenerational increase in pregnancy abnormalities and female adult onset disease states are promoted (Nilsson et al., 2008).

The existing information supports the hypothesis that Vz steroid-mediated actions in vertebrates have similar sub-lethal effects in invertebrates. In *Daphnia magna* it induces a decrease in the number of newborn males (Haeba et al., 2008). In molluscs Vz was shown to cause female virilisation (imposex development) and reduction of accessory sex organ expression in the fresh water snail *Marisa cornuarietis* and two marine prosobranchs *N. lapillus* and *Nassarius reticulatus* (Tillmann et al., 2001). Snails were exposed to nominal concentrations of Vz ranging from 0.03 to 1 µg/L for up to 5 months. In exposed juvenile *Marisa*, males had a slight decrease in the male accessory sex organ, particularly the penis and penis sheath. This response was only detected during the first 2-3 months of exposure for the lowest concentrations (0.03 and 0.1 µg of Vz/L), and was reversible once they attained puberty. Adult male *N. lapillus* exposed to the fungicide developed shorter penis, smaller prostate gland, and there were less males with ripe sperm stored in the seminal vesicle. Compared to the reported effects of estrogens and androgens on these two same species, these anti-androgenic responses seem to be less drastic, and might not have any biological effect at the population level (Tillmann et al., 2001). The immobilising effect (EC50, 48h) of Vz on the American oyster (*Crassostrea virginica*) and *D. magna* was reported to be 3.2 mg/L and 3.65 mg/L, respectively. For the opossum shrimp (*American bahia*) the LC50 (96h) was 1.5 -2.1 mg Vz/L (U.S. EPA, 2000).

Studies of Vz effects in the soil compartment are scarce. Vinclozolin has been reported as non-toxic to earthworms (Tomlin, 2003), but the most complete and extensive research was performed in a terrestrial isopod (Lemos et al. 2009, 2010a, 2010b, 2010c). In *Porcellio scaber*, Vz exhibited ecdysteroidal activity in ecdysone endogenous levels in a concentration-dependent way (Lemos et al., 2009). The results from this study demonstrated that the fungicide caused endocrine disruption in the isopod with an ecdysteroid up-regulation resulting in molting disturbances being further related to developmental and reproductive toxicity which enabled to suggest a causal link to ED in this class of organisms.

Since a sharp rise of ecdysteroid followed by a decrease of the hormone level triggers the ecdysis process (Bodar et al., 1990), when basal concentration of 20-hydroxyecdysone (20E) is maintained at high levels (hyperecdysonism) the shedding of the old cuticle is impaired and mortality due to incomplete ecdysis occurs (Fig. 4). Therefore, molting behaviour was associated with hyperecdysonism, delaying molt and in many cases impairing the molting process, and death surmounting at higher toxicant concentrations (1 g of a.i. Vz/kg of soil) (Lemos et al., 2009).

Fig. 4. Isopod *Porcellio scaber* and shedded cuticle after posterior half molt (A), and dead animal within the unshedded old cuticle after failed posterior half molt, after exposure to vinclozolin contaminated soil (B).

The same authors (Lemos et al. 2010a) also reported that young isopods respond with reduced growth at lower concentrations (LOEC of 100 mg a.i. Vz/kg soil) compared to adults (LOECs 300 mg a.i. Vz/kg soil). The increased sensitivity of the juvenile life stage may be either due to easier absorption of the toxicant through their relatively larger body surface/volume ratio and thin cuticle, or due to their lower capacity to metabolize the contaminants (Fischer et al., 1997; Lemos et al., 20010a) but another possibility pointed was the differential life-stage endocrinology.

Vinclozolin also elicited overall reproductive toxicity to *P. scaber* (Lemos et al., 2010b) decreasing the reproductive allocation for exposed females, the number of pregnancies, and the number of juveniles, while increasing the percentage of abortions. It induced a decrease of the brood period, with the isopods releasing juveniles almost 43 hours earlier at 100 mg a.i./kg dry soil and five days earlier at 300 mg a.i./kg dry soil. These two factors together considerably reduced the total juvenile output.

One of the reasons attributed to this impairment was the increased 20E titres (Lemos et al., 2009) that have been previously correlated with increased vitellogenin - a key protein of extreme importance in crustacean reproduction - synthesis and uptake in developing oocytes (Gohar &d Souty, 1984).

The molecular effect of Vz was assessed by differential protein expression of Vz exposed *P. scaber* in the gut, hepatopancreas, and gonads (Lemos et al., 2010c). In this study it was possible to detected up-regulated proteins at concentrations as low as 10 mg a.i./kg soil in the testes while for hepatopancreas this was only possible at concentrations equal to 1000 mg a.i./kg soil. Proteins from the heat shock protein family, Hsp70 (known as ubiquitous stress response proteins, anti-apoptotic, and protects cells from cytotoxicity and inhibiting cell death induced by several agents) were over-expressed at the lower concentration of Vz in the testes of organisms exposed to Vz (around 160% increase after exposure to 10 mg a.i./kg and around 130% at 30 mg a.i./kg).

Vinclozolin up-regulated arginine kinase (around 150% at 1000 mg a.i./kg for Vz) in the isopods' hepatopancreas. This enzyme is involved in the cellular energy metabolism, suggesting an increase of resources allocated to the activation of metabolic processes related to detoxification and the metabolisation of energy reserves to provide for these processes (Lemos et al., 2010c).

The fact that the gonads showed increased protein expression at concentrations higher than 10 mg a.i./kg soil (Lemos et al., 2010c), suggests that testes are more susceptible to these compounds than other organs. Thus, male isopod reproductive traits may therefore be especially susceptible and sensitive to this fungicide. Previous studies have also stressed the gonads and reproductive traits as preferential targets by EDCs in vertebrates (Navas and Segner, 2006).

6. Conclusions

Although TBT levels have been decreasing in the last decades, mainly due to restrictions in its use, it is still present at ng/L levels in the environment (Fent, 2006). Additionally, marine prosobranch gastropods and other invertebrates are extremely sensitive to TBT contamination, and imposex can be elicited in some species at concentrations of < 1 ng TBT/L (Gibbs et al. 1988). Moreover, TBT is very persistent in sediments being is considered as the most toxic substance ever introduced into the marine environment and a major threat to the environment for many years (e.g. Fent, 2006) since its sediment concentrations (up to several mg/kg sediment) are still highly toxic to benthic fauna while organotins resuspension is possible through storms or dredging leading to an increase of organotin levels in the water column (Unger et al., 1988). Additionally one should not forget that organotins residues such as TBT can also be transferred to humans via dietary uptake. Sousa and co-workers predict that in traditional seafood-consuming countries the estimated Sn daily intake is high enough to cause damage to humans (Sousa et al., 2009b), making TBT contamination more than "just" an environmental issue of concern.

Vinclozolin is classified as very toxic (EC50 < 1 mg/L), and it has been shown to have adverse effects in the laboratory on aquatic snail at concentrations below 1 µg/L. Nevertheless, Vz is non-persistent in the environment, and degrades rapidly, particularly under alkaline conditions (Tomlin, 2003; Ueoka et al., 1997). Moreover, a survey of German ground and surface waters found only 1% of water samples to have detectable levels (>0.1 µg/L) of Vz (Funari et al., 1995). Mediterranean estuarine waters have also been investigated for fungicide contamination, and from the sites examined only the River Po in Italy had detectable levels of vinclozolin (Readman et al., 1997).

In the soil compartment, despite these reported severe effects in isopods, the concentrations that elicit effects are far from ecological relevant concentrations.

So far there is no clear evidence suggesting effects of Vz on the terrestrial or aquatic ecosystems. Nevertheless, in Europe, the use of this substance is no longer authorized according to EC Directive 91/414.

Despite the examples showing that invertebrates are susceptible to ED, endocrine changes following exposure to certain compounds may be missed due to scarce knowledge of invertebrate endocrinology or due to very low concentrations of certain compounds (below detection limits) eliciting ED effects. In fact there is no reason to suppose that the array of endocrine changes such as the ones demonstrated for TBT and Vz are in any sense unique and most certainly similar effects are plausible and to be expected in most invertebrate species.

Through small biochemical and molecular changes, these contaminants may however interfere with different systems (e.g., reproductive, endocrine, immunological, and nervous) in different life stages of non-target species, causing medium and long term effects at the population level. Due to the nature of the ED mode of action, the consequences of the exposure to such class of compounds in the communities' structure and function are thus not always immediate and can be extremely hard to predict.

As said before many fungicides can behave like EDC, and because ED effects at the population and community levels might be only detected after several generations exposed to sub-lethal levels of pollutants, there is a need to develop and validate quantifiable tools for the identification of ED effects of fungicides and pesticides in general. It is our conviction that this has to be achieved with a mechanistic approach where alterations of hormonal vital processes in particular species are assessed. For that it is vital to conduct long term chronic assays where biochemical, organismal and reproductive parameters are measured and compared with changes in hormone levels of invertebrate species. It is not an easy task but due to the expected increase in fungicide production and utilization in the next decades, it is critical to adopt this kind of integrative approaches to better understand the mechanisms that link endocrine level responses to population- and community-level processes and to improve environmental risk assessment of these compounds

7. Acknowledgements

The authors are grateful to Professor Toshihiro Horiguchi and Dr. Ana Sousa for kindly providing the photos of imposex-exhibiting female gastropods used in this book chapter. AC Esteves and JLT Pestana wish to acknowledge the financial support given by Fundação para a Ciência e Tecnologia (SFRH/BPD/38008/2007 and SFRH/BPD/45342/2008).

8. References

Antizar-Ladislao, B. (2008). Environmental levels, toxicity and human exposure to tributyltin (TBT)-contaminated marine environment: a review. *Environment International*, 34, pp. 292-308.

Anway, M.D., Cupp, A.S., Uzumcu, M., & Skinner, M.K. (2005). Epigenetic transgenerational actions of endocrine disruptors and mate fertility. *Science*, 308, pp. 1466-1469.

Anway, M.D., Skinner, M.K. (2008). Transgenerational effects of the endocrine disruptor vinclozolin on the prostate transcriptome and adult onset disease. *Prostate*, 68, pp. 517-529.

Baker, M.E., Ruggeri, B., Sprague, L.J., Eckhardt-Ludka, C., Lapira, J., Wick, I., Soverchia, L., Ubaldi, M., Polzonetti-Magni, A.M., Vidal-Dorsch, D., Bay, S., Gully, J.R., Reyes, J.A., Kelley, K.M., Schlenk, D., Breen, E.C., Sasik, R., & Hardiman, G. (2009). Analysis of endocrine disruption in Southern California Coastal fish using an aquatic multispecies microarray. *Environmental Health Perspectives*, 117, pp. 223-230.

Barata, C., Porte, C., & Baird, D.J. (2004). Experimental designs to assess endocrine disrupting effects in invertebrates - A review. *Ecotoxicology*, 13, pp. 511-517.

Bettin, C., Oehlmann, J., & Stroben, E. (1996). Induced imposex in marine neogastropods is mediated by an increasing androgen level. Helgoländer Meeresunters 50, pp. 299-317

Blaber, S.J.M. (1970). The occurrence of a penis-like outgrowth behind the right tentacle in spent females of *Nucella lapillus* (L.). *Journal of Molluscan Studies*, 39, pp. 231-233.

Bodar, C.W.M., Voogt, P.A., & Zandee, D.I. (1990). Ecdysteroids in *Dapnhia magna* - Their role in molting and reproduction and their levels upon exposure to cadmium. *Aquatic Toxicology*, 17, pp. 339-350.

Bryan, G.W., Gibbs, P.E., Hummerstone, L.G., & Burt, G.R. (1986). The decline of the gastropod *Nucella lapillus* around south-west England: evidence for the effects of tributyltin from antifouling paints. *Journal of the Marine Biological Association of the United Kingdom*, 66, pp. 611-640.

Bursztyka, J., Debrauwer, L., Perdu, E., Jouanin, I., Jaeg, J.P., & Cravedi, J.P. (2008). Biotransformation of vinclozolin in rat precision-cut liver vices: Comparison with in vivo metabolic pattern. *Journal of Agricultural and Food Chemistry*, 56, pp. 4832-4839.

Carfi, M., Croera, C., Ferrario, D., Campi, V., Bowe, G., Pieters, R., & Gribaldo, L. (2008). TBTC induces adipocyte differentiation in human bone marrow long term culture. *Toxicology*, 249, pp. 11-18.

Cruz, A., Caetano, T., Suzuki, S., & Mendo, S. (2007). *Aeromonas veronii*, a tributyltin (TBT)-degrading bacterium isolated from an estuarine environment, Ria de Aveiro in Portugal. *Marine Environmental Research*, 64, pp. 639-50.

Damstra, T., Barlow, S., Bergman, A., Kavlock, R., & Van Der Kraak, G. (2002). Global Assessment of the State-of-the Science of Endocrine Disruptors. WHO/PCS/EDC/02.2. World Health Organisation, International Programme on Chemical Safety, Geneva, Switzerland.

De Falco, M., Sciarrillo, R., Capaldo, A., Russo, T., Gay, F., Valiante, S., Varano, L., & Laforgia, V. (2007). The effects of the fungicide methyl thiophanate on adrenal gland morphophysiology of the lizard, *Podarcis sicula*. *Archives of Environmental Contamination and Toxicology*, 53, pp. 241-248.

deFur, P.L., Crane, M., Ingersoll, C., & Tattersfield, L. (1999). Endocrine disruption in invertebrates: endocrinology, testing, and assessment. *Proceeding of the Workshops on Endocrine Disruption in Invertebrates*, 12-15 December 1998, Noordwijkerhout, The Netherlands. SETAC Press, Pensacola, U.S.

European Commission (1999). Communication from the Commission to the Council and the European Parliament - Community Strategy for Endocrine Disrupters. COM(1999)706. Commission of the European Communities, Brussels, Belgium.

European Commission (2000). Towards the establishment of a priority list of substances for further evaluation of their role in endocrine disruption - preparation of a candidate list of substances as a basis for priority setting. BKH-RPS Group, The Netherlands. P. 29.

European Commission (2001). Communication to the Council and the European Parliament on the implementation of the Community Strategy for Endocrine Disrupters - a range of substances suspected of interfering with the hormone systems of humans and wildlife. COM(2001)262. Commission of the European Communities, Brussels, Belgium.

European Commission (2004). Commission Staff Working Document on implementation of the Community Strategy for Endocrine Disrupters - a range of substances suspected of interfering with the hormone systems of humans and wildlife (COM (1999) 706). SEC(2004) 1372. Commission of the European Communities, Brussels, Belgium.

European Commission (2007). Commission Staff Working Document on the implementation of the "Community Strategy for Endocrine Disrupters" - a range of substances suspected of interfering with the hormone systems of humans and wildlife (COM (1999) 706), (COM (2001) 262) and (SEC (2004) 1372). SEC(2007) 1635. Commission of the European Communities, Brussels, Belgium.

European Commission DG ENV (2002). Study on gathering information on 435 substances with insufficient data. BKH-RPS Group, The Netherlands. p. 52.

Fent, K. (2004). Ecotoxicological effects at contaminated sites. *Toxicology*, 205, pp. 223-240.

Fent, K. (2006). Worldwide occurrence of organotins from antifouling paints and effects in the aquatic environment. *The Handbook of Environmental Chemistry*, 5, pp. 71-100.

Fischer, E., Farkas, S., Hornung, E., & Past, T. (1997). Sublethal effects of an organophosphorous insecticide, dimethoate, on the isopod *Porcellio scaber* Latr. *Comparative Biochemistry and Physiology C-Pharmacology Toxicology & Endocrinology*, 116, pp. 161-166.

Funari, E., Donati., Sandroni, D., & Vighi, M. (1995). Pesticide levels in groundwater: value limitations of monitoring. In: *Pesticide risk in groundwater*, Vighi, M. & Funari, E. (eds.), pp- 3-33. CRC press. Boca Raton, FL, U.S.

Gadd, G. (2000). Microbial interactions with tributyltin compounds: detoxification, accumulation, and environmental fate. *The Science of The Total Environment*, 258, pp. 119-127.

Gianessi, L.P., Reigner, N. (2005). The Value of Fungicides In U.S. Crop Production. CropLife Foundation - Crop Protection Research Institute. Washington, DC, U.S.

Gibbs, P.E., Bryan, G.W. (1986). Reproductive failure in populations of the dog-whelk, *Nucella lapillus*, caused by imposex induced by tributyltin from antifouling paints. *Journal of the Marine Biological Association of the United Kingdom*, 66, pp. 767-777.

Gibbs, P.E., Bryan, G.W., Pascoe, P.L., & Burt, G.R. (1987). The use of the dog-whelk, *Nucella lapillus*, as an indicator of tributyltin (TBT) contamination. *Journal of the Marine Biological Association of the United Kingdom*, 67, pp. 507 – 523.

Gibbs, P.E., Pascoe, P.L., & Burt, G.R. (1988). Sex change in the female dogwhelk, *Nucella lapillus*, induced by tributyltin from antifouling paints. *Journal of the Marine Biological Association of the United Kingdom*, 68, pp. 715-731.

Gibbs, P.E., Bryan, G.W., & Pascoe, P.L. (1991). Tributyltin-induced imposex in dogwhelk, *Nucella lapillus*: Geografical uniformity of the response and effects. *Marine Environmental Research*, 32, pp. 79-87.

Gohar, M., Souty, C. (1984). The temporal action of ecdysteroids on ovarian protein-synthesis invitro in the terrestrial Crustacean Isopoda, *Porcellio ditatatus* (Brandt). *Reproduction Nutrition Development*, 24, pp. 137-145.

Gonzalez-Rodriguez, R.M., Rial-Otero, R., Cancho-Grande, B., & Simal-Gandara, J. (2008). Determination of 23 pesticide residues in leafy vegetables using gas chromatography-ion trap mass spectrometry and analyte protectants. *10th International Symposium on Advances in Extraction Techniques,* Bruges, Belgium, pp. 100-109.

Gooding, M.P., Wilson, V.S., Folmar, L.C., Marcovich, D.T., & LeBlanc, G.A. (2003). The biocide tributyltin reduces the accumulation of testosterone as fatty acid esters in the mud snail (*Ilyanassa obsoleta*). *Environmental Health Perspectives,* 111, pp. 426-430.

Guo, S.Z., Qian, L.J., Shi, H.H., Barry, T., Cao, Q.Z., & Liu, J.K. (2010). Effects of tributyltin (TBT) on *Xenopus tropicalis* embryos at environmentally relevant concentrations. *Chemosphere,* 79, pp. 529-533.

Haeba, M.H., Hilscherova, K., Mazurova, E., & Blaha, L. (2008). Selected endocrine disrupting compounds (vinclozolin, flutamide, ketoconazole and dicofol): Effects on survival, occurrence of males, growth, molting and reproduction of *Daphnia magna. Environmental Science and Pollution Research,* 15, pp. 222-227.

Halldin, K., Berg, C., Bergman, A., Brandt, I., & Brunstrom, B. (2001). Distribution of bisphenol A and tetrabromobisphenol A in quail eggs, embryos and laying birds and studies on reproduction variables in adults following in ovo exposure. *Archives of Toxicology,* 75, pp. 597-603.

Harding, A.K., Daston, G.P., Boyd, G.R., Lucier, G.W., Safe, S.H., Stewart, J., Tillitt, D.E., & Van der Kraak, G. (2006). Endocrine disrupting chemicals research program of the US Environmental Protection Agency: Summary of a peer-review report. *Environmental Health Perspectives,* 114, pp. 1276-1282.

International Maritime Organization (2001). Final Act. International Conference on the Control of Harmful Anti-Fouling Systems for Ships, 1–5 October 2001. Report No. AFS/CONF/25. London: International Maritime Organization.

IUPAC, 2006. *Global availability of information on agrochemicals: vinclozolin* (Ref: BAS 352F), accessed in 20 of July 2011, available from: http://sitem.herts.ac.uk/aeru/iupac/680.htm

Jansch, S., Garcia, M., & Rombke, J. (2005). Acute and chronic isopod testing using tropical *Porcellionides pruinosus* and three model pesticides. *European Journal of Soil Biology,* 41, pp. 143-152.

Kaneko, M., Okada, R., Yamamoto, K., Nakamura, M., Moscom, G., Polzonetti-Magni, A.M., & Kikuyama, S. (2008). Bisphenol A acts differently from and independently of thyroid hormone in suppressing thyrotropin release from the bullfrog pituitary. *General and Comparative Endocrinology,* 155, pp. 574-580.

Kang, I.H., Kim, H.S., Shin, J.H., Kim, T.S., Moon, H.J., Kim, I.Y., Choi, K.S., Kil, K.S., Park, Y.I., Dong, M.S., & Han, S.Y. (2004). Comparison of anti-androgenic activity of flutamide, vinclozolin, procymidone, linuron, and p,p '-DDE in rodent 10-day Hershberger assay. *Toxicology,* 199, pp. 145-159.

Kannan, K., Guruge, K.S., Thomas, N.J., Tanabe, S., & Giesy, J.P. (1998). Butyltin residues in southern sea otters (*Enhydra lutris nereis*) found dead along California coastal waters. *Environmental Science and Technology,* 32, pp. 1169-1175.

Kavlock, R.J., Daston, G.P., DeRosa, C., FennerCrisp, P., Gray, L.E., Kaattari, S., Lucier, G., Luster, M., Mac, M.J., Maczka, C., Miller, R., Moore, J., Rolland, R., Scott, G.,

Sheehan, D.M., Sinks, T., & Tilson, H.A. (1996). Research needs for the risk assessment of health and environmental effects of endocrine disruptors: A report of the US EPA-sponsored workshop. *Environmental Health Perspectives*, 104, pp. 715-740.

Kavlock, R., Cummings, A. (2005). Mode of action: Inhibition of androgen receptor function - Vinclozolin-induced malformations in reproductive development. *Critical Reviews in Toxicology*, 35, pp. 721-726.

Kelce, W.R., Lambright, L.R., Gray, L.E., & Roberts, K.P. (1997). Vinclozolin and p,p'-DDE alter androgen-dependent gene expression: In vivo confirmation of an androgen receptor-mediated mechanism. *Toxicology and Applied Pharmacology*, 142, pp. 192-200.

Lemos, M.F.L., van Gestel, C.A.M., & Soares, A.M.V.M. (2009). Endocrine disruption in a terrestrial isopod under exposure to bisphenol A and vinclozolin. *Journal of Soils and Sediments*, 9, pp. 492-500.

Lemos, M.F.L., van Gestel, C.A.M., & Soares, A.M.V.M. (2010a). Developmental toxicity of the endocrine disrupters bisphenol A and vinclozolin in a terrestrial isopod. *Archives of Environmental Contamination and Toxicology*, 59, pp. 274-281.

Lemos, M.F.L., van Gestel, & C.A.M., Soares, A.M.V.M. (2010b). Reproductive toxicity of the endocrine disrupters vinclozolin and bisphenol A in the terrestrial isopod *Porcellio scaber* (Latreille, 1804). *Chemosphere*, 78, pp. 907-913.

Lemos, M.F.L., Esteves, A.C., Samyn, B., Timperman, I., van Beeumen, J., Correia, A., van Gestel, C.A.M., & Soares, A.M.V.M. (2010c). Protein differential expression induced by endocrine disrupting compounds in a terrestrial isopod. *Chemosphere*, 79, pp. 570-576.

Matthiessen, P., Gibbs, P.E. (1998). Critical appraisal of the evidence for tributyltin-mediated endocrine disruption in molluscs. *Environmental Toxicology and Chemistry*, 17, pp. 37-43.

McLachlan, J.A. (2001). Environmental signaling: what embryos and evolution teach us about endocrine disrupting chemicals. *Endocrinology Reviews*, 22, pp. 319-341.

Michel, P., Averty, B., Andral, B., Chiffoleau, J.F., & Galgani, F. (2001). Tributyltin along the coasts of Corsica (Western Mediterranean): a persistent problem. *Marine Pollution Bulletin*, 42, pp. 1128-1132.

Mimura, H., Sato, R., Furuyama, Y., Taniike, A., Yagi, M., Yoshida, K., & Kitamura, A. (2008). Adsorption of tributyltin by tributyltin resistant marine *Pseudoalteromonas sp.* cells. *Marine Pollution Bulletin*, 57, pp. 877-82.

Navas, J. M., Segner, H. (2006). Vitellogenin synthesis in primary cultures of fish liver cells as endpoint for in vitro screening of the (anti)estrogenic activity of chemical substances. *Aquatic Toxicology*, 80, pp. 1-22.

Nilsson, E.E., Anway, M.D., Stanfield, J., & Skinner, M.K. (2008). Transgenerational epigenetic effects of the endocrine disruptor vinclozolin on pregnancies and female adult onset disease. *Reproduction*, 135, pp. 713-721.

Oehlmann, J., Schulte-Oehlmann, U. (2003). Endocrine disruption in invertebrates. *Pure and Applied Chemistry*, 75, pp. 2207-2218.

Pinder, L.C.V., Pottinger, T.G. (1998). Endocrine Function in Aquatic Invertebrates and Evidence for Disruption by Environmental Pollutants. Draft report to the United

Kingdom Environmental Agency and the CEFIC Endocrine Modulator Steering Group, 178p.

Porte, C., Janer, G., Lorusso, L.C., Ortiz-Zarragoitia, M., Cajaraville, M.P., Fossi, M.C., & Canesi, L. (2006). Endocrine disruptors in marine organisms: Approaches and perspectives. *Comparative Biochemistry and Physiology C-Toxicology & Pharmacology*, 143, pp. 303-315.

Price, T.M., Murphy, S.K., & Younglai, E.V. (2007). Perspectives: The possible influence of assisted reproductive technologies on transgenerational reproductive effects of environmental endocrine disruptors. *Toxicological Sciences*, 96, pp. 218-226.

Readman, J.W., Albanis, T.A., Barcelo, D., Galassi, S., Tronczynski, J., & Gabrielides, G.P. (1997). Fungicide contamination of Mediterranean estuarine waters: Results from a MED POL pilot survey. *Marine Pollution Bulletin*, 34, pp. 259-263

Reddy, P.S., Sarojini, R., & Nagabhushanam, R. (1991). Impact of tributyltin oxide (TBTO) on limb regeneration of the prawn, *Caridina rajadhari*, after exposure to different time intervals of amputation. *Journal of Tissue Research*, 1, pp. 35-39.

Rodriguez, E.M., Medesani, D.A., & Fingerman, M. (2007). Endocrine disruption in crustaceans due to pollutants: A review. *Comparative Biochemistry and Physiology A - Molecular and Integrative Physiology*, 146, pp. 661-671.

Ruppert, E.E., Fox, R.S., & Barnes, R.D. (Eds.). (2003). *Invertebrate Zoology: A Functional Evolutionary Approach*. Brooks Cole Thomson. Belmont, CA, U.S..

Santos, M.M., Ten Hallers-Tjabbes, C.C., Santos, A.M., & Vieira, N. (2002). Imposex in *Nucella lapillus*, a bioindicator for TBT contamination: re-survey along the Portuguese coast to monitor the effectiveness of EU regulation. *Journal of Sea Research*, 48, pp. 217- 223.

Santos, M.M., Enes, P., Reis-Henriques, M., Kuballa, J., Castro, L.F.C., & Vieira, M.N. (2009). Organotin levels in seafood from Portuguese markets and the risk for consumers. *Chemosphere*, 75, pp- 661-666.

Short, J.W., Thrower, F.P. (1987). Toxicity of tri-n-butyl-tin to chinook salmon, *Oncorhynchu tshawytscha*, adapted to seawater. *Aquaculture*, 61, pp. 193-200.

Shulte-Oehlmann, U., Bettin, C., Fioroni, P., Oehlmann J., & Stroben, E. (1995). *Marisa cornuarietis* (Gastropoda, Prosobranchia): a potential TBT bioindicator for freshwater environments. *Ecotoxicology*, 4, pp. 372-384.

Shulte-Oehlmann, U. (1997). *Fortpflanzungsstörungen bei süß- und brackwasserschnecken - Einfluß der umweltchemikalie tributylzinn*. Wissenschaft und Technik Verlag. Berlin, Germany.

Sidharthan, M., Young, K.S., Woul, L.H., Soon, P.K., & Shin, H.W. (2002). TBT toxicity on the marine microalga Nannochloropsis oculata. *Marine Pollution Bulletin*, 45, pp. 177-180

Smith, B.S. (1981). Tribultyltin compounds induce male characteristics on female mud snails *Nassarius obsoletus = Ilyanassa obsolete. Journal of Applied Toxicology*, 1, pp. 141-144.

Sonak, S. (2009). Implications of organotins in the marine environment and their prohibition. *Journal of Environmental Management*, 90, pp. 1-3.

Sousa, A., Ikemoto, T., Takahashi, S., Barroso, C., & Tanabe, S. (2009a). Distribution of synthetic organotins and total tin levels in *Mytilus galloprovincialis* along the Portuguese coast. *Marine Pollution Bulletin*, 58, pp. 1130-1136.

Sousa, A., Laranjeiro, F., Takahashi, S., Tanabe, S., & Barroso, C.M. (2009b). Imposex and organotin prevalence in a European post-legislative scenario: Temporal trends from 2003 to 2008. *Chemosphere*, 77, pp. 566-573.

Spooner, N., Gibbs, P.E., Bryan, G.W., & Goad, L.J. (1991). The effects of tributyltin upon steroid titres in the female dogwhelk, *Nucella lapillus*, and the development of imposex. *Marine Environmenal Research*, 32, pp. 37-49.

Szeto, S.Y., Burlinson, N.E., Rahe, J.E., Oloffs, P.C. (1989). Persistence of the fungicide vinclozolin on pea leaves under laboratory conditions. *Journal of Agricultural and Food Chemistry*, 37, pp. 529-534.

Tabuchi, M., Veldhoen, A., Dangerfield, N., Jeffries, S., Helbing, C.C., & Ross, P.S. (2006). PCB-related alteration of thyroid hormones and thyroid hormone receptor gene expression in free-ranging Harbor seals (*Phoca vitulina*). *Environmental Health Perspectives*, 114, pp. 1024-1031.

Takeda N. (1980). Hormonal control of head-wart development in the snail *Euchadra peliomphala*. *Journal of Embryology and Experimental Morphology*, 60, pp. 57-69.

Takeuchi, I., Takahashi, S., Tanabe, S., & Miyazaki, N. (2004). Butyltin concentrations along the Japanese coast from 1997 to 1999 monitored by *Caprella spp.* (Crustacea: Amphipoda). *Marine Environmental Research*, 57, pp. 397-414.

Tillmann, M., Schulte-Oehlmann, U., Duft, M., Markert, B., & Oehlmann, J. (2001). Effects of endocrine disruptors on prosobranch snails (Mollusca: Gastropoda) in the laboratory. Part III: Cyproterone acetate and vinclozolin as antiandrogens. *Ecotoxicology*, 10, pp. 373-388.

Thomas, K.V. (2001). Antifouling paint booster biocides in the UK coastal environment and potential risks of biological effects. *Marine Pollution Bulletin*, 42, pp. 677-688

Tomlin, C.D.S. (ed). (2003). *The pesticide manual: a world compendium*. British Crop Protection Council. Farnham, U.K.

Ueoka, M., Allinson, G., Kelsall, Y., Graymore, M., & Stagnitti, F. (1997). Environmental fate of pesticides used in Australian viticulture: Behaviour of dithianon and vinclozolin in the soils of the South Australian Riverland. *Chemosphere*, 35, pp. 2915-2924.

Unger, M.A., MacIntyre, W.G., & Huggett, R.J. (1988). Sorption behavior of tributyltin on estuarine and freshwater sediments. *Environmental Toxicology and Chemistry*, 7, pp. 907-915.

U.S. EPA (1991). *Pesticide environmental fate one liner summaries: vinclozolin*. Environmental Fate and Effects Division, Washington, DC, U.S.

U.S. EPA (1998). *Endocrine Disruptor Screening and Testing Advisory Committee (EDSTAC): final report*. Environmental Protection Agency, Washington, DC, U.S.

U.S. EPA (2000). *Reregistration eligibility decision for vinclozolin*. Office for prevention, pesticides and toxic substances, Washington, DC, U.S.

Vinggaard, A.M., Joergensen, E.C.B., & Larsen, J.C. (1999). Rapid and sensitive reporter gene assays for detection of antiandrogenic and estrogenic effects of environmental chemicals. *Toxicology and Applied Pharmacology*, 155, pp. 150-160.

Waite, M. E., Evans, K. E., Thain, J. E., & Waldock, M. J. (1989). Organotin concentrations in the Rivers Bure and Yare, Norfolk Broads, England. *Appllied. Organometal. Chemistry*, 3, pp. 383–391.

Waldock, M.J., Thain, J.E. (1983). Shell thickening in *Crassostrea gigas*: Organotin antifouling or sediment induced? *Marine Pollution Bulletin*, 14, pp. 411-415

Weis, J.S, Gottlieb, J., & Kwiatkowski, J. (1987). Tributyltin retards regeneration and produces deformities in the limbs in the fiddler crab, Uca pugilator. *Archives of Environmental Contamination and Toxicology*, 16, pp. 321-326

13

Evaluation of Soybean (*Glycine max*) Canopy Penetration with Several Nozzle Types and Pressures

Robert N. Klein, Jeffrey A. Golus and Greg R. Kruger
University of Nebraska WCREC, North Platte, NE,
USA

1. Introduction

Fungicides, when applied in the most effective way, can greatly improve the efficacy of the fungicide, reduce the risk of resistance, and potentially increase yields or preserve crops. When making fungicide applications, there are several things that must be considered. Most sprayers use hydraulic nozzles with pressure against an orifice. The applicator must consider which type of nozzle to use (both orifice size and nozzle type) as well as operating pressure.

Soybean rust is a foliar disease which has for many years been found mainly in Asian countries such as Taiwan, Japan, India, and more recently South Africa, Paraguay, Brazil, and Argentina (Dorrance, et al, 2009). *Phakopsora pachyrhizi* is one of the fungal species known to cause soybean rust and is the most aggressive. This species was identified in US soybean production fields in November of 2004. US cultivars are thought to be highly susceptible to this fungus, and efforts are underway to identify partial resistance or slow rusting traits. Fungicides have proven to be very effective in managing this disease and this will be the primary means of management for the first several years.

Soybean rust in the early days following infection can be found on the lower, first leaves of soybean plants (Geiseler, 2009). Therefore, to obtain control with fungicides one must penetrate the soybean canopy and get the fungicide down to where the infection occurs.

The objective of this study is to determine the optimum spray particle size that delivers the greatest coverage at lower levels of the soybean canopy.

Since the key to this research is the spray nozzle tip let's discuss spray nozzle tip technology.

The spray nozzle tip is important because it:

1. Controls the amount applied - GPA
2. Determines the uniformity of application
3. Affects the coverage
4. Affects the spray drift potential
5. Breaks the mix into droplets
6. Forms the spray pattern
7. Propels the droplets in the proper direction

Where does one start in choosing a spray nozzle tip? The two important factors are pesticide efficacy and spray drift management.

Applicators want to use low water volumes to save time which makes coverage more of a concern. One needs knowledge of the product being used, whether it is a systemic or contact pesticide. Contact pesticides, such as paraquat, need more thorough coverage than systemic pesticides like glyphosate. Most fungicides and insecticides require thorough coverage which requires a smaller spray droplet size. Will this smaller droplet size penetrate the canopy? Also, what is the target? Is it soil, or grass, or broadleaf plants. Are the plant surfaces smooth, or hairy, or waxy? Leaf orientation and even the time of day of the application can affect the coverage needed and hence pesticide efficacy.

2. Nozzle description

Nozzle types commonly used in low-pressure agricultural sprayers include: flat-fan, flood, raindrop, hollow-cone, full-cone, and others. Special features, or subtypes such as extended range, low pressure, drift guard, venturi-type and turbos are available for some nozzle types.

2.1 Flat-fan

Flat-fan nozzles, used for broadcast spraying, produce a tapered-edge spray pattern. These nozzles are also available for band spray. These nozzles are called even flat-fans, which produce a pattern with the same amount applied across the entire spray pattern. Other flat-fan nozzle subtypes include the standard flat-fan, even flat-fan, low pressure flat-fan, extended range flat-fan, drift guards, Turbo TeeJet, and some special types such as off-center flat-fan and twin-orifice flat-fan.

The **standard** flat-fan normally is operated between 2.07 and 4.14 bar, with an ideal range 2.07 and 2.76 bar. The **even** (E) flat-fan nozzles (nozzle number ends with E) apply uniform coverage across the entire width of the spray pattern. They are used for **banding** pesticide over the row and should not be used for broadcast applications. The band width can be controlled with the nozzle height, spray angle, and the orientation of the nozzles.

The **low pressure** (LP) flat-fan develops a normal flat-fan angle and spray pattern at operating pressures between 1.03 and 1.72 bar. Lower pressures result in larger droplets and less drift, but a LP nozzle produces smaller droplets than a standard nozzle at the same pressure.

The **extended range** (XR or LFR) flat-fan provides excellent drift control when operated between 1.03 and 1.72 bar. This nozzle is ideal for an applicator who likes the uniform distribution of a flat-fan nozzle and desires lower operating pressure for drift control. Since extended range nozzles have an excellent spray distribution over a wide range of pressures 1.03 to 4.14 bar, they are ideal for sprayers equipped with flow controllers if spray particle drift is not a problem.

The **Turbo TeeJet** has the widest pressure range of the flat-fan nozzles - 1.03 to 6.21 bar. It produces larger droplets for less drift and is available only in 110 degree spray angle.

The **drift guard** flat-fan has a pre-orifice which controls the flow. The spray tip is approximately one nozzle size larger than the pre-orifice and therefore produces larger droplets and reduces the small drift prone droplets.

The **venturi-type** nozzle produces large air-filled drops through the use of a venturi air aspirator for reducing drift. These include the Delavan Raindrop Ultra, Greenleaf TurboDrop, Lurmark Ultra Lo-Drift, Spraying Systems AI Teejet, ABJ Agri. Products Air Bubble Jet, and Wilger's Combo-Jet. Some of these nozzle tips are available in extended range for pressures.

Flat-fan nozzles also include the **off-center** (LX) flat-fan which is used for boom end nozzles so a wide swath projection is obtained and the **twin-orifice** (TJ) flat-fan which produces two spray patterns -- one angled 30 degrees forward and the other directed 30 degrees backwards. The TJ droplets are small because the spray volume is passing through two smaller orifices instead of one larger one. The two spray directions and smaller droplets improve coverage and penetration, a plus when applying postemergence contact herbicides. To produce fine droplets, the twin-orifice usually operates between 2.07 and 4.14 bar.

Flat-fan nozzles are available in several spray angles. The most common spray angles are 65, 73, 80, and 110 degrees. Recommended nozzle heights for flat-fan nozzles during broadcast application are given in Table 1.

| | Spray height (cm) | | | |
| | 51 cm Spacing With Overlap of | | 76 cm Spacing With Overlap of | |
Nozzle angle°	30%	100%	30%	100%
65	56	-NR-	-NR-	-NR-
73	51	-NR-	73.7	-NR-
80	43	66	66	97
110	25	38	36	64

-NR- Not recommended because of drift potential.

Table 1. Suggested minimum spray heights.

The correct nozzle height is measured from the nozzles to the target, which may be the top of the ground, growing canopy, or stubble. Use 110 degree nozzles when boom heights are less than 76 cm and 80 degree nozzles when the booms are higher.

Although wide-angle nozzles produce smaller droplets that may be more prone to drift, the reduction of boom height reduces the overall drift potential. The net reduction in drift potential more than offsets the effect of smaller droplet size. The nozzle spacing and orientation should provide for 100 percent overlap at the target height. Nozzles should not be oriented more than 30 degrees from vertical.

Most nozzle manufacturers identify their flat-fan nozzles with a four or five digit number. The first numbers are the spray angle and the other numbers signify the discharge rate at rated pressure. For example, an 8005 has an 80 degree spray angle and will discharge 1.9 liters per minute (LPM) at the rated pressure of 2.75 bar. A 11002 nozzle has a 110 degree spray angle and will discharge 0.8 LPM at the rated pressure of 2.75 bar. Additional designations are: " BR" - brass material; "SS" - stainless steel; "VH" - hardened stainless steel; and "VS" - color codes. See Table 2 for nozzle type and discharge rates.

Nozzle Type	Discharge (lpm)	Rated Pressure (bar)	Operating Range	
			Min (bar)	Max (bar)
Regular flat-fan 8006	2.3	2.76	2.07	4.14
Regular flat-fan 11008	3.0	2.76	2.07	4.14
Low pressure flat-fan 8006LP	2.3	1.03	1.03	2.76
Low pressure flat-fan 11008LP	3.0	1.03	1.03	2.76
Extended range flat-fan 8006XR	2.3	2.76	1.03	4.14
Extended range flat-fan 11008XR	3.0	2.76	1.03	4.14
Turbo TeeJet TT11002VP	0.8	2.76	1.03	6.21
Turbo TeeJet TT11005VP	1.9	2.76	1.03	6.21
Turbo TeeJet Induction TTI11002	0.8	1.03	2.76	6.89
Turbo TeeJet Induction TTI11005	1.9	1.03	2.76	6.89
Drift Guard DG8002VS	0.8	2.76	2.76	4.14
Drift Guard DG11005VS	1.9	2.76	2.07	4.14
AI TeeJet AI11002-VS	0.8	2.76	2.07	6.89
AI TeeJet AI11005-VS	1.9	2.76	2.07	6.89
AIXR TeeJet 11002VS	0.8	2.76	1.03	6.21
AIXR TeeJet 11005VS	1.9	2.76	1.03	6.21
Flood TKSS 6	2.3	0.69	0.69	2.76
Flood TKSS 8	3.0	0.69	0.69	2.76
Turbo FloodJet TF-VS2	0.8	0.69	0.69	2.76
Turbo FloodJet TF-VS10	3.8	0.69	0.69	2.76
Raindrop RA-6	2.3	2.76	1.38	3.45

Table 2. Nozzles types and discharge rates at rated pressure.

Delavan flat-fan nozzles are identified by LF or LF-R, which reflect the standard and extended range flat-fan nozzles. The first numbers are the spray angle followed by a dash, and then the discharge rate at rated pressure. For example, an LF80-5R is an extended range nozzle with an 80 degree spray angle, and will apply 1.9 LPM at the rated pressure of 2.75 bar.

2.2 Flood

Flood nozzles are popular for applying suspension fertilizers where clogging is a potential problem. These nozzles produce large droplets at pressures of 0.69 to 1.72 bar. The nozzles should be spaced less than 152 cm apart. The nozzle height and orientation should be set for 100 percent overlap.

Nozzle spacing between 76 and 102 cm produces the best spray patterns. Pressure influences spray patterns of flood nozzles more than flat-fan nozzles. However, the spray pattern is not as uniform as with the flat-fan nozzles, and special attention to nozzle orientation and correct overlap is critical.

Flooding nozzles are designated "TK" by Spraying Systems and "D" by Delavan. The value following the letters is the flow rate at the rated pressure of 0.69 bar. For example, TK-SS2 or D-2 are flood nozzles that apply 0.8 LPM at 0.7 bar.

The Turbo FloodJet incorporates a pre-orifice which controls the flow plus a turbulence chamber. The tip design more closely resembles a flat fan nozzle, which greatly reduces the surface area and the result is a much improved pattern with tapered edges. Use the turbo flood for pesticide application for incorporated herbicides because of the improved pattern. The turbo flood nozzles are a good choice if drift is a concern because they produce larger droplets than standard flood nozzles. Because of their large droplet size do not use the turbo flood nozzle where good coverage is needed.

2.3 Raindrop

Raindrop nozzles produce large drops in a hollow-cone pattern at pressures from 1.38 to 3.45 bar. The "RA" Raindrop nozzles are used for pre-plant incorporated herbicide and are usually mounted on tillage implements. When used for broadcast application, nozzles should be orientated 30 degrees from the horizontal. The spray patterns should be overlapped 100 percent to obtain uniform distribution. These nozzles are not satisfactory for postemergence or non-incorporated herbicides because the small number of large droplets produced would not provide satisfactory coverage.

2.4 Cone

Hollow-cone - Hollow cone nozzles generally are used to apply insecticides or fungicides to field crops when foliage penetration and complete coverage of the leaf surface is required. These nozzles operate in a pressure range from 2.76 to 6.89 bar. Spray drift potential is higher from hollow-cone nozzles than from other nozzles due to the small droplets produced.

Full-cone - Full-cone nozzles usually are recommended over flood nozzles for soil-incorporated herbicides. Full-cone nozzles operate between a pressure range of 1.03 to 2.76 bar. Optimum uniformity is achieved by angling the nozzles 30 degrees and overlapping the spray coverage by 100 percent.

Fine Hollow-cone - The ConeJet (Spray Systems) and WRW-Whirl Rain (Delavan) are wide-angle (80 to 120 degrees), hollow-cone nozzles. These nozzles are used for postemergence contact herbicides where a finely atomized spray is used for complete coverage of plants or weeds under a hood for band spraying. Drift potential is great for these nozzles.

2.5 Nozzle material

Nozzles can be made from several materials. The most common are brass, nylon, stainless steel and hardened stainless steel and ceramic. Stainless steel nozzles last longer than brass or nylon and generally produce a more uniform pattern over an extended time period. Nylon nozzles with stainless steel or hardened stainless steel inserts offer an alternative to solid stainless steel nozzles at a reduced cost. Thermoplastic nozzles have good abrasion resistance but swelling can occur with some chemicals, and they are easily damaged when cleaned. Ceramic has superior wear life and is highly resistant to abrasive and erosive chemicals. Where available ceramic is usually the best choice.

Do not mix nozzles of different materials, types, spray angles, or spray volumes on the same spray boom. A mixture of nozzles produces uneven spray distribution.

2.6 Combination nozzles

These are where the spray tip is built right into the cap as one. These keep the spray tip and cap from separating and when available is usually the best choice.

2.7 When to replace nozzles

- Spray pattern distorted
- Nozzles show irregular wear
- Nozzle flow rates is 10% greater than new nozzles

Note: Each nozzle's flow rate on spray boom needs to be within 5% of the average nozzle flow rate.

Spray particle size affects both pesticide efficacy (coverage) and spray drift. Cutting the droplet size in half results in eight times the number of spray droplets, see Figure 1.

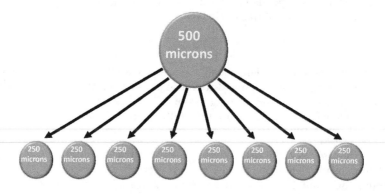

Fig. 1. Reducing droplet size by 50% results in eight times the number of droplets

The origin of standardization of spray droplet sizes started with the British Crop Protection Council in 1985 with droplet size classification, primarily designed to enhance efficacy. It uses the term, *SPRAY QUALITY* for droplet size categories. In 2000 the ASAE Standard S572 established the droplet size classification in the U.S. primarily designed to control spray drift and secondarily efficacy. It uses the term *DROPLET SPECTRA CLASSIFICATION* for droplet size categories. In March 2009 ANSI/ASAE approved S572.1 as the American National Standard. This added extremely fine and ultra coarse to the classification categories.

The specifics of the Standard ASAE S572.1.

- Based on spraying water through reference nozzles
- Nozzle manufacturers can conduct the tests
- Must use a set of reference tips and a laser-based instrument
- Droplet Spectra measurements must be with the same instrument, measuring method, sampling technique, scanning technique, operator, and in a similar environmental condition.

Important Droplet Characteristics:

Dv0.1(μm) - 10% of spray volume is of droplet sizes less than this number
Dv0.5(μm) - 50% of spray volume is of droplet sizes less than this number; volume median diameter (VMD)
Dv0.9(μm) - 90% of spray volume is of droplet sizes less than this number

Though not part of the standard, the percent of spray volume less than 200 microns identifies the particle sizes most prone to spray drift.

Figures 2 and 3, and Table 3 illustrate volume median diameter, the ASAE Standard and a Turbo TeeJet spray tip droplet spectra.

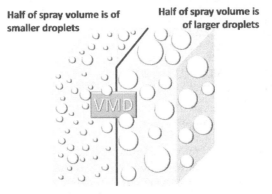

Fig. 2. Volume Median Diameter (VMD)

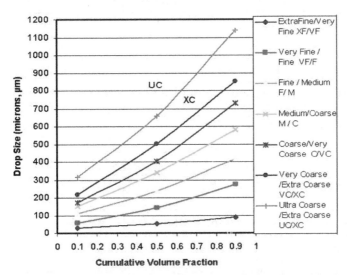

Fig. 3. ANSI/ASAE S572.1 March 2009; Approved as an American National Standard

Nozzle	Bar										
	1.03	1.38	1.72	2.07	2.41	2.76	3.45	4.14	4.83	5.52	6.21
TT11001	C	M	M	M	M	M	F	F	F	F	F
TT110015	C	C	M	M	M	M	M	M	F	F	F
TT11002	C	C	C	M	M	M	M	M	M	M	F
TT11003	VC	VC	C	C	C	C	M	M	M	M	M
TT11004	XC	VC	VC	C	C	C	C	C	M	M	M
TT11005	XC	VC	VC	VC	VC	C	C	C	C	M	M
TT11006	XC	XC	VC	VC	VC	C	C	C	C	C	M
TT11008	XC	XC	VC	VC	VC	VC	C	C	C	C	M

Table 3. Turbo TeeJet® Nozzle Droplet Spectra

Three nozzle tips were evaluated for the control of triazine resistant kochia and green foxtail in winter wheat stubble. The treatment parameters used to evaluate the nozzle tips and percent control are shown in Table 4 and Figures 4 and 5.

Trt	Nozzle	Spray Particle Size	Volume		Speed	
			(gpa)	(L/ha)	(mph)	(km/h)
1	XR11005	Coarse	10	94	8.6	14
2	DG11005	Coarse	10	94	8.6	14
3	TF-VS2.5	Extremely Coarse	10	94	8.6	14
4	XR11004	Medium	7.5	70	9.2	15
5	DG11004	Coarse	7.5	70	9.2	15
6	TF-VS2	Extremely Coarse	7.5	70	9.2	15
7	XR11003	Fine	5.0	47	10.3	17
8	DG11003	Coarse	5.0	47	10.3	17
9	Untreated Check			---		---

*All treatments applied at 2 bar
*Herbicide applied was Paraquat + Atrazine (0.35 + 0.56 Kg ai/ha)

Table 4. Treatment parameters used to evaluate three nozzle types

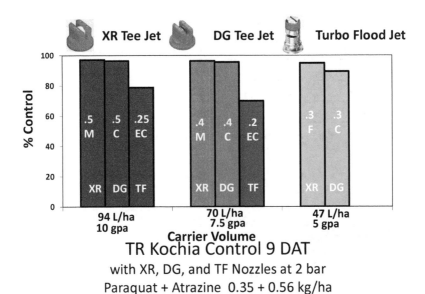

Fig. 4. Triazine resistant kochia control with several nozzles and carrier volumes

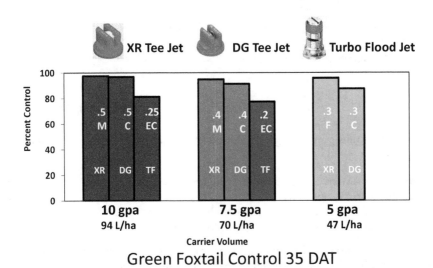

Fig. 5. Green foxtail control with several nozzles and carrier volumes

The results of a study on the retention and total uptake and root translocation of glyphosate of glyphosate resistant corn by Paul Feng of Monsanto are in Table 5 and Figure 6.

Droplet Size	% Retention (Actual over Calculated)
Fine	47 ± 2
Medium	37 ± 7
Coarse	38 ± 8

Table 5. Effect of droplet size on glyphosate retention in glyphosate resistant corn

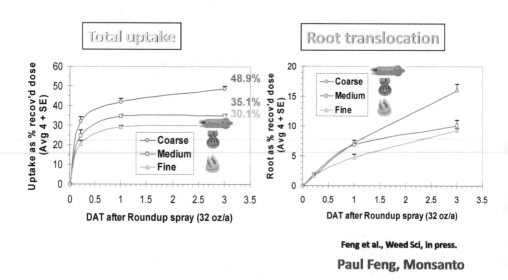

Feng et al., Weed Sci, in press.

Paul Feng, Monsanto

Fig. 6. Uptake and translocation in glyphosate resistant corn with fine, medium and coarse spray droplets.

In the glyphosate resistant corn study even though there was less glyphosate retained on the plants with the coarse droplet size as compared to the fine, the coarse droplet size resulted in increased uptake and root translocation. It is thought that there must be enough glyphosate present in the spray droplet to translocate in the plant.

A study across Nebraska in 2004 with five nozzle tips resulted in almost identical weed control. One should therefore select those tips which produce the smallest amount of fines which are the Turbo TeeJet, Turbo Flood and Air Induction nozzle tips. See Figure 7.

Figure 8 gives the particle sizes of five nozzle tips with water and water with Roundup WeatherMax + 2% Ammonium Sulfate and three additives: Array, In-Place and Interlock.

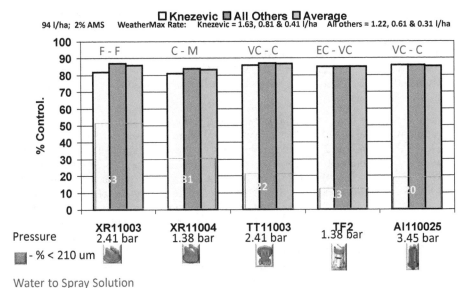

R. Klein, S. Knezevic, A. Martin, R. Wilson, B. Kappler and F. Roeth, 2004

Fig. 7. Percent visual control of corn, oil sunflower, velvetleaf, green foxtail and watermemp with glyphosate.

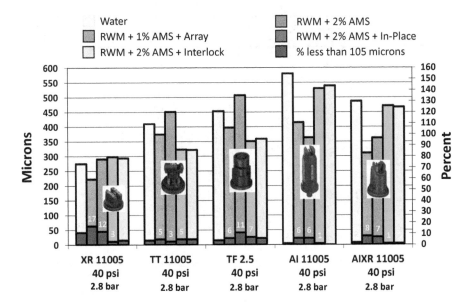

Fig. 8. Volume median diameter of several nozzles and spray solutions.

The volume median diameter of the spray particle sizes are greatly reduced as was the amount of spray volume under 210 microns with the addition of Roundup WeatherMax and 2% AMS. The additive Array increased the volume median diameter and reduced the amount of spray volume under 210 microns with the extended range, Turbo TeeJet and Turbo Flood nozzles. The additives In-Place and Interlock performed with similar results with the air induction and extended range air induction tip. All additives do not work with all nozzles as evident in Figure 9.

To evaluate the nozzle tips, pressure, nozzle spacing and angle in getting penetration into the soybean canopy, research was conducted over several years. Soybeans were planted in 76.2 cm rows in May 2006, 2007, and 2008. Field applications were conducted in August and September of these years. Six different nozzles were included, all Spraying Systems Co: XRC11003, XRC11006, TT11003, TT11006, AIC110025, and AIC11005. Each nozzle was used at three pressures, and two different nozzle setups were included. Nozzles were set on 76.2 cm spacing, and 190 l ha-1 was the carrier volume (Table 6). For the smaller nozzle size of each type, two nozzles were used: one directed 45 degrees forward from vertical and the other directed 45 degrees back from vertical. Boom height was 43.2 cm above the canopy. White indicating cards were set into a row of soybeans. The cards were attached to an electric fencepost at heights of 14 cm (low), 42 cm (middle) and 70 cm (high), with the soybeans being 84 cm tall. A pull type sprayer was used to apply the treatments. Water dyed with Garrco Products Vision Pink indicating dye was sprayed over the cards. Four sets of cards were placed for each treatment to create four replications. A nozzle setup ran directly over the row of soybeans containing the cards. The cards were allowed to dry and placed in Ziploc bags. The cards were then scanned with the program DropletScan, which determines the number of drops, volume median diameter (VMD) and percent coverage for each card. VMD is the micron size of which half the spray volume in made of smaller droplets and half is of larger droplets.

Treatment	Nozzle(s)	Pressure bar	Total Nozzle Output L min-1	Speed km hr-1
1		1.03	1.36	05.8
2	XRC 11003 (2)	2.07	1.97	08.2
3		4.14	2.80	11.8
4		1.03	1.36	05.8
5	XRC 11006	2.07	1.97	08.2
6		4.14	2.80	11.8
7		1.03	1.36	05.8
8	TT 11003 (2)	2.07	1.97	08.2
9		4.14	2.80	11.8
10		1.03	1.36	05.8
11	TT 11006	2.07	1.97	08.2
12		4.14	2.80	11.8
13		2.07	1.67	05.8
14	AIC 110025 (2)	4.14	2.35	08.2
15		6.21	3.88	11.8
16		2.07	1.67	05.8
17	AIC 11005	4.14	2.35	08.2
18		6.21	3.88	11.8

Table 6. Nozzles and pressures used in soybean canopy penetration study.

The nozzles were also analyzed with a Sympatec Helos Vario KF particle size analyzer. With the R6 lens installed, it is capable of detecting particle sizes in a range from 0.5 to 1550 microns. This system uses laser diffraction to determine particle size distribution. Each treatment was replicated three times. The width of the nozzle plume was analyzed by moving the nozzle across the laser by means of a linear actuator. Information obtained includes VMD.

3. Results

The results of coverage on the cards scanned with the program DropletScan are reported in Table 7. This includes the three pressures, VMD and the percent coverage on the lower, middle and upper cards. Also listed in the table are the laser VMD for each nozzle and pressure. Results are from 2006, 2007 and 2008 combined.

Nozzle(s)	Pressure	Lower Card VMD[a]	Pct Cov[b]	Middle Card VMD[a]	Pct Cov[b]	Upper Card VMD[a]	Pct Cov[b]	Laser VMD[a]
	Bar	μm	%	μm	%	μm	%	μm
XRC11003 (2)	1.03	267	0.92	293	4.11	342	13.14	382
	2.07	231	0.86	239	4.06	316	10.45	244
	4.14	230	0.56	237	3.50	285	8.81	189
XRC11006	1.03	395	1.01	374	3.44	453	7.85	465
	2.07	314	0.70	338	4.18	394	12.56	316
	4.14	309	0.68	330	4.47	390	10.95	262
TT11003 (2)	1.03	352	1.46	408	4.37	506	16.27	516
	2.07	314	1.09	394	4.16	448	13.63	380
	4.14	268	1.05	344	2.50	394	12.67	254
TT11006	1.03	392	1.21	454	3.98	572	11.10	640
	2.07	379	0.81	417	5.20	504	11.46	406
	4.14	326	1.43	334	3.25	400	11.14	347
AIC110025 (2)	2.07	438	0.82	519	3.70	582	10.18	677
	4.14	376	0.37	376	2.79	480	8.73	532
	2.28	333	0.64	351	2.21	429	9.09	462
AIC11005	2.07	454	0.94	448	3.95	550	10.94	662
	4.14	403	0.85	433	3.80	485	10.43	508
	2.28	349	1.23	407	4.77	461	11.22	436
	LSD 5%	70	0.58	52	2.14	38	3.48	29

[a]Volume median diameter
[b]Percent coverage

Table 7. Volume median diameter (VMD) and percent coverage for various nozzles, pressures and card placement in soybean canopy.

4. Discussion

The figures 9, 10 and 11 contain results of VMD and percent coverage of the card in the three card levels in soybean canopy for the 3 years, 2006 to 2008. Data in the figures represents the average of the three pressures used, while Table 7 includes data for each pressure. With the lower card, the TT11003 had the best coverage closely followed by the TT11006. With the XRC nozzles, as the pressure increased the coverage decreases in the lower canopy. This was also true with the smaller TT (2-TT11003 in opposite directions) but the larger TT spraying straight down and the AIC nozzle coverage decreased with the medium pressure but increased again with the highest pressure. The two TT11003 had the highest amount of coverage of the three nozzle types in the study.

For the lower card, laser VMD was larger than card VMD for each nozzle. As particle size increased (comparing nozzle types), the difference between the VMD's increased, especially at laser VMD sizes of 375 μm and greater. For the middle card, the two VMD's were equal up to 375 μm. Above this point, laser VMD once again becomes larger. For the upper card, card VMD was greater than laser VMD up to laser VMD being 450 μm. Above that, card VMD was smaller. This suggests larger particles landed in the upper canopy, especially for nozzles producing a smaller VMD (smaller particle size). The nozzles producing the highest percent coverage of the lower card were the TT11003 and TT11006. These nozzles produced a laser VMD of 383 μm and 464 μm respectively, suggesting a micron size of around 400 may be optimal.

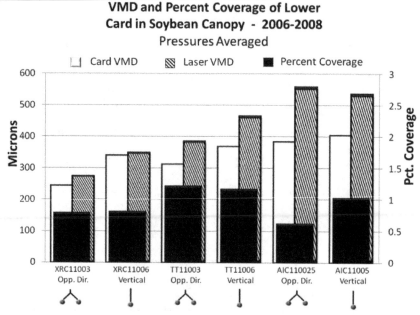

Fig. 9. VMD and percent coverage of lower card in soybean canopy

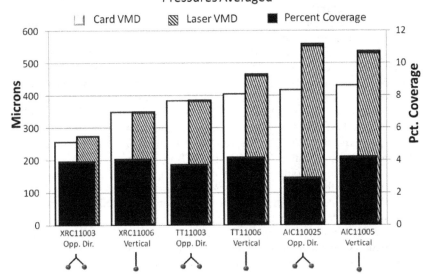

Fig. 10. VMD and percent coverage of middle card in soybean canopy

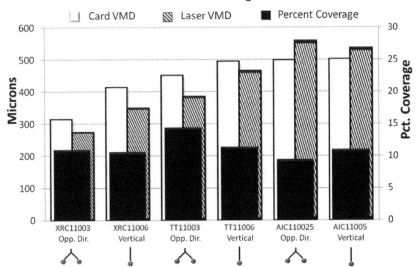

Fig. 11. VMD and percent coverage of upper card in soybean canopy

5. References

Dorrance, Anne E., Patrick E. Lipps, Dennis Mills and Miguel Vega-Sanchez. 2009. Soybean Rust. Ohio State University Extension Fact Sheet AC-0048-04. Ohio State University, Columbus Ohio.

Geiseler, Loren J. 2009. Asian Soybean Rust. University of Nebraska-Lincoln Plant Disease Central. University of Nebraska-Lincoln, Lincoln NE.

Klein, Robert N., Jeffrey A. Golus, Alexander R. Martin, Fred W. Roeth, and Brady F. Kappler. 2006. Glyphosate Efficacy With Air Induction, Extended Range, Turbo FloodJet and Turbo TeeJet Nozzle Tips. pp. 453-460. International Advances in Pesticide Application 2006, Robinson College, Cambridge, UK, January 10-12, 2006.

Klein, Robert, Jeffrey Golus and Amanda Cox. 2008. Spray Droplet Size and How It Is Affected by Pesticide Formulation, Concentrations, Carrier, Nozzle Tips, Pressure and Additives. pp. 232-238. International Advances in Pesticide Application 2006, Robinson College, Cambridge, UK, January 9-11, 2008.

Klein, Robert and Jeffrey Golus. 2010. Evaluation of Soybean (*Glycine max*) Canopy Penetration With Several Nozzle Types and Pressure. pp. 35-39. 2010. International Advances in Pesticide Application 2006, Robinson College, Cambridge, UK, January 5-7, 2010

Klein, Robert N. 2010. Spray Nozzle Tip Technology. pp. 195-200. Proceedings 2010 Crop Production Clinics, University of Nebraska-Lincoln-Extension.

Feng, Paul C.C., Tommy Chiu, R. Douglas Sammons, and Jan S. Ryerse. 2003. Droplet Size Affects Glyphosate Retention, Absorption, and Translocation in Corn. Weed Science, 51:443-448.

Permissions

The contributors of this book come from diverse backgrounds, making this book a truly international effort. This book will bring forth new frontiers with its revolutionizing research information and detailed analysis of the nascent developments around the world.

We would like to thank Dr. Nooruddin Thajuddin, for lending his expertise to make the book truly unique. He has played a crucial role in the development of this book. Without his invaluable contribution this book wouldn't have been possible. He has made vital efforts to compile up to date information on the varied aspects of this subject to make this book a valuable addition to the collection of many professionals and students.

This book was conceptualized with the vision of imparting up-to-date information and advanced data in this field. To ensure the same, a matchless editorial board was set up. Every individual on the board went through rigorous rounds of assessment to prove their worth. After which they invested a large part of their time researching and compiling the most relevant data for our readers. Conferences and sessions were held from time to time between the editorial board and the contributing authors to present the data in the most comprehensible form. The editorial team has worked tirelessly to provide valuable and valid information to help people across the globe.

Every chapter published in this book has been scrutinized by our experts. Their significance has been extensively debated. The topics covered herein carry significant findings which will fuel the growth of the discipline. They may even be implemented as practical applications or may be referred to as a beginning point for another development. Chapters in this book were first published by InTech; hereby published with permission under the Creative Commons Attribution License or equivalent.

The editorial board has been involved in producing this book since its inception. They have spent rigorous hours researching and exploring the diverse topics which have resulted in the successful publishing of this book. They have passed on their knowledge of decades through this book. To expedite this challenging task, the publisher supported the team at every step. A small team of assistant editors was also appointed to further simplify the editing procedure and attain best results for the readers.

Our editorial team has been hand-picked from every corner of the world. Their multi-ethnicity adds dynamic inputs to the discussions which result in innovative outcomes. These outcomes are then further discussed with the researchers and contributors who give their valuable feedback and opinion regarding the same. The feedback is then collaborated with the researches and they are edited in a comprehensive manner to aid the understanding of the subject.

Apart from the editorial board, the designing team has also invested a significant amount of their time in understanding the subject and creating the most relevant covers. They scrutinized every image to scout for the most suitable representation of the subject and create an appropriate cover for the book.

The publishing team has been involved in this book since its early stages. They were actively engaged in every process, be it collecting the data, connecting with the contributors or procuring relevant information. The team has been an ardent support to the editorial, designing and production team. Their endless efforts to recruit the best for this project, has resulted in the accomplishment of this book. They are a veteran in the field of academics and their pool of knowledge is as vast as their experience in printing. Their expertise and guidance has proved useful at every step. Their uncompromising quality standards have made this book an exceptional effort. Their encouragement from time to time has been an inspiration for everyone.

The publisher and the editorial board hope that this book will prove to be a valuable piece of knowledge for researchers, students, practitioners and scholars across the globe.

List of Contributors

Snježana Topolovec-Pintarić
Department for Plant Pathology, Faculty of Agriculture, University of Zagreb, Croatia

Angel Rebollar-Alviter
Centro Regional Morelia, Universidad Autonoma Chapingo, Mexico

Mizuho Nita
Virginia Polytechnic Institute and State University, Alson H. Smith Jr. Agricutural Research and Extension Center, Winchester, VA, USA

Roland W. S. Weber and Alfred-Peter Entrop
Esteburg Fruit Research and Advisory Centre, Jork, Germany

A. Billard, S. Fillinger, P. Leroux, J. Bach, C. Lanen and D. Debieu
INRA UR 1290 BIOGER-CPP Thiverval-Grignon, France

H. Lachaise
Bayer SAS, Bayer CropScience, Research Center La Dargoire, Lyon, France

R. Beffa
Bayer CropScience AG, Frankfurt, Germany

Kris Audenaert, Sofie Landschoot, Adriaan Vanheule and Geert Haesaert
Associated Faculty of Applied Bioscience Engineering, Ghent University College, Ghent, Belgium
Laboratory of Phytopathology, Faculty of Bioscience Engineering, Ghent University, Ghent, Belgium

Willem Waegeman and Bernard De Baets
KERMIT, Department of Mathematical Modelling, Statistics and Bioinformatics, Ghent University, Ghent, Belgium

Simona Marianna Sanzani and Antonio Ippolito
Department of Environmental and Agro-Forestry Biology and Chemistry, University of Bari "Aldo Moro", Italy

Suzana Kristek, Andrija Kristek and Dragana Kocevski
University of J. J. Strossmayer, Faculty of Agriculture, Croatia

Maurizio Mulas
University of Sassari, Italy

Emese D. Nagygyörgy, László Hornok and Attila L. Ádám
Agricultural Biotechnology Center and Department of Plant Protection, Szent István University, Mycology Group of the Hungarian Academy of Sciences, Gödöllő, Hungary

S. Gomathinayagam
Faculty of Agriculture and Forestry, University of Guyana, Berbice Campus, Tain, Guyana

N. Balasubramanian
CIRN and Department of Biology, University of Azores, Ponta Delgada, Azores, Portugal

V. Shanmugaiah
Department of Microbial Technology, School of Biological Sciences, Madurai Kamaraj University, Madurai, Tamil Nadu, India

M. Rekha
Department of Biotechnology, Kalasalingam University, Krishnankovil, Tirunelveli, Tamil Nadu, India

P. T. Manoharan
Department of Botany, Vivekananda College, Thiruvedakam, Madurai, Tamil Nadu, India

D. Lalithakumari
Research and Development, Bio-Soil Enhancers, Hattiesburg, MS, USA

Selim Sameh, Roisin-Fichter Céline, Andry Jean-Baptiste and Bogdanow Boris
Platform Biotechnology and Plant Pathology, Institut Polytechnique LaSalle Beauvais, GIS PhyNoPi: Groupement d'Intérêt Scientifique, Phytopathologie Nord-Picardie, France

Marco F. L. Lemos
ESTM and GIRM, Polytechnic Institute of Leiria, Portugal

Marco F. L. Lemos, Ana C. Esteves and João L. T. Pestana
Department of Biology and CESAM, University of Aveiro, Portugal

Robert N. Klein, Jeffrey A. Golus and Greg R. Kruger
University of Nebraska WCREC, North Platte, NE, USA

Printed in the USA
CPSIA information can be obtained
at www.ICGtesting.com
JSHW011442221024
72173JS00004B/902